Linguistics and the Study of Comics

# Linguistics and the Study of Comics

Edited by
Frank Bramlett
*University of Nebraska at Omaha, USA*

Selection and editorial matter © Frank Bramlett 2012
Individual chapters © their respective authors 2012

All rights reserved. No reproduction, copy or transmission of this publication may be made without written permission.

No portion of this publication may be reproduced, copied or transmitted save with written permission or in accordance with the provisions of the Copyright, Designs and Patents Act 1988, or under the terms of any licence permitting limited copying issued by the Copyright Licensing Agency, Saffron House, 6–10 Kirby Street, London EC1N 8TS.

Any person who does any unauthorized act in relation to this publication may be liable to criminal prosecution and civil claims for damages.

The authors have asserted their rights to be identified as the authors of this work in accordance with the Copyright, Designs and Patents Act 1988.

First published 2012 by
PALGRAVE MACMILLAN

Palgrave Macmillan in the UK is an imprint of Macmillan Publishers Limited, registered in England, company number 785998, of Houndmills, Basingstoke, Hampshire RG21 6XS.

Palgrave Macmillan in the US is a division of St Martin's Press LLC,
175 Fifth Avenue, New York, NY 10010.

Palgrave Macmillan is the global academic imprint of the above companies and has companies and representatives throughout the world.

Palgrave® and Macmillan® are registered trademarks in the United States, the United Kingdom, Europe and other countries.

ISBN 978–0–230–36282–6

This book is printed on paper suitable for recycling and made from fully managed and sustained forest sources. Logging, pulping and manufacturing processes are expected to conform to the environmental regulations of the country of origin.

A catalogue record for this book is available from the British Library.

A catalog record for this book is available from the Library of Congress.

10  9  8  7  6  5  4  3  2  1
21 20 19 18 17 16 15 14 13 12

Printed and bound in Great Britain by
CPI Antony Rowe, Chippenham and Eastbourne

# Contents

*List of Tables and Figures* vii

*Acknowledgements* x

*Notes on Contributors* xi

Introduction 1
*Frank Bramlett*

1  Image Schemas and Conceptual Metaphor in Action Comics 13
   *Elisabeth Potsch and Robert F. Williams*

2  Creating Humor in Gary Larson's *Far Side* Cartoons Using
   Interpersonal and Textual Metafunctions 37
   *Richard Watson Todd*

3  Metaphors and Topoi of H1N1 (Swine Flu) Political
   Cartoons: A Cross-cultural Analysis 59
   *Jill Hallett and Richard W. Hallett*

4  Comics, Linguistics, and Visual Language: The Past and
   Future of a Field 92
   *Neil Cohn*

5  Constructing Meaning: Verbalizing the Unspeakable in
   Turkish Political Cartoons 119
   *Veronika Tzankova and Thecla Schiphorst*

6  Plurilingualism in Francophone Comics 142
   *Miriam Ben-Rafael and Eliezer Ben-Rafael*

7  To and Fro Dutch Dutch: Diachronic Language Variation
   in Flemish Comics 163
   *Gert Meesters*

8  Linguistic Codes and Character Identity in *Afro Samurai* 183
   *Frank Bramlett*

9  Pocho Politics: Language, Identity, and Discourse
   in Lalo Alcaraz's *La Cucaracha* 210
   *Carla Breidenbach*

| | | |
|---|---|---|
| 10 | The Use of English in the Swedish-Language Comic Strip *Rocky*<br>*Kristy Beers Fägersten* | 239 |
| 11 | 'Ah, laddie, did ye really think I'd let a foine broth of a boy such as yerself get splattered...?' Representations of Irish English Speech in the Marvel Universe<br>*Shane Walshe* | 264 |

Conclusion  291
*Frank Bramlett*

*Index of Language Varieties*  294

*Subject Index*  296

# List of Tables and Figures

## Tables

| | | |
|---|---|---|
| 2.1 | Summary of the original and adapted Gary Larson cartoons | 44 |
| 2.2 | Frequency of choices between original and adapted versions | 49 |
| 2.3 | Examples of reasons for themes | 49 |
| 2.4 | Frequencies of themes associated with original and adapted versions | 51 |
| 2.5 | Frequencies of themes associated with individual cartoons | 52 |
| 3.1 | Metaphors used in swine flu cartoons by country (raw numbers) | 74 |
| 3.2 | Fears exploited in swine flu cartoons by country (raw numbers) | 81 |
| 5.1 | A seal's protest | 128 |
| 5.2 | A worker has a conversation with his boss | 132 |
| 5.3 | Prime Minister and President exchange pleasantries | 135 |
| 5.4 | Valentine's Day in politics | 137 |
| 6.1 | Examples of *schtroumpf* expressions and translation | 156 |
| 7.1 | Lexical regionalisms in *Suske en Wiske* and in *Jommeke* | 171 |
| 7.2 | Grammatical regionalisms in *Suske en Wiske* and in *Jommeke* | 172 |
| 7.3 | Distribution of grammatical regionalisms in *Suske en Wiske* and *Jommeke*, in order of decreasing type presence in subsections of the corpus | 175 |
| A.1 | Comics in the corpus | 181 |
| 8.1 | Adult Afro's utterances in *AS–anime* | 202 |
| 8.2 | Adult Afro's utterances in *AS–manga*, Volumes 1 and 2 | 204 |
| 8.3 | Excerpts from Ninja Ninja's utterances in *AS–anime* | 205 |
| 8.4 | Excerpts from Ninja Ninja's utterances in *AS–manga*, Volumes 1 and 2 | 206 |
| 9.1 | Seven main themes in *La Cucaracha* 2007–10 | 222 |

| 9.2 | Spanish used for place names | 225 |
|---|---|---|
| 9.3 | Mock Spanish words or phrases in *La Cucaracha* | 226 |
| 11.1 | The distribution of supposedly typical Irish English features in the corpus | 285 |

## Figures

| 0.1 | How language and image complement each other | 6 |
|---|---|---|
| 1.1 | Billy's dotted line path in Bil Keane's *Family Circus* | 21 |
| 1.2 | Example of a ribbon path | 23 |
| 1.3 | Example of motion lines | 24 |
| 1.4 | Example of an impact flash | 26 |
| 1.5 | A panel depicting complex action | 30 |
| 3.1 | Overarching metaphors employed in cartoons across all countries | 75 |
| 3.2 | Overarching metaphors employed in Indian cartoons | 77 |
| 3.3 | Overarching metaphors employed in US cartoons | 79 |
| 3.4 | Overarching metaphors employed in cartoons of other countries | 80 |
| 3.5 | Overarching fears employed in cartoons across all countries | 81 |
| 3.6 | Overarching fears exploited in Indian cartoons | 84 |
| 3.7 | Overarching fears exploited in US cartoons | 86 |
| 3.8 | Overarching fears exploited in other countries' cartoons | 88 |
| 4.1 | Ungrammatical line junctions in a child's drawing of a kangaroo | 100 |
| 4.2 | Variation in photology | 101 |
| 4.3 | Various levels of photological constructions | 103 |
| 4.4 | Morphological processes in graphic form | 105 |
| 4.5 | Inference generated by an action star | 107 |
| 4.6 | Panel transitions analyzing a sequence of images | 108 |
| 4.7 | Sequential images with center-embedded clause analyzed with a tree diagram highlighting the narrative structure | 108 |
| 4.8 | 'Convergence' construction from a sequence in *Usagi Yojimbo* | 109 |

| | | |
|---|---|---|
| 5.1 | Inseparability between semantics and semiotics within social coherence of verbal meaning | 123 |
| 5.2 | Shattered plate: Saussurian fragmented language system attributed to Turkish political cartoons | 124 |
| 5.3 | Language as a circular connection between object, author, and reader forming an intertextual narrative | 130 |
| 7.1 | A strip from the 1955 *Suske en Wiske* album *De dolle musketiers* | 166 |
| 7.2 | Two panels from the 1960 *Jommeke* album *Purpere pillen* | 168 |
| 9.1 | Bug Blog – Lou Dobbs Quits CNN | 224 |
| 9.2 | Lopez! We need a new comic strip! | 227 |
| 9.3 | Cinco de Marcho | 228 |
| 9.4 | Self-hate mail | 229 |
| 9.5 | The Latino Vote | 230 |
| 9.6 | Viva Obama: Barack Obama as Emiliano Zapata | 230 |
| 9.7 | An early request for a birth certificate | 232 |
| 9.8 | Anglicizing Spanish-language names | 233 |
| 9.9 | Stuff Latinos Like #2 | 234 |
| 10.1 | Rocky calls New York | 246 |
| 10.2 | Rocky meets an American rooster | 247 |
| 10.3 | Rocky speaks of serious topics | 249 |
| 10.4 | Rocky resists the police | 252 |
| 10.5 | Rocky proves he knows about country life | 255 |
| 10.6 | Rocky attends a hip-hop music festival | 256 |
| 11.1 | Shamrock | 269 |
| 11.2 | Siryn in *X-Force* #58 | 286 |

# Acknowledgements

To the volume contributors: you took a chance on a project that was experimental and amorphous at the start, and I appreciate your hard work and perseverance. I am very happy with this collection, and I hope that for all of you it is worth the wait. Thanks also to Jill Lake at Palgrave Macmillan for all her help with copyediting and finalizing the script.

I am deeply indebted to several organizations, colleagues, and friends for their support and encouragement as I learn more about linguistics and comics. In 2003, I presented my very first comics and linguistics paper at Lavender Languages and Linguistics in Washington, DC. In subsequent years, both the International Comic Arts Forum and the Graphic Novels and Comics Conference have provided an intellectually stimulating and remarkably facilitative environment for thinking about comics in broad, cross-disciplinary ways. I especially appreciate refreshing, supportive conversations with colleagues such as Roy T. Cook, Cécile Danehy, Charles Forceville, Charles Hatfield, Ben Little, Simon Locke, Joan Ormond, Julia Round, Roger Sabin, and Qiana J. Whitted, among many others.

Closer to home, my research has been supported by the UNO English Department and College of Arts and Sciences. Time for reading and writing is very precious, and I am fortunate to work in an environment that promotes scholarship.

For food, drink, laughter, and love: Joanie Latchaw, Barbara Robins, Kathy Radosta, Crystal Edwards and Tom Sanchez, Sekhmet Ra Em Kht Maat, Eduardo Millán, Diane and Gary Grobeck, and Sara and Darin Jensen.

FRANK BRAMLETT

*Images*

We would like to extend a special thanks to all the artists and publishers who allowed us to use images in this collection. Details of publication and permissions for use in this book are given alongside the individual illustrations in the text.

# Notes on Contributors

**Kristy Beers Fägersten** is an Assistant Professor of English Linguistics at Södertörns högskola, Sweden. Her current research interests include the non-native use of English swear words and the discourse of Swedish contemporary newspaper comics.

**Eliezer Ben-Rafael** is Professor Emeritus in Sociology of the Tel-Aviv University. He is past president of the International Institute of Sociology, and past incumbent of the Weinberg Chair of Political Sociology. He has authored and edited numerous books about language, identity, ethnicity, the Israeli society, and globalization. His recent works include: *Is Israel One? Religion, Nationalism and Ethnicity Confounded* (2006), *Transnationalism: Diasporas and the Advent of a New (Dis)order* (2009) and *World Religions and Multiculturalism: A Dialectic Relation* (2010).

**Miriam Ben-Rafael** has taught French for many years and is an independent researcher in sociolinguistics. She has investigated and published in areas that include Hebrew-French language contact, sociolinguistic variation under the influence of globalization, the linguistic landscape of ethnic communities. Among her recent publications are 'The linguistic landscape of transnationalism: The divided heart of Europe' (2010), 'English in French comics' (2008), 'Language attrition and ideology: Two groups of immigrants in Israel' (2007).

**Frank Bramlett** is an associate professor at the University of Nebraska at Omaha. He has published research on discourse styles in interviews at a social service agency; conversation analysis and the short fiction of Raymond Carver; and the interaction of gender and anti-gay prejudice and its impact on social stigma. Most recently, his work on verbal camp in *The Rawhide Kid* appeared in the on-line comics journal *ImageText*.

**Carla Breidenbach**, a Chicana originally from California, is now an Assistant Professor of Spanish at the College of Charleston in South Carolina. Her areas of study include Chicano sociolinguistics, identity, ethnicity, and sexuality, racial humor, immigration, and pop culture linguistics. When not working, Carla can be found on the beach with her dogs.

**Neil Cohn** is a scholar of the cognition underlying the visual language of comics. He is the author of *Early Writings on Visual Language* and

*Meditations*, and the illustrator of *We the People* (with Thom Hartmann) and *A User's Guide to Thought and Meaning* (by Ray Jackendoff).

**Jill Hallett** is a graduate student in linguistics at the University of Illinois and a high school teacher in Chicago. Her research interests include sociolinguistics, specifically American and world Englishes, linguistic identity in literature, inner city pedagogical discourse, and second language acquisition. Her dissertation research analyzes teacher accommodation to student language.

**Richard W. Hallett** is the coordinator of the Linguistics Program at Northeastern Illinois University. During the 2009-2010 academic year, he was a Fulbright-Nehru Senior Scholar based in New Delhi researching the sociolinguistic aspects of the Incredible India tourism campaign. His research interests include the language of tourism, discourse analysis, world Englishes, and second language vocabulary acquisition. He is the co-author of *Official Tourism Websites: A Discourse Analysis Perspective* (2010).

**Gert Meesters** holds a PhD in Dutch Linguistics from the University of Leuven and teaches linguistics at the University of Liege, Belgium. In addition to his academic work, he has written for many years for the general press as a comics critic, currently for the Flemish news weekly *Knack*.

**Elisabeth Potsch** currently works as a library specialist at the University of Illinois at Urbana-Champaign, where she also attends the Graduate School of Library and Information Science. When not combing the bookstacks for comics, she enjoys working on her constructed language.

**Thecla Schiphorst** is a Media Artist/Designer and Faculty Member in the School of Interactive Arts and Technology at Simon Fraser University in Vancouver, Canada. Her background in performance and computing forms the basis for her research, which focuses on embodied interaction, sense-making, and the aesthetics of interaction.

**Veronika Tzankova** is a PhD student at the Communication & Culture program at York University in Toronto, Canada. Her background education in civil law was obtained in Turkey, where she spent seven years of her life exploring the Oriental culture and its influence on moral values and language. Her current interests include Islam and visual culture, Islam and virtual reality, and discrepancies in representation between physical reality and electronic environments in the context of Muslim societies.

**Shane Walshe** is a lecturer in the English Department at the University of Zurich, Switzerland. His primary research interests are perceptual dialectology, the depiction of varieties of English in popular culture, and the notion of linguistic stereotyping.

**Richard Watson Todd** has worked for nearly 20 years at King Mongkut's University of Technology Thonburi where he is Associate Professor and Head of the Centre for Research and Services in the School of Liberal Arts. He is the author of *Much Ado about English* and *Classroom Teaching Strategies*, and he has published numerous articles in the areas of text linguistics, computer-based analyses of language, and curriculum innovation.

**Robert F. Williams** is Associate Professor of Education at Lawrence University in Appleton, Wisconsin. He teaches courses in education, cognitive linguistics, distributed cognition, and gesture studies. His research explores conceptual and bodily aspects of meaning construction in everyday cognition and instruction.

# Introduction

*Frank Bramlett*

This collection of essays is a book for anybody who wants to know more about how language and comics go together. It begins to construct a space in which the scientific and literary study of language may productively combine with the burgeoning scholarship in comics.

## 0.1 'Language' as metaphor, language through linguistics

Many comics scholars who use the word 'language' in their writing tend to use it metaphorically, as a vessel to stand in for something other than a linguistic system. McCloud (1994: 47) helps set the stage for this approach: 'Words, pictures, and other icons are the vocabulary of the language called comics. A single unified language deserves a single, unified vocabulary. Without it, comics will continue to limp along as the 'bastard child' of words and pictures.' Of course, McCloud takes his cue from Eisner (1985: 8): 'comics employ a series of repetitive images and recognizable symbols. When these are used again and again to convey similar ideas, they become a language – a literary form, if you will. And it is this disciplined application that creates the "grammar" of Sequential Art.' On the one hand, Eisner's metaphor of 'the language of comics' is brilliant because it gives scholars and artists alike some common ground for discussing their research and their art. This metaphor has been used in a great deal of scholarship, serving in some form as the title or subtitle of books, journal articles, and conference papers, and these works proved to be very important for furthering comics scholarship. On the other hand, the metaphor facilitates the neglect of comics scholarship from a linguistic point of view.

Thus, in writing about the language of comics, Eisner and McCloud may have accidentally gotten in the way of understanding comics through the study of language itself: they may have interfered with the study of language *in* comics because they called for a language *of* comics. However, this is not to say that they utterly fail to address language in comics. Early in *Comics & Sequential Art*, Eisner (1985: 13) identifies the nature of word and image directly:

> 'Comics' deal with two major communicating devices, words and images. Admittedly, this is an arbitrary separation. But, since in the modern world of communication they are treated as independent disciplines it seems valid. Actually, they are derivatives of a single origin and in the skillful employment of words and images lies the expressive potential of the medium.

Even though Eisner equates the linguistic notion of language with the notion of a writing system, he later separates them appropriately. He demonstrates the non-visual nature of speech and the requirements of representing speech visually in three successive panels (p. 26). The character Eisner draws says 'My words cannot be seen!' – which is exactly correct. Spoken words cannot be seen; only words or utterances that are written or produced through sign language are visible. What comics artists do is use a writing system to represent speech, conversation, narration, and thoughts of the characters. Whether the writing system is an alphabet, syllabary, or logographic system, it is a visual representation of a cognitive/social construct, and writing plays an immensely important role in comics.

But the 'language of comics' metaphor seems to have inspired many other scholars in the quest to theorize the nature of the verbal and visual construct. Groensteen (1999[2007]) attempts to articulate those parts, relationships, and processes that form the system of comics, and he refers to this system as 'the language of the ninth art' (p. 23). (See also Bongco 2000; Varnum and Gibbons 2001; and Saraceni 2003.) Some, like Bongco, sense that the metaphor is problematic, however: 'A close look at comicbooks reveals an ingenious form, with a highly developed grammar and vocabulary based on a unique combination of verbal and visual elements. [...] Reading a comicbook is as a complex semiotic process [...]. The appreciation of [comics] is not possible without the recognition that its language and grammar consist of not one but two elements: words and images' (p. 46). The key word in Bongco's text for me is *semiotic*. It is absolutely true that comics consist of codes, and that the codes function in certain ways that make the relationship between visual and verbal possible.

Harvey (1996: 3), like other scholars, calls for 'a vocabulary and a critical perspective forged expressly in the image of the form,' though he seems to avoid calling this perspective a *language*. He agrees in spirit, however, and specifies 'an analysis of the verbal-visual blending' which may 'give us a way of approaching comics, of getting into the art form, and of seeing how it does what it does' (p. 4). Most of Harvey's analysis proceeds through the visual element rather than the verbal. He employs the metaphor of weaving, the warp and woof of a tapestry, to get at the medium's visual nature. The four threads Harvey identifies are: (1) narrative breakdown: panel units; (2) composition: elements inside a panel; (3) layout: arrangement of panels on a page; and (4) style: the way an artist handles pen, brush (1996: 9, 10). Clearly, an analysis of the visual representations of language alights squarely in these four categories. Harvey's call for a unified approach to understanding comics invokes those both before and after him, but his metaphor of weaving avoids confusing the linguistic material in comics with the systems of forms and functions, the semiotic of the visual-verbal construct. Like Harvey, Hatfield (2005: 37) attempts to bring focus to the tension of comics, the relationship between words on the one hand and images on the other. As an example, he gives readers a case study on Chris Ware's 'I Guess,' explaining that the artist 'experiments with a radically disjunctive form of verbal/visual play' in which the 'parallel verbal and pictorial narratives […] tell two different tales.'

One of the most sustained treatments of language in comics is David William Foster's (1989) book, *From Mafalda to Los Supermachos: Latin American Graphic Humor as Popular Culture*. His chapter on text production principles includes narrative patterns (p. 14), verbal images (p. 17), and disjunction (p. 20), among others. The analysis in the book is wide-ranging. For instance, Foster discusses linguistic and rhetorical devices in Fontanarrosa's *Las aventuras de Indoro Pereyra*, arguing that those language devices make up a highly exaggerated stylistic register spoken by characters in the strip, which 'becomes functionally significant for the ironic role it plays in the characterization of a rural way of life derided and degraded in spite of the vacuous myths of Argentine gauchomania' (p. 41). These agonistic ideologies – of the romanticized Argentine cowboy (*gaucho*) and the urban denigration of the rural ways of life – become instantiated by the linguistic play of the characters in the comic.

Foster also discusses language in another Argentinian comic, *Mafalda*, by Joaquin Salvador Lavador (pen-name 'Quino'):

> Quino's superb ear [finds] that special combination of pretentious verbosity and gritty urban turns of speech that characterize the dominant Buenos Aires sociodialect as represented in a mainline of

twentieth-century narrative and theater, and [zeroes] in on patterns of behavior that reveal significant underlying social [...] values. (1989: 55)

Foster's discussion stretches sometimes toward the physical representation of characters and the social and cultural values present in the comics, but it in general includes a concrete use of linguistic detail, ranging from vocabulary like slang and regionalisms to patterns of conversation among characters.

## 0.2 Sampling extant scholarship in linguistics and comics

Many different linguistic approaches are evident across the field of comics research. See Cohn (this volume) for a survey of cognitive research. Early twentieth-century scholars valued dialect research as well as what we now term *register variation*. One of the earliest scholarly articles in English on the linguistics of comics appears in the journal *American Speech* (Tysell 1935). It focuses on naming practices, nonstandard spelling, slang, and jargon in newspaper comic strips and examines those linguistic features that may indicate dialect or accent. Thirty years later, but also in *American Speech*, Malin (1965) published his essay on 'eye dialect' in *Li'l Abner*. This study focuses on spelling deviations that stray from 'prescriptive pronunciation norms' and phonetic spellings that reveal differences in education or social status among the characters (p. 230). Inge (1990) devotes a chapter in his book to a discussion of American English in comics, focusing in large measure on vocabulary.

Petersen (2007) discusses the function of sound in manga. In particular, he invents the term 'narrative erotics' to describe 'those moments when the narrative becomes embodied through a sensual presence' and 'create[s] an animated interior for the story to live within, allowing it to become more evocative and memorable' (p. 580). Salgueiro (2008) proposes a system for understanding the functions of synesthesia in comics and discusses the place of word (p. 590) and visible sound (p. 591), among other concepts.

Rosen (1995: 257) explores the way that the English language functions metaphorically in Spiegelman's *Maus*, arguing for example that the foreign quality of Vladek's English was designed to 'convey the foreignness of the Holocaust itself.' Barker (2009 [1989]: 199) explores semiotics in general, including language signs, to critique Angela McRobbie's 'definitive study of *Jackie*.' Rauch (2004) employs a Foucauldian approach to explaining power and control in the language of Morrison's *The Invisibles*. He argues that, '[d]espite all the power and control over the world that language has given us, ultimately it blinds us to higher truths' (p. 350). In another linguistic nod to Morrison, Manning (2008: 37) argues that the world of *The Invisibles* is 'bound together by language, and it ends in language.' Language in superhero comics, according to Devarenne (2008: 52), instantiates power, ideology, and hegemony via nationalism, and even though 'the American superhero genre is compatible with nationalist ideology in some respects, its vernacular linguistic format, restrained and regulated, both complies with this ideology and represents the potential for its subversion.'

Exploring lesbian language in comics, Queen (1997: 254) finds that comics authors 'not only draw on linguistic stereotypes in molding their characters but also combine various [stereotypical] linguistic features' from other stereotype categories, like heterosexual women and men. In writing about *The Rawhide Kid: Slap Leather*, Bramlett (2010) argues that the Rawhide Kid 'expresses same-sex sexual attraction […] and instantiates a celebration of queerness' by blending sociocultural codes of gun fighters, sissies, cowboys, and gay men. The Kid uses the linguistic discourse style of 'verbal camp as a means of both protecting and preserving other characters in the story yet simultaneously overturning heteromasculinist ideologies.'

## 0.3  The ways that language and comics go together

How a comic book artist renders words may be more or less realistic, more or less iconic, more or less abstract. This is the idea behind discussions of font choices: size, bold, or majuscule/minuscule letters. See Kannenberg (2001) for an extended discussion of font choice, etc. in Chris Ware's comics. See also Khordoc (2001) for a discussion of 'photostylistic elements' like balloons. Forceville (2005) explores cognitive models of anger in the Astérix album *La Zizanie* and how anger is represented visually. He argues that pictorial runes 'are not arbitrary signs, but signs metonymically motivated' by cognitive models (p. 74). Generalizing away from models of anger, Forceville (2011) catalogues all pictorial runes in *Tintin and the Picaros*, arguing 'that it is possible to assign each of these runes a more or less specific meaning, and that runes are to a considerable extent motivated signs' (p. 876). See also Forceville, Veale and Fayaerts (2010) for a quantitative analysis of balloonics.

An instructive example of how artists manipulate fonts, speech balloons, and narrative captions can be found in Mark Haven Britt's graphic novel *Full Color* (2007). Some of Britt's speech balloons are round or elliptical, some are square, some are angular and irregular; some speech occurs outside balloons altogether. Fonts for speech tend to be small, thin and light, and upper and lower case. When characters like Boom and David yell, the size of the words grows enormously and becomes bold and all-capital. Narrative captions are rendered in a variety of ways, in narrative boxes distributed in different parts of the panels (not always at the bottom or top), and the boxes are double-bordered. The use of these diverse representations of language index both character development and narrative arc in the text.

*Figure 0.1* Boom says 'no' to David. How language and image complement each other. Copyright Mark Haven Britt 2007. Used by permission.

In chapter one of *Full Color*, 'That's When Mister Nonsense Showed Up,' David stands naked in Boom's apartment, explaining to Boom how he spent his evening covered in Vaseline, trying to evade police officers (pp. 20–21):

David: 'Can you help me?'
Boom: No. No. I can't. I broke up with my girlfriend yesterday. An hour ago, I quit my job. I just couldn't fucking take it anymore. The drama, the politics, the nonsense. No fucking way. I'm making it right, David. I've given myself one day to get it all right or I'm going to kill myself. What do you think of that?
David: Chaos it is then. We'll need coffee.

As Boom articulates her refusal, her posture and gestures speak powerfully to her emotional state. In a three-quarter page panel, Boom is drawn below the waist as a single person with one pair of legs. From the waist up, the reader sees five arms and three heads, almost as if there are three Booms, one superimposed on the other. In addition, we see the reflection of her back in the mirror behind her. Boom's physical movement, suggested to be a kind of swaying from left to right above the waist, reflects the inner turmoil, the emotional chaos she experiences.

The content of Boom's words is divided over five speech balloons. The balloon on the far left is probably the first balloon in time; it has a tail pointed toward Boom's mouth. However, even though the other four balloons are connected to each other, none has a tail pointing toward Boom, so the reader has to infer that all the speech belongs to her. This is possible through attending to the content of the speech. While it is not unusual for a speaker's turns to be rendered in this way, the visual design of the language is significant in that it reflexively supports and is supported by the image representing Boom's movement. In a single panel, Boom's language and body movement parallel each other, but even more important, the verbal and the visual blend together to reinforce the inner turmoil Boom feels.

Arguably, all of Boom and David's adventures that night border on chaos: sexual harassment, fist fights, breaking-and-entering, kidnapping, and murder. The chaos of the narrative and the emotional chaos the characters feel are indexed in part by Britt's representations of speech and, of course, are triggered by David's use of the lexeme *chaos* in his response to Boom. It is this interweaving reliance of word and image that make *Full Color* such a powerful story, and it is these interdependencies that Hatfield identifies as the *tension* in comics and the verbal-visual blending that Harvey explores.

## 0.4 What this book will do for the reader

This volume contains chapters in which researchers blend linguistic scholarship with comics scholarship to help propel us into a deeper and clearer understanding of what language is from a disciplinary sense, what comics is from a disciplinary sense, and how productively the two fields contribute to each other. Some of the chapters look at language more than at the visual, some look at the visual more than language, and some tread in the space where the verbal and the visual come together. The first four chapters cohere in their attempts to peer into the minds of readers and artists, accessing linguistic and visual codes through cognitive linguistics, especially cognitive metaphor, but also how the metafunctions of systemic functional linguistics operate in creating humor.

**Elisabeth Potsch and Robert F. Williams** use cognitive linguistics as their starting point but analyze the graphic, i.e., non-verbal, representations of speed and direction lines in action comics. They analyze common visual conventions used in superhero comics and how these function conceptually to create the perception of whole action events. The analysis is based on two foundations of cognitive linguistics: image schemas and conceptual metaphor, relying on the idea that the way humans conceptualize motion events is structured by the SOURCE-PATH-GOAL image schema (Johnson 1987). An analysis of a single comic panel deciphers how the reader parses the visual cues to conceptualize the sequence of events. The authors conclude that conceptual mechanisms proposed by cognitive linguists as essential to meaning-making in language play an equivalent role in making meaning from the static images of comics.

**Richard Watson Todd** uses systemic functional linguistics to examine how humor arises in Gary Larson's single-panel *Far Side* cartoons. He argues that it occurs in part through the manipulation of the interpersonal metafunction of language. In particular, Watson Todd proposes that Larson's cartoons also involve creation of atmosphere through the uses of both the interactional metafunction in the caption and the textual metafunction to more closely link the caption and the cartoon. Using proven methodologies in humor research, Watson Todd relies on audience survey responses to test the linkages of the metafunctions and their relative weight in creating humor. While he does not question the centrality of semantic or ideational incongruity in creating humor, his findings highlight the importance of the interactional and textual metafunctions in making what may already be humorous even more so.

**Jill Hallett and Richard W. Hallett** write about representations of H1N1 (swine flu) in international political cartoons, especially those in

India and the United States. By building a corpus, they are able to see trends in political cartoons via the lens of cognitive metaphors. They situate their analysis by employing the concept of *topos* and locating the predominant cognitive metaphors expressed in the cartoons. Their analysis elucidates how fears are addressed through language and media cross-culturally. Further, they argue that these kinds of cartoons can play on associations and fears relevant to each particular culture. Their findings implicate the reader in the predictive nature of the image and text relationship.

**Neil Cohn** employs research in psycholinguistics and cognitive linguistics in an ambitious attempt to further articulate a theory of visual language. His chapter reviews previous work in cognitive linguistics and comics scholarship, and it advances his prior research in identifying structures and functions that arise from the mental capacity that humans have for creating comic art. The essence of Cohn's chapter is not to reinforce the notion of the language of comics. Instead, he argues that humans have the capacity to create visual language and that artists create their comics using visual language.

The remaining seven chapters take as their point of departure the notion that language is a sociocultural phenomenon, that speakers of languages – and characters in comics – engage in shaping the world around them through code choice and that the presence of a linguistic code plays an important role in shaping the sociocultural landscape.

**Veronika Tzankova and Thecla Schiphorst** analyze political cartoons from Turkey and trace artists' attempts to communicate with their readers on multiple levels. Artists in Turkey employ their linguistic resources to communicate different messages simultaneously, engendering enough ambiguity that they can speak the unspeakable, that they can express their opinions at the same time that they appear not to be doing so. Tzankova and Schiphorst discuss the use of Turkish grammar, especially inflectional suffixation and pronominals, as well as the use of English, and they conclude that these artists successfully resist the political atmosphere they live in despite powerful forces of censorship arrayed against them.

**Miriam Ben-Rafael and Eliezer Ben-Rafael** provide a vast corpus of French-language comics (*bandes dessinées*) and measure the extent to which the language in them remains essentially French despite its invasion by youth vernacular and English. Their investigation surveys a sample of popular comics of different epochs, among them, *Tintin*, *Astérix*, *Le Chat du Rabbin*, and *Les Schtroumpfs*. They find that even though English plays a significant role in French language comics, it is just one language among many, including Spanish, German, Russian, and more. Their findings indicate not that English harms the French language or *bandes dessinées* but in fact that English and many other languages play a central role in francophone youth culture.

**Gert Meesters** builds a corpus of Flemish comics to trace how grammatical and lexical features diverge and converge in two varieties of Dutch. His chapter describes the language situation in the Netherlands and Flanders, the Dutch-speaking region in Belgium. In recent centuries, Dutch from the Netherlands and Flemish Dutch have undergone different evolutions, resulting mostly in differences in pronunciation and vocabulary, much like British and American English. Using corpus linguistics methods, the study surveys two comics, *Suske en Wiske* by Willy Vandersteen and *Jommeke* by Jef Nys. The study determines that mainstream children's comics can be perfect witnesses of the evolution of language variants. Meesters concludes by discussing how comics research can benefit from a number of quantitative techniques adapted from corpus linguistics.

**Frank Bramlett** maps out the role of English varieties in *Afro Samurai* by Takashi Okazaki. Identity politics revolving around race and ethnicity in the US are often vexed by an equally powerful politics of language, and this relationship reveals itself in the characters of Afro Samurai and Ninja Ninja, in which the sociocultural codes of Japanese identity, samurai warrior culture, 1970s R&B, and twenty-first century hip-hop in the US blend together. Analyzing both the anime and the manga versions, Bramlett argues that the characters engage in linguistic and social struggle, both with each other and with the other characters. At the same time, linguistic and cultural codes vie for dominance, mirroring the social realities of race, ethnicity, stereotype, and forces of appropriation.

**Carla Breidenbach** examines from a discourse analysis perspective how racial and ethnic identity plays a significant role in the US comic strip *La Cucaracha* by Lalo Alcaraz. This chapter enumerates the linguistic tools Alcaraz uses to create and comment on Latino identities in the US, especially in light of political discussion about English as an official language, immigration policy, and racism. Among others, these tools include Spanish, English, Spanglish, and bilingual code switching. Breidenbach argues that Alcaraz extends the Chicano identity movement into his art by creating a 'pocho' identity for his characters and for himself as a Latino living in the US.

**Kristy Beers Fägersten** studies the interplay of English and Swedish in the comic strip *Rocky*, with particular attention paid to code switching and code crossing. Since Swedish is the dominant language of *Rocky*, switches to English are discursively significant, and not only reflect the in-group linguistic norms shared by the *Rocky* characters and their real-life counterparts, but can also reflect or even introduce a similar linguistic behavior among the wider Swedish reading public. The use of English among the *Rocky* characters reflects a stylistic choice, serving to reflect cultural alignment, in particular with the popular culture of the United States, notably hip-hop. Beers Fägersten argues that a familiarity with and affinity for US popular culture, and especially African American culture, is clearly valued by the comic strip characters, and the use of English is a hallmark of cultural appropriation.

The volume ends with an examination of language and identity in Marvel superhero comics. **Shane Walshe** surveys representations of Irish English and finds that comics artists rely on linguistic stereotype for the portrayal of speech by such heroes as Banshee, Siryn, Tom Cassidy, and Shamrock. Walshe begins his essay by discussing representations of the Irish people in popular culture, especially representations of how Irish people speak. Using a corpus compiled from Marvel comic books, he examines how accurate the portrayals of Irish speech in these publications really are. It compares the dialect and accent features rendered in the comic books with those which are generally regarded as being typical of spoken Irish English. In keeping with similar studies on representations of Irish English in films and television, Walshe also examines whether the writers rely predominantly on grammatical, lexical, or discourse features to create the impression of Irishness or whether accent carries the greater functional load.

## References

Barker, M. (2009[1989]) *Jackie* and the problem of romance. In J. Heer and K. Worcester (eds.) *A Comics Studies Reader*, pp. 190–206. Jackson: University Press of Mississippi.

Bongco, M. (2000) *Reading Comics: Language, Culture, and the Concept of the Superhero in Comic Books*. New York and London: Garland Publishing.

Bramlett, F. (2010) The confluence of heroism, sissyhood, and camp in *The Rawhide Kid: Slap Leather*. *ImageTexT: Interdisciplinary Comics Studies* 5 (1). Retrieved on 4 August 2010 from http://www.english.ufl.edu/imagetext/archives/v5_1/bramlett/

Britt, M. H. (2007) *Full Color*. Berkeley, CA: Image Comics.

Devarenne, N. (2008) 'A language heroically commensurate with his body': nationalism, fascism, and the language of the superhero comic. *International Journal of Comic Art* 10 (1): 48–54.

Eisner, W. (1985) *Comics & Sequential Art*. Expanded edition: print and computer. Tamarac, FL: Poorhouse Press.
Forceville, C. (2005) Visual representations of the idealized cognitive model of anger in the Asterix album *La Zizanie*. *Journal of Pragmatics* 37 (1): 69–88.
Forceville, C. (2011) Pictorial runes in *Tintin and the Picaros*. *Journal of Pragmatics* 43 (3): 875–90.
Forceville, C., Veale, T., and Feyaerts, K. (2010) Balloonics: the visuals of balloons in comics. In J. Goggin and D. Hassler-Forest (eds.) *The Rise and Reason of Comics and Graphic Literature: Critical Essays on the Form*, pp. 56–73. Jefferson, NC: McFarland & Co.
Foster, D. W. (1989) *From Mafalda to Los Supermachos: Latin American Graphic Humor as Popular Culture*. Boulder: L. Rienner.
Groensteen, T. (1999[2007]) *The System of Comics*. [Trans. B. Beaty and N. Nguyen, 2007.] Jackson: University Press of Mississippi.
Hatfield, C. (2005) *Alternative Comics: An Emerging Literature*. Jackson: University Press of Mississippi.
Harvey, R. (1996) *The Art of the Comic Book: An Aesthetic History*. Jackson: University Press of Mississippi.
Inge, M. T. (1990) *Comics as Culture*. Jackson: University Press of Mississippi.
Johnson, M. (1987) *The Body in the Mind: The Bodily Basis of Meaning, Imagination, and Reason*. Chicago: University of Chicago Press.
Kannenberg, Jr., G. (2001) The comics of Chris Ware. In R. Varnum and C. T. Gibbons (eds.) *The Language of Comics: Word and Image*, pp. 174–97. Jackson: University Press of Mississippi.
Khordoc, C. (2001) Comic book's soundtrack: visual sound effects in *Asterix*. In R. Varnum and C. T. Gibbons (eds.) *The Language of Comics: Word and Image*, pp. 156–73. Jackson: University Press of Mississippi.
Malin, S. D. (1965) Eye dialect in 'Li'l Abner.' *American Speech* 40 (3): 229–32.
Manning, S. (2008) Language and fiction in the creation of reality in *The Invisibles*. *International Journal of Comic Art* 10 (1): 32–8.
McCloud, S. (1994 [1993]) *Understanding Comics: The Invisible Art*. New York: HarperPerennial.
Petersen, R. S. (2007) The acoustics of manga: narrative erotics and the visual presence of sound. *International Journal of Comic Art* 9 (1): 578–90.
Queen, R. M. (1997) 'I don't speak spritch': locating lesbian language. In A. Livia and K. Hall (eds.) *Queerly Phrased: Language, Gender, and Sexuality*, pp. 233–56. Oxford: Oxford University Press.
Rauch, S. (2004) 'We have all been sentenced': language as means of control in Grant Morrison's *Invisibles*. *International Journal of Comic Art* 6 (2): 350–63.
Rosen, A. (1995) The language of survival: English as metaphor in Spiegelman's *Maus*. *Prooftexts* 15 (3): 249–62.
Salgueiro, J. (2008) Synesthesia and onomatopoeia in graphic literature. *International Journal of Comic Art* 10 (2): 581–97.
Saraceni, M. (2003) *The Language of Comics*. Intertext series. London and New York: Routledge.
Tysell, H. T. (1935) The English of the comic cartoons. *American Speech* 10 (1): 43–55.
Varnum, R. and Gibbons, C. T. (2001) *The Language of Comics: Word and Image*. Jackson: University Press of Mississippi.

# 1
# Image Schemas and Conceptual Metaphor in Action Comics

*Elisabeth Potsch and Robert F. Williams*[1]

## 1.1 Introduction

Comics is cinema without motion or sound. Like films, comics tell stories through a sequence of images. Likewise, they incorporate sound effects, spoken dialogue, and voice-over narration, all rendered as text so that the sound emerges in the reader's mind rather than from sound waves impinging on the ear. Unlike film, comics present images simultaneously, in durable form, rather than in rapid succession, to produce the illusion of moving images. This means that the comics reader must add motion and dynamics to the story conceptually, mentally animating the narrated events. The static, soundless nature of comics poses problems of representation for the comics artist and of interpretation for the comics reader. These problems are acute in the popular genre of superhero comics – with characters like Superman, Batman, Wonder Woman, Spider-Man, Green Lantern, Captain America, Iron Man, X-Men, and so on – where complex, fast-paced action is central to the story. How do the static images of action comics become dynamic events in the mind of the reader? What representational conventions prompt these interpretations, and what is the conceptual basis for these representations and their functions? The present chapter addresses these questions of depiction and meaning-making from the perspective of cognitive linguistics. Specifically, it draws upon studies of image schemas and conceptual metaphors to explain the conceptual basis for several key conventions for representing dynamic action in contemporary superhero comics, and it illustrates how these conventions function together through detailed analysis of a single comic panel depicting complex action.

## 1.2 From static images to dynamic events

When employed representationally, visual media such as photography and painting depict individual moments in time. Skilled photographers and artists capture precisely those key moments that, together with visual cues for context, imply whole events that are part of a larger narrative. Comics gain narrative power by presenting depicted moments in a visual array, where the reader's habituated strategy of reading (viewing) the images from left to right produces a succession of moments, and bridging inferences link these moments into a coherent story. In this way, comics substitute space for time (McCloud 2000: 2). Within that space, artists can manipulate the size, shape, and juxtaposition of panels to affect the consideration a reader gives to each part of the page, guiding the reader's selective attention to each depicted moment and generating a sense of pacing for the action. This level of dynamics suffices for simple drama or for the four-panel jokes that populate the comics page of daily newspapers, but for action comics like those of the superhero genre, this panel-to-panel pacing is too slow to render the experience of rapid, often simultaneous action, impacts and collisions, and other complex events. For action comics, motion and force are vital to the story and to the storyteller's art, and the artist must overcome the constraints of the medium to show movement and impact in the work, even and especially within the constraints of individual panels. While depiction of movement was rudimentary in early comics history, comics art has progressed over the last half century to render motion and force with greater vividness, maximizing the impact of panels that portray action. The composition of such panels will be our primary focus.

Comics tell stories through the juxtaposition of images and text for speech and sounds, but as Will Eisner observes in *Graphic Storytelling and Visual Narrative* (1996: 1–2), 'the major dependence for description and narration is on universally understood images.' To be universally understood (or nearly so), comics images employ conventions of representation that are readily interpretable by the reader and that prompt for the construction of particular meanings. In this respect, the images function somewhat like language. With respect to language, Talmy (2000) has argued that 'the basic function of grammatical forms is to structure conception while that of lexical forms is to provide conceptual content' (p. 24). Similarly, in comics images, the visual representational conventions structure conception while the rendered characters, objects, and settings provide conceptual content. Like grammatical forms, the visual conventions have a schematic quality and conceptual structuring function. There are some notable differences, of course. Language is sequential, segmented, and hierarchically structured, and it must use words (and gestures in spoken discourse) to prompt for the spatial composition of scenes as well as for their dynamic qualities. Because comics images directly depict the visual composition of scenes (albeit in two dimensions, using artistic conventions for representing visual perspective that are not the primary focus here), the grammar of comics consists not of patterned constructions for speaking but of an inventory of stylized symbols, a kind of 'visual shorthand' (Wolk 2007: 120) for depicting qualities of experience such as emotions and processes (changing relations through time). Comics artists draw from a collective pool of visual symbols (McCloud 1994: 128), symbols that rely on the reader's 'stored memory of experience' and that 'require readers to participate in the acting out of the story' (Eisner 1996: 17, 57). Readers use conceptual structure derived from embodied, cultural, and linguistic experience to construct the meaning of each panel of comics art, while they rely upon pragmatic abilities, such as bridging inferences, to string these panels together into a story.

In this chapter we explore the conceptual basis for three stylized symbols commonly used in action comics to represent the dynamics of events: ribbon paths, motion lines, and impact flashes. *Ribbon paths* indicate movement within a comic panel from one location to another, emphasizing the path traveled by the character or object that moves; the reader views this action from an observer's (a hidden spectator's) perspective. In the years since the creation of modern comics, artists have experimented with different ways of representing movement within a single panel, and ribbon paths are a modern stylization from earlier techniques. *Motion lines* emphasize motion without regard to path (to starting and ending locations) and are used to place the reader in the center of action as if moving with the characters, providing a participant's perspective to heighten the drama. *Impact flashes* represent the application or exchange of forces: they indicate sites where movements are initiated or terminated and, in particular, collisions between characters or objects in motion. In action comics today, these symbols are widespread – nearly universal – and readers understand them without explanation or study. To understand how, we need to examine the conceptual structures that readers employ to make meaning. Our analysis will focus on two aspects of meaning construction that have been the subject of extensive study in cognitive linguistics: image schemas and conceptual metaphors.

## 1.3 Image schemas and conceptual metaphors

An image schema is a mental representation of a pattern that people frequently encounter as embodied beings experiencing a physical world. As originally defined by Johnson (1987: xiv), an image schema is 'a recurring dynamic pattern of our perceptual interactions and motor programs that gives coherence and structure to our experience.' Common examples are PART-WHOLE, CENTER-PERIPHERY, SUPPORT, BALANCE, PROXIMITY, and CONTAINMENT. Image schemas related to motion include ANIMATE (OR SELF-) MOTION, CAUSED MOTION, and PATH (SOURCE-PATH-GOAL), while those related to force include COMPULSION, ATTRACTION, RESTRAINT, BLOCKAGE, and DIVERSION, among others. As the examples show, a specific image schema, such as SUPPORT or CAUSED MOTION, can integrate aspects of spatial organization with force or motion dynamics as these occur as patterned gestalts in our experience.

In her introduction to a volume on image schema research, Hampe (2005: 1–2) provides a succinct summary of the characteristics of image schemas as originally described by Johnson (1987) and Lakoff (1987):

- Image schemas are *directly meaningful* ('experiential'/'embodied'), *preconceptual* structures, which arise from, or are grounded in, human recurrent bodily movements through space, perceptual interactions, and ways of manipulating objects.
- Image schemas are highly *schematic* gestalts which capture the structural *contours* of sensory-motor experience, integrating information from multiple modalities.
- Image schemas exist as *continuous* and *analogue* patterns *beneath* conscious awareness, prior to and independently of other concepts.
- As gestalts, image schemas are both *internally structured*, i.e., made up of very few related parts, and highly *flexible*. This flexibility becomes manifest in the numerous transformations they undergo in various experiential contexts, all of which are closely related to perceptual (gestalt) principles. (pp. 1–2, Hampe's emphasis)

The 'image' portion of the term 'image schema' refers not just to visual perception but to 'all types of sensory-perceptual experience' (Evans and Green 2006: 179), including visual, auditory, haptic (touch), and vestibular (balance/movement), all of which generate what psychologists call 'images' in the mind. The 'schema' portion of the term is meant to distinguish image schemas from rich visual images: what an image schema describes is not a picture in the mind's eye but a schematized pattern that recurs in such images and that gives them their meaningful (relational/processual) structure. Image schemas are an embodied, emergent alternative to an innate mental calculus, language of thought, or other source of propositional structure rooted in disembodied logic or universal rationality.

An example that illustrates this point is the UP-DOWN schema described by Johnson (1987) and discussed in Evans and Green (2006: 178). From a purely logical point of view, UP and DOWN are merely opposite directions along a vertical axis, but from an embodied point of view, they are experienced quite differently. Unsupported objects fall downward while stationary objects require support to maintain their elevation and rising objects must be propelled upward by an applied force. For embodied beings in a world with gravity, space and force are entwined, so that we experience the vertical axis as functionally asymmetric. This asymmetry structures the way we perceive and conceptualize motion events, eliciting surprise when something appears to be inconsistent with this pattern. The example of UP-DOWN shows how image schemas become associated with 'broad classes of concepts or experiences' (Grady 2005: 36), providing what cognitive linguists consider to be the embodied foundation for the human conceptual system.

The functional asymmetry of UP-DOWN inheres in other conceptual domains via conceptual metaphor, a fixed set of correspondences or 'mappings' across domains that enables us to 'conceptualize one mental domain in terms of another' (Lakoff 1993: 203). A common example is the conceptual metaphor MORE IS UP. In everyday experience, adding items to a pile makes the pile higher and adding liquid to a container makes the level rise; these directly perceived correspondences are the basis for a mapping that can be exploited in non-spatial domains of experience, such as economics, in which we can say 'prices are rising' or 'wages are falling' though no actual motion is present. Here we conceptualize increases or decreases in quantity as movements upward or downward along a vertical axis, an axis on which things fall naturally unless supported or boosted upward by an applied force. As human beings, we stand and walk upright, with our head at the top, and we maintain this posture through alertness, wellness, and effort. These experiential associations provide the basis for a series of metaphors that exploit the asymmetry of the vertical axis, including HAPPY IS UP ('My spirits rose' / 'I'm feeling down'), CONSCIOUS IS UP ('I'm waking up' / 'He sank into a coma'), HEALTH AND LIFE ARE UP ('He rose from the dead' / 'He fell ill'), and CONTROL IS UP ('I'm on top of the situation' / 'It's under [my] control'). Other metaphors that derive from the orientation of the human body and positive associations with verticality include STATUS IS UP ('She rose to the top' / 'He fell from power'), VIRTUE IS UP ('She has high standards' / 'That was a low thing to do'), and, quite generally, GOOD IS UP ('Things are looking up' / 'Things are at an all-time low'), among others described in Lakoff and Johnson (1980: 14–17). The systematicity of these metaphors is no accident: all incorporate the functional asymmetry of the UP-DOWN image schema with mappings (patterns of correspondences or neural connections) across domains of experience. These mappings preserve image-schematic structure (Lakoff 1993: 215), equipping us to conceptualize abstract domains like economics in terms of concrete experiences like objects rising or falling. The combination of image-schematic structure and conceptual metaphor makes it possible for the entire conceptual system to be grounded, directly or indirectly, in embodied experience.

Cross-domain mappings not only link the abstract with the concrete, but they also link different domains of sensory experience (Kogan *et al.* 1980: 1). Some metaphors conflate the senses through a kind of synesthesia, equating one sense with another, so that a person can 'look sharp' or wear a 'loud shirt,' a musical note can 'sound flat,' and a food can 'taste dull.' These metaphorical expressions characterize sensations cross-modally, providing apt descriptions where words might otherwise fail us. Comics use visual cues in a similar way, exploiting synesthetic mappings in the conceptual system to make visual symbols stand for other sense perceptions. In a medium that can portray only pictorial or textual information, the ability to map one type of sensory perception onto another is invaluable. Artists may use bright colors in onomatopoeic sound effects and large bold letters for loud noises or shouting, where the shape and scale of letters on the page represents the quality and magnitude of the sound as it would be perceived by the auditory system. This metaphor based on conflated sensory perceptions has a wide range of applicability in the world of comics (McCloud 1994: 128). In action comics, visual representations of collisions combine sensory conflation (a primary form of metaphor) with image-schematic structure to render the images interpretable as dynamic happenings in the mind of the reader.

The example of representing sound magnitude by letter scale illustrates another key feature of metaphor in comics: its multimodality. While Lakoff and Johnson (1980) identified conceptual metaphors based on patterns in language, metaphorical expressions in comics – the means through which conceptual metaphors are expressed – can consist of words, images, or (especially) both in combination. The multimodality of conceptual metaphor has been noted by comics artists as well as by metaphor researchers. In *The Language of Comics*, Varnum and Gibbons (2001: xi) write: 'In comics, words take on some of the properties of pictures, and conversely, pictures take on some of the properties of words.... Comics is a system of signification in which words and pictures are perceived in much the same way.' McCloud makes a similar observation in *Understanding Comics* (1994: 128): 'Not really a picture anymore, these lines are more a visual metaphor – a symbol. And symbols are the basis of language!' In a recent academic volume on multimodal metaphor, Yus (2009: 167–68) argues that the interpretation of visual metaphor 'does not differ substantially' from the interpretation of verbal metaphor: the initial perception delivers information concerning a subject which the reader must subsequently interpret through encyclopedic knowledge of the subject or through the subject's associated metonymic relationships. From a cognitive linguistic point of view, meaning is conceptualization, so language, gesture, image, and social action all engage common conceptual structures and operations in the act of meaning creation, which is not to deny differences in the format, patterns, affordances, and apprehension of these different modes of expression and the roles they play. See, for example, the contrasts between linguistic and imagistic realizations of metaphor described by Forceville (2008).

With regard to motion events, early evidence of the metaphorical nature of motion representations in comics comes from experiments by Kennedy (1982) on the interpretation of speed lines, which are parallel black lines drawn behind moving figures to represent movement at different speeds. As with sound effects, we find again a conceptual metaphor linking size to magnitude: longer lines represent faster motion. Kennedy found that hearing children, who have greater exposure to metaphor in language, more readily understand speed lines as symbolic of motion than do deaf children. Kennedy argues that the children's exposure to metaphor correlates with their understanding of the visual motion symbols because those symbols are metaphorical in nature (Kennedy 1982: 593). From a cognitive linguistics point of view, we argue that metaphor resides primarily in thought, i.e., in conceptualizing one domain in terms of another, but that experience with metaphorical expressions, primarily linguistic but also pictorial, facilitates the interpretation of symbols that rely upon metaphorical mappings for their intended meaning.

With this brief introduction to some fundamental concepts in cognitive linguistics, we turn now to analyzing specific conventions in action comics for visually representing motion and force events: ribbon paths, motion lines, and impact flashes. Our focus here is on representing action within a single panel – a static image – such that the reader can interpret the dynamics of the depicted event. Once these conventions have been explicated, we examine how they function together, with time and pacing, to render the larger-than-life action familiar to fans of superhero comics.

## 1.4 Ribbon paths for movement

The rapid pace and drama of action comics demands that action events unfold in a single panel or short series of panels, which presents an immediate problem of depicting characters' movements as they interact. Comics artists have experimented with various ways of exhibiting movement since the medium's rise in popularity (McCloud 1994: 110). A character's movement through the space of the panel could be depicted by a series of drawings showing the character in different poses reflecting its changing configuration as it moves; this would create an effect reminiscent of Marcel Duchamp's famous painting *Nude Descending a Staircase*. While comics artists occasionally use this technique to depict high-speed actions in rapid sequence (for Superman or the Flash, for example), the technique fails as a general means of depicting motion because of its inefficiency (due to repeated drawing) and because it clutters the panel, obscuring the other contents of the scene. A more economical approach is to distill the visual representation of motion to its essential elements: those that depict the basic image-schematic structure of the motion event with just enough visual perspective to add three-dimensionality to the interpretation of motion.

The elegant solution to be described below is a nearly direct depiction of SOURCE-PATH-GOAL image-schematic structure. The SOURCE-PATH-GOAL image schema is the basic conceptual structure of a motion event: a moving object (which cognitive linguists call the *trajector*) begins its motion at one location (the source), travels through a series of contiguous locations in space (the path), and ends its motion at another location (the goal). At any given moment, the trajector occupies some position along the path from source to goal. In our everyday experience, we frequently travel along real physical paths, such as sidewalks, as we travel to a destination. Real, visible paths can also be formed by our movements through the world, as when a boat leaves behind a wake or a vehicle leaves ruts in the mud (Kennedy 1982: 592). Conceptually, we form a path whenever we move through space, even when no physical trace of the path remains; we can, for example, retrace our steps across a room despite the fact that there is no discernible difference between the parts of the floor we crossed and those we did not. We can visualize the path because it is conceptually real: it is the route we traveled between two locations. Lakoff and Johnson (1980: 90) argue that, conceptually, A JOURNEY DEFINES A PATH and THE PATH OF A JOURNEY IS A SURFACE. In their words, 'paths are conceived of as surfaces (think of a carpet unrolling as you go along, thus creating a path behind you).' This elemental structure of a journey along a path is the basis for many conceptual metaphors, including LIFE IS A JOURNEY, A CAREER IS A JOURNEY, A RELATIONSHIP IS A SHARED JOURNEY, and so on, which are reflected in the typical ways we talk about these phenomena.

For our purposes, the issue is not so much metaphorical paths as it is how to depict the basic conceptual structure of literal, though fictional, movement events in still images. Here the answer is to reify the SOURCE-PATH-GOAL image-schematic structure in the visual representation: in other words, to draw the path defined by the journey of the object in motion. One of the most recognizable examples of this approach is the dotted line path used repeatedly by Bil Keane in *Family Circus* to depict young Billy's circuitous route of travel through a complex visual scene; a sample comic panel is shown in Figure 1.1. Here Billy is the trajector, and viewers have no trouble interpreting the dotted line as Billy's path

*Figure 1.1* Billy's dotted line path in Bil Keane's *Family Circus*. From Bil Keane, *The Family Circus Memories*, Ballantine Books, 1989. Used with permission. FAMILYCIRCUS © 1986 BIL KEANE, INC. KING FEATURES SYNDICATE.

of travel from his starting point (the source, marked here by an 'x') and his present position. In this example the shape of the dotted line path also depicts certain aspects of the manner of motion, as Billy has apparently jumped on or over many objects (a bed, a trashcan, a football, and so on), circumnavigated others (a potted plant), and even climbed a tree, all in contrast with his mother's request for direct, goal-directed action, thereby producing the humor of the scene. The image is static; it gains significance from the reader visually tracing the line of Billy's path and interpreting the various events that appear to have happened along the way. While this provides a pleasant diversion in a Sunday comic, the tracing of a dotted line path proceeds at far too slow a pace for the high-speed action of superhero comics.

In action comics, the preferred way to depict SOURCE-PATH-GOAL image-schematic structure is to draw a ribbon path behind the object in motion, as shown by the example in Figure 1.2. A ribbon path is a swath of light color (white, yellow, or the predominant color of the moving object) edged by lines that diverge or converge, taking advantage of visual perspective to add apparent depth to the depicted motion. The drawn path looks like a segment of ribbon oriented horizontally (or sometimes tipped to align with the long axis of the trajectory), depicting precisely the extended flat surface identified by Lakoff and Johnson as the path defined by a journey. Unlike Billy's dotted line path on the ground, ribbon paths commonly depict objects swinging or flying through the air, so the path appears as a strand of ribbon arcing through space where no path would normally be visible. Readers have no trouble interpreting a ribbon path drawn 'in empty air, rather than on an actual surface' as standing for the virtual or conceptual path traversed by the drawn object (Kennedy 1982: 593). In particular, readers understand that the ribbon indicates the path the object has already traversed (past tense) because the SOURCE-PATH-GOAL schema implies that the depicted object, drawn in its present position, must 'already have been at the source and path locations' (Dodge and Lakoff 2005: 59). The juncture of space, motion, and time, those elements which are inseparable in the physical world, helps the artist introduce an impression of the passage of time into the comics panel (a topic explored in a later section). The thin lines and continuous swath of color also provide a sense of smooth, rapid motion – fast, fluid visual scanning – that gives speed to the action of the scene. The conceptual path creates the illusion of movement, as though the panel were a moment frozen in time, a snapshot of an object in motion. The drawn path represents a concrete, visible form of the idea that there is motion in such images. In this way, the

*Image Schemas and Conceptual Metaphor* 23

*Figure 1.2* Example of a ribbon path. From *Green Lantern 80-Page Giant #2*, DC Comics, 1999. Used with permission.

artist tricks the reader into concluding that time passed as the character 'moved' through the conceptualized space.

Artists who create action comics draw a traveled path as a visible surface, a ribbon path, in depictions of environments that would not ordinarily exhibit such paths, directly portraying this essential but invisible aspect of motion. This might seem like a self-evident way to depict movement, but that is only because we as human beings with bodily experience in the physical world of moving objects have the necessary patterns in our minds, the image schemas that structure our conceptualization, to enable us to look at a stripe of color drawn on paper and interpret it as an object's journey through space and time.

## 1.5 Motion lines for participant viewpoint

Every comics panel depicts its scene from a particular vantage point. The reader is typically positioned as a viewer outside the action, viewing it as a kind of hidden spectator, whether near or far. Occasionally, the reader is positioned inside the action for startling effect, viewing it as if somehow a co-participant. Comics artists manipulate point of view to shape the reader's experience of the events and degree of emotional engagement. While the outside perspective (hidden spectator viewpoint) is pervasive in all comics, the inside perspective (participant viewpoint) appears in action comics at dramatic moments, drawing the viewer into the action. In these situations, comics artists use motion lines, thin lines radiating from a central point that is the source (or goal) of movement, to simulate the effect of optic flow: the expansion or contraction of the visual scene as the observer moves toward or away from the focal center. Figure 1.3 provides an example of how motion lines create the effect of a character oriented directly toward the reader and moving with the reader through the space, which flows inward toward the source of motion. For this type of effect, a ribbon path would fail as a representational device: it would be obscured by the character's body or, for motion away from the viewer, would itself obscure the body in motion. On the other hand, the complete omission of motion

*Figure 1.3* Example of motion lines. From *Ultimate Spider-Man #81*, Marvel Comics, 2005. Used with permission.

symbols would render an apparently static scene rather than a motion event. Motion lines add the dynamic of motion in the z-axis without representing the path structure of a complete movement.

Here it is worth returning to the comparison of comics to cinema to consider similarities in the depiction of motion events. In the typical depiction of movement, a comics panel employs a film-like composition, with the trajector's movement carrying it from one viewable location in the panel to another. This is the planar component of the object's trajectory. Depth of motion out of or into the plane is suggested by the object's increased or reduced size in relation to other recognizable objects in the panel, making it appear nearer or more distant, and by the tapering or spreading of the lines outlining the ribbon path, emphasizing visual perspective. In Figure 1.2, for example, the trajector is drawn quite small with a noticeably tapering ribbon path, making it appear to have receded far into the distance. In contrast, panels with motion lines, such as the example shown in Figure 1.3, deviate from theatrical convention by breaking the 'fourth wall': orienting the moving object directly toward (or away from) the viewer, with both appearing to move together through space; this is similar to the cinematic effect created by the camera moving with the hero through the setting. The excitement elicited by this apparent joint motion comes at a cost: it tends to disrupt the hidden observer/spectator viewpoint so important to the voyeuristic pleasure of cinema.

Panels containing motion lines elicit a stronger first-person perspective, a perspective that replicates the point of view of a person directly in the middle of the action. Motion lines encourage the reader to '[call] up personal references to action – blurred mental pictures of objects in motion' (Taylor 2001: 46) or, more properly, of the background scenery in motion when focusing on an object moving with the viewer through space. In comics, motion lines lead the reader's eye to the focal object, precluding perception of a background. In these respects, motion lines are a visual metaphor for a first-person embodied perspective, supporting the link between real-world experiences of motion and their depiction on the comics page, thus 'imitating [and] exaggerating reality' (Eisner 1996: 1–2). The reader is placed momentarily in the center of the action, before returning to the more comfortable position as hidden spectator of the unfolding events.

## 1.6  Impact flashes for force events

Up to this point we have discussed motion events without regard to the forces that propel objects into motion or deflect objects already in motion. To add force dynamics (Talmy 2000) to their representations, artists employ two devices: impact flashes and sound effects rendered as text in exaggerated typefaces. An example of an impact flash is shown in Figure 1.4. Here a 'flash' or spot of bright color with radiating points marks a collision between objects. Flashes are also often used to mark the source of movement: the place where a ribbon path originates when force has been exerted or applied to launch an object into motion. In both uses, flashes mark the sites of force-dynamic events, bursts of energy that initiate or modify movements through space.

Together, ribbon paths and impact flashes create a visual map for an entire action event. Flashes capture visual attention while ribbon paths lead the eye across the page, so the reader's own eye movements produce a sense of motion in the art. Time and cause-and-effect are compressed such that a single still image with ribbon paths and flashes comes to represent a rapid sequence of connected events that lead directly to the depicted moment, the endpoint, captured in the art of the panel. As the reader's eye is attracted to flashes and scans along paths, the reader conceives of a sequence of happenings, creating dynamic action from static art.

*Figure 1.4* Example of an impact flash. From *Ultimate Spider-Man #84*, Marvel Comics, 2005. Used with permission.

Conventionally, the visual representation of an impact in action comics is straightforward: a sunburst-shaped bright spot in the image appears to be a flash of light captured at the moment of bursting into radiation. From a conceptual point of view, however, a mystery emerges: just how is the depiction of a flash of light (a visual phenomenon) so readily interpreted as a physical impact (a non-visual event)? Partly, the answer is learned representational convention, but there is naturalness to the mapping that needs a conceptual explanation. We argue that the impact flash functions so effectively because it is based on a primary conceptual metaphor, that is, a mapping between different modalities of sensory experience. The bright spot on the page depicts a flash of light, a visual perception of bursting energy that is associated with perceptions of bursting energy in other sensory modalities: pressure impacting the body and a percussive sound in the ear. An explosion is a prototypical example of this kind of burst: a flash of light, felt pressure, and a bang. A collision provides pressure and a bang without a flash of light. Because comics are still and silent (audible only through text), the visual flash metaphorically invokes the embodied experience of sudden percussive force. The temporal dynamics of a visual flash – sudden onset, brief high energy or power, and immediate release – correspond to the dynamics of a felt or heard impact, so that, conceptually speaking, the image-schematic structure of a flash structures the reader's conceptualization of the force-dynamic event. The characteristic sound associated with the substance of the colliding objects is rendered onomatopoetically in text next to the flash, completing the sensory image of impact. In Figure 1.4, the text reads 'FUNK,' representing the sound of pointed metal becoming lodged in wood.

Evidence for this cross-modal mapping appears in the scalar rendering of intensity: a more intense (higher-force) collision is associated with a brighter flash and a louder sound. In general, bright colors and loud noises are highly stimulating, whereas dark colors and soft noises are less stimulating (Gibbs and Colston 1995: 361). The metaphoric mapping preserves this directionality, so that a brightly colored shirt is 'loud' while dark colors are 'muted.' On the comics page, an impact flash is drawn larger to represent a higher-force collision while the text for the sound effect is written larger to represent a higher-volume noise. This invokes the familiar metaphoric mapping linking intensity to physical size: MORE IS BIGGER. A larger flash is a more powerful exertion or exchange of force, while a larger font size is a louder sound. The mapping is so natural that we hardly discern its metaphoric nature.

## 1.7 Time and pacing of action

In comics, time is elastic: it can be stretched or compressed for dramatic effect in rendering the events of the story. Pacing emerges partly in the reader's experience of taking in panel after panel, so artists can exert control over pacing through the size and placement of panels on the comics page, affecting how readers shift their gaze from image to image. Artists can also manipulate the contents of a series of panels so that they appear to depict happenings in a sequence, happenings separated by time or space (often with a textual cue such as 'later...' or 'meanwhile...'), closely spaced moments in a single happening (like frames of slow motion), or even a single moment rendered from multiple vantage points or in increasing close-up or pull-back. The combination of gaze-shifting from panel to panel and apparent temporal spacing of moments depicted in subsequent images gives the reader an experience of pacing that is more dramatic than real, including fast motion, slow motion, frozen moments, and leaps in time. Here again, comics show a kinship to cinema: the composition of images, shifts in vantage point, and temporal manipulations are key visual techniques of film, but in comics the reader retains greater control over pacing by deciding when and where to shift gaze to take in new information and how quickly to move from panel to panel. While the frames of a film are displayed to the viewer sequentially at a fixed pace in a single location, the panels in a comic book are available to the viewer simultaneously, spread across the page, with the sequence and pace determined by the action of reading. The reader has the ability to 'roam, to peek at the ending, or dwell' on a particular image (Eisner 1985: 71). Part of the joy of reading comics is exerting control over how one experiences the story.

More relevant to our discussion is how comics artists manipulate the experience of time within a single panel. On a comics page, each panel represents a certain length of time within the story. Janson (2002) argues that readers interpret similarly-sized panels as representing similar lengths of time and that any action within the panels is read as occurring within the same general amount of time (p. 111). As McCloud observes in *Understanding Comics* (1994: 99), time and space tend to be 'defined more by the contents of the panel than by the panel itself.' The within-panel experience of time is affected by dialogue or monologue rendered as text which must be read, by sound effects taking the form of onomatopoetic words placed near the objects meant to produce the sounds, and by symbols representing motion (McCloud 1994: 110). Speech and other sounds in real life are perceived over time, so readers interpret the reading of text and sound effects in a panel as representing a comparable amount of time passing in the depicted story. In the absence of speech or sound effects, however, the within-panel sense of time passing comes primarily from the depiction of movement, which is our focus in the present chapter.

In the physical world, every movement takes place through time, so time must pass during the course of any movement (Eisner 1985: 25). Motion and time are conceptually linked: whenever we conceive of motion, we conceive of time passing. Ribbon paths and impact flashes not only help the reader decipher what type of event produced the illustrated moment (a punch, a kick, or a throw, for example), they also instill time into the image. Interpreting an object as having moved along a ribbon path to its present position entails interpreting a corresponding amount of time as having passed in leading to that moment. How much time depends on the reader's encyclopedic knowledge of different types of events; a bullet flies faster than a bird, for example. Relative time durations are also reinforced by visual clues: a thin, straight path appears faster than a curved or meandering path; longer or shorter speed lines trailing from a fast-moving object suggest faster or slower speed; and so on. These guide the interpretation of speed in the depicted motion and therefore the sense of a certain interval of time passing. Even impact flashes convey temporal information. In the real world, a flash of light, impact, or bang is sudden and brief. A reader viewing an impact flash interprets it as standing for a fraction of a second in the action. Thus, a comics reader simultaneously interprets how the action unfolds and how the time passes. Visual devices like ribbon paths and impact flashes add temporal information to the image, shaping readers' impressions of the internal structure of portrayed events, whether action happens quickly or slowly, and how much time passes in the space of a single panel.

In an action sequence portrayed across a series of panels, such as a fight between a hero and a villain, the amount of time understood to pass for the sequence depends on how many movements are depicted, while the pacing of action depends largely on how the depictions of movement are spread across the panels. One movement per panel tends to be read as a metered, steady pace of action; this approach was standard in the early days of comics, when each panel portrayed a single action in the sequence of the story. Today, artists achieve a more dynamic, faster-paced sense of action by layering multiple movements in a single panel. If one panel of equivalent size to another represents roughly the same amount of time, several movements in one panel seem to happen more quickly than one movement per panel. This is an advantage in action comics, as a fast-paced battle is more exciting and engaging to the reader. It does, however, present the reader with a greater conceptual challenge: how to assemble the multiple depictions of movement into a meaningful action event within the context of the story. The movements depicted in a panel might be simultaneous, overlapping, or sequential, and they might be independent or interdependent. Determining their sequence and relations requires the reader to draw on knowledge of action types, durations, and interrelations to add CAUSE-EFFECT structure to the conceptualization. This, together with the image-schematic structure provided by ribbon paths and impact flashes, equips the reader to build a coherent understanding of what the panel portrays. In the next section, we illustrate this idea with analysis of a single panel portraying complex action.

## 1.8 Interpreting the action in a comic panel

As an example of multiple symbols in a single image, consider this panel (Figure 1.5) from *The Brave and the Bold #13*, published by DC Comics. Readers easily understand the sequence of events represented here despite the complexity of the composition and its layering of symbols. All of the motion symbols appear concurrently and are simultaneously available to perception, yet readers interpret the movements as occurring one after another through time. Because so many movements are drawn within the same panel, the reader deciphers the movements as occurring in rapid succession over a brief interval. How do readers parse this complex image to produce the intended meaning?

A cursory evaluation of the panel reveals the events as the artist likely intended them to be interpreted: Batman throws a batarang

*Figure 1.5* A panel depicting complex action. From *The Brave and the Bold #13*, DC Comics, 2008. Used with permission.

across the space between himself and the android, and the android blocks the batarang with his sword. In order to recognize the nature of the events and understand their sequence, the reader must do several things together: classify the actions depicted (as throwing an object or swinging a sword), extract image-schematic structure from the visual symbols (ribbon paths and impact flashes) to simulate paths of motion and force-dynamic interactions, and add CAUSE-EFFECT structure to conceptualize the sequence of connected events.

Interpretation of the panel begins with the reader's encyclopedic knowledge. Reader knowledge includes familiarity with objects in the world, like the window, the cape, the sword, and their properties; familiarity with actions, such as throwing an object or swinging a stick or sword; and familiarity with characteristic sounds and their associations, such as the sound of the word 'KTANG' emulating the sound of metal striking metal. Readers of action comics also have familiarity with comics in general and with the superhero genre in particular, including its stereotypical characters, storylines, and representational conventions. Additionally, readers of Batman comics will already know much about this superhero and his history, personality, and behavior, and readers of the present comic will know the events leading up to the depicted moment and thus have expectations about what will happen next. All of this knowledge shapes the construction of a particular meaning from this panel, yet what remains to be added is the image-schematic structure of the cognitive representation: the conceptual structure needed to support mental simulation of the action.

Paths of motion and force dynamics form the conceptual basis for piecing together the events of the panel. Here a combination of ribbon paths depicts four distinct phases of movement, and the reader must employ SOURCE-PATH-GOAL image-schematic structure to understand the illustration, orientation, and direction of each. By visually tracing these paths, the reader's scanning creates a dynamic sense of motion. The composition of the panel helps the reader determine the sequence of actions by taking advantage of the reader's entrenched habit of reading from left to right. Beginning on the left, the reader first encounters the ribbon path that symbolizes the swinging of Batman's hand and arm from the top left corner through an arc toward the bottom of the panel. Because the ribbon path is narrower near the top of the panel than at the bottom, the reader understands Batman's hand as having moved from the background into the foreground, closer to the reader's vantage point. Batman's arm is depicted in a position near the end of the movement, after having released the batarang (whose path is discussed below), this being the critical element that defines the movement as a throwing action. The ribbon path provides the source, a nearly complete path, and the direction of movement, with the ribbon itself marking the portion of the path already traversed, and with the object at the end of the ribbon, Batman's hand, marked as the trajector (the object in motion). Without actually moving, the image makes Batman's arm appear to have swung around his body, flinging the batarang away from him.

After tracing this ribbon path, the reader's eye then follows the batarang's path across the panel. The orientation of this second ribbon path with respect to the first indicates that the batarang started to travel away from Batman while his arm was in mid-motion, consistent with a throwing action. Interestingly, though, the ribbon path for the batarang does not begin at the ribbon path for Batman's swinging hand, which must have released it on its flight; instead, the ribbon path for the batarang starts closer to Batman's body, appearing to cross the path for Batman's hand. This apparent logical inconsistency provides two conceptual advantages. First, it makes use of the Gestalt principle of continuation to visually separate the paths, helping the reader see two distinct paths crossing rather than a unitary object with branching arms. Second, it uses proximity (incorporating the PROXIMITY image schema) to encourage the viewer to see Batman as the originator or source of the batarang's motion. This example clearly accentuates the conceptual function of ribbon paths: rather than functioning as objects in the scene, they supply image-schematic structure to guide the reader's conceptualization.

The ribbon path for the batarang moves horizontally across the panel in the familiar rightward reading direction, so that the reader's eyes scan smoothly and effortlessly across the page, creating a sensation of speed, until they encounter the jagged impact flash where the batarang collides with the sword. Here the brightly colored flash suggests a forceful impact, using the primary conceptual metaphor described above, while the sound effect (the text 'KTANG' read subvocally, with the enlarged 'A' expanding the central vowel sound) provides the onomatopoetic sensation of metal striking metal. The ribbon path deflects, as shown by the upward shift in the angle of the path and by the change from tapering to expanding outlines indicating a shift in motion toward the viewer. The angle of deflection is consistent with the reader's experience of real-world moving objects colliding with one another, so the artist avoids violating the reader's expectations. Following the new direction of the path leads the reader's eye directly to the batarang itself, the object in motion (the trajector) which becomes visibly identifiable for the first time just before it exits the frame, an implication of continuing motion. Behind the batarang, the swinging of the sword into blocking position is represented by a latticed ribbon path drawn in silver to match the sword; this effectively conveys the movement of this elongated trajector without obscuring the background or dominating the panel with color. More importantly from a conceptual point of view, the latticed ribbon path allows for the layering of movement symbols, so that the ribbon path for the batarang can be drawn on top of, and thus appear in front of, the latticed path signifying motion of the sword. In this way, four distinct movements – throw, fly, swing, deflect – emerge from the constellation of ribbon paths and an impact flash in this single panel.

While the ribbon paths and impact flash provide important image-schematic support to the reader's conceptualization of dynamics, they do not in themselves provide the cause-effect structure needed to link these dynamic actions into a coherent event. Talmy (2000) argues that we interpret cause and effect via force-dynamic image schemas and that abstract causes and effects are conceptualized metaphorically using image-schematic structure derived from physical events. Readers know that an effect must have a cause, and that a cause must result in an effect. Each of the four movements in the panel must be parsed to produce the correct order and relation of cause and effect, or they will not amount to a logical sequence of action. This panel has been carefully composed so that the visual symbol for each movement implies the movement's orientation and direction toward a goal, as well as the span of time through which it unfolds. The reader interprets the depicted entities, salient features, spatial relations, and conceptual symbols against a backdrop of encyclopedic knowledge to determine the sequence and causal connections. In order for the batarang to fly across the room, it must have been propelled. In order for its path to change, it must have been deflected by a force. The swinging of Batman's arm and his placement at the start of the batarang's path imply that he threw it. At the same time as he threw it, the android must have been bringing up its sword to block the batarang, or the sword would not have arrived in time to deflect it, and so on. The left-to-right layout of the panel helps the viewer read the motion events in sequence, while the depiction of these movements in a single panel preserves the simultaneity of overlapping action. Even in the absence of alignment with the conventional direction of reading, the events could be reconstructed using knowledge of the world and force-dynamic schemas of cause and effect to determine or explain how objects move. The artwork itself has no motion, sequence, or time, yet the reader's visual scanning, interpretation of schematic structure in the ribbons paths and impact flash, and incorporation of world knowledge create a dynamic sense of meaningful action unfolding through a brief interval of time.

## 1.9 Conclusion

As a genre brimming with motion and force dynamics, action comics pose a considerable representational challenge to artists trying to tell action-filled stories in still images. To meet this challenge, comics artists have developed stylized symbols to prompt the reader's conceptualization, symbols which are tied to established patterns in the human mind. Ribbon paths depict routes traveled by moving objects, motion lines provide an embodied participant perspective on action, and impact flashes (with sound effects) denote the sites and magnitudes of collisions. All of these function best if readers have the conceptual apparatus to interpret them naturally and effortlessly.

Approaching the study of these visual symbols from the perspective of cognitive linguistics shows that they are not arbitrary, nor does the readers' understanding of them depend solely on the conventionality of their use. Though not necessarily predictable, the form of the symbols is clearly motivated. A ribbon path encapsulates SOURCE-PATH-GOAL image-schematic structure, while its light color and tapering lines attract visual attention and add apparent depth to motion. Motion lines act as an analogue to the optic flow we experience when focusing on an object while moving with it through the environment. Flashes have the proper temporal dynamics and synesthetic associations with pressure waves and percussive sounds to stand as symbols for impacts. While these symbols do become familiar to readers through repetition so that their interpretation becomes automatic, they are nevertheless readily decipherable by novices precisely because they are yoked to these familiar patterns in the human mind.

In action comics, artists use visual symbols of movement and force to evoke basic conceptual patterns in readers' minds: image schemas derived from bodily experience in the physical world and conceptual metaphors linking different domains of experience. Through ordinary processes of meaning construction, readers add time, motion, and event structure to the panels on the page, generating the fast pace and thrilling action of superhero stories, thus turning comics into cinema in the mind.

## Note

1. The order of authorship is alphabetical. Send correspondence to: Robert F. Williams, Lawrence University, 711 E. Boldt Way SPC 22, Appleton, WI 54911 (robert.f.williams@lawrence.edu).

## References

Dodge, E. and Lakoff, G. (2005) Image schemas: from linguistic analysis to neural grounding. In B. Hampe with J. E. Grady (eds.) *From Perception to Meaning: Image Schemas in Cognitive Linguistics*, pp. 57–92. Berlin: Walter de Gruyter.
Eisner, W. (1985) *Comics & Sequential Art*. Tamarac, FL: Poorhouse Press.
Eisner, W. (1996) *Graphic Storytelling and Visual Narrative*. Tamarac, FL: Poorhouse Press.
Evans, V. and Green, M. (2006) *Cognitive Linguistics: An Introduction*. Edinburgh: Edinburgh University Press.
Forceville, C. (2008) Metaphor in pictures and multimodal representations. In R. W. Gibbs, Jr. (ed.) *The Cambridge Handbook of Metaphor and Thought*, pp. 462–82. Cambridge: Cambridge University Press.
Gibbs, Jr., R. and Colston, H. L. (1995) The cognitive psychological reality of image schemas and their transformations. *Cognitive Linguistics* 6 (4): 347–78.
Grady, J. E. (2005) Image schemas and perceptions: refining a definition. In B. Hampe with J. E. Grady (eds.) *From Perception to Meaning: Image Schemas in Cognitive Linguistics*, pp. 35–56. Berlin: Walter de Gruyter.
Hampe, B. (2005) Image schemas in cognitive linguistics: introduction. In B. Hampe with J. E. Grady (eds.) *From Perception to Meaning: Image Schemas in Cognitive Linguistics*, pp. 1–12. Berlin: Walter de Gruyter.
Janson, K. (2002) *The DC Comics Guide to Pencilling Comics*. New York: Watson-Guptill.
Johnson, M. (1987) *The Body in the Mind: The Bodily Basis of Meaning, Imagination, and Reason*. Chicago: University of Chicago Press.
Kennedy, J. M. (1982) Metaphor in pictures. *Perception* 11 (5): 589–605.
Kogan, N., Connor, K., Gross, A., and Fava, D. (1980) Understanding visual metaphor: developmental and individual differences. *Monographs of the Society for Research in Child Development* 45 (1): 1–78.
Lakoff, G. (1987) *Women, Fire, and Dangerous Things: What Categories Reveal about the Mind*. Chicago: University of Chicago Press.
Lakoff, G. (1993) The contemporary theory of metaphor. In A. Ortony (ed.) *Metaphor and Thought*, 2nd edition, pp. 202–51. Cambridge: Cambridge University Press.
Lakoff, G. and Johnson, M. (1980) *Metaphors We Live By*. Chicago: University of Chicago Press.
Lakoff, G. and Johnson, M. (1999) *Philosophy in the Flesh: The Embodied Mind and Its Challenge to Western Thought*. New York: Basic Books.
McCloud, S. (1994) *Understanding Comics: The Invisible Art*. New York: HarperCollins.
McCloud, S. (2000) *Reinventing Comics*. New York: Perennial.
Talmy, L. (2000) *Toward a Cognitive Semantics. Vol. 1: Concept Structuring Systems*. Cambridge, MA: MIT Press.

Taylor, T. (2001) If he catches you, you're through: coyotes and visual ethos. In R. Varnum and C. Gibbons (eds.) *The Language of Comics: Word and Image*, pp. 40–59. Jackson: University of Mississippi Press.

Varnum, R. and Gibbons, C. T. (eds.) (2001) *The Language of Comics: Word and Image*. Jackson: University of Mississippi Press.

Wolk, D. (2007) *Reading Comics: How Graphic Novels Work and What They Mean*. Cambridge: Da Capo Press.

Yus, F. (2009) Visual metaphor versus verbal metaphor: a unified account. In C. Forceville and E. Urios-Aparisi (eds.) *Multimodal Metaphor*, pp. 147–72. Berlin: Mouton de Gruyter.

# 2
# Creating Humor in Gary Larson's *Far Side* Cartoons Using Interpersonal and Textual Metafunctions

*Richard Watson Todd*

## 2.1 Introduction

Cartoons, one of the most frequently encountered forms of humor, have a venerable history. Eighteenth-century Britain saw the establishment and widespread acceptance of cartoons as a medium for expressing political and moral opinions, most notably in Hogarth's *The Rake's Progress*. Such cartoons had, and still have in the case of editorial cartoons in newspapers, an overtly political purpose achieved primarily through satire and irony. The nineteenth century saw less politically focused cartoons emerge as a common genre. Gag cartoons initiated in magazines such as *Punch* are designed primarily for entertainment with the picture and accompanying caption creating humor. The analysis of such cartoons therefore falls within the domain of humor research.

Research into humor has tended to encompass two categories. First, there is linguistic and psychological research that looks at how humor is created, in other words, what makes something funny. Second, there is sociological research into the effects of humor (see Norrick 2003), such as the role of humor in moods (e.g. Martin and Lefcourt 1983) and in leadership (e.g. Priest and Swain 2002). While most of the research into the creation of humor has focused on verbal jokes, it is generally acknowledged that the findings also apply broadly to cartoons. However, since cartoons are multimodal, combining images and text, the creation of humor in cartoons may be more complex than in purely verbal jokes, and this chapter investigates such complexities by examining through systemic functional linguistics the relationship between the caption and the picture in creating humor in cartoons.

The multimodal nature of cartoons also means that the linguistics of humor can be investigated through comparing variations in linguistic elements of captions while retaining the same contexts of the pictures. If, as in systemic functional linguistics, language is viewed as a series of choices in using linguistic resources to achieve intended functions and to express intended meanings, are some of these choices perceived as funnier than others? In this chapter, choices concerning whether to emphasize interpersonal functions and meanings through language are investigated for how they influence perceptions of humor.

## 2.2 Theories of humor and cartoons

The three main theories of humor – incongruity theory, superiority theory and psychic release theory – largely explain what makes things funny in terms of semantic content (Cook 2000; Ross 1998). Incongruity theory emphasizes clashes between two competing semantic scripts or schemata; superiority theory focuses on semantic or experiential blunders by a group of others; and psychic release theory stresses the role of taboo semantic domains. Of the three theories, incongruity theory is both the most influential theory and the theory that places the greatest emphasis on semantic content in humor.

Incongruity theory, and its most detailed statement, the general theory of verbal humor (Attardo 2001; Attardo and Raskin 1991), argue that humor is created when schemata either overlap or are broken (Cook 2000; Minsky 1985; Raskin 1985, 2008). Schemata, also termed scripts and frames, refer to cognitive structures of knowledge of ordinary events (Nassaji 2002) which enable participants to generate expectations of what is likely to be heard or read next. Generally, people find things funny under two broad circumstances: first, when an incongruity is created either by breaking a script through the inclusion of content at odds with the script or by activating a mix of two conflicting scripts; and second, when the incongruity is resolved (Deckers and Avery 1994). In such cases, the most likely expectations generated by the original schema at the start of a joke are not fulfilled; rather, they are replaced by expectations from a second, less highly activated, schema which may be applied retrospectively. Where no resolution is possible, the supposed joke may just be perceived as illogical. This process can be seen most clearly in children's jokes such as 'What do you call a cow lying on the earth?' 'Ground beef.' Here, the word 'ground' can be interpreted in two ways based on two conflicting schemata associated with 'earth' and 'beef,' and the joke is resolved by accepting that 'ground beef,' the most likely meaning of which is minced meat, can also be interpreted literally as a 'cow lying on the earth.' Incongruity theory, then, clearly foregrounds semantic content as the central and necessary factor in how people understand jokes (Scovel 2001).

This centrality of semantic content to humor has been supported by numerous research studies. For instance, early humor research by Brownell *et al.* (1983) highlighted the importance of clashes between the content of the body of a joke and the content of the punchline as being key to perceptions of the funniness of jokes, while Deckers and Buttram (1990) identified the importance of content schemata in both incongruity and incongruity resolution theories of humor.

From such research, the importance of semantic content to humor is now generally accepted (Raskin 2008). However, while semantic content creating incongruity may be essential to humor, it seems likely that a range of other factors may come into play either in specific types of humor or in increasing the funniness of humor based on certain semantic content.

Researchers have also highlighted the centrality of semantic content to creating humor and the importance of other factors as well. Most cartoons require multimodal semiotic interpretations integrating at least two modes: text and image (Hempelmann and Samson 2008). These two modes may interact in three possible ways: (i) the humor is in the text with the picture simply providing an illustration; (ii) the picture itself is the joke and, in such cases, there may be no caption; and (iii) the humor relies on an interaction between the text and the image, one of which could complement or contradict the other (Tsakona 2009).

Being multimodal, cartoons allow more than two script oppositions to be created, and thus may be more complex in how they create humor than purely text-based jokes. For instance, there may be opposition between two schemata in the text, between two schemata in the image, and between one or two schemata in the text and one or two schemata in the image. Even with so many potential script incongruities, and despite some claims that some cartoons have no resolution, there is evidence that humor in cartoons follows the two-stage process of creating incongruity and then resolving it (Paolillo 1998).

## 2.3 Humor and the metafunctions of language

Systemic functional linguistics proposes that there are three main purposes of language, termed metafunctions: the *ideational* concerning content, the *interpersonal* concerning social purposes and interaction, and the *textual* concerning organization and linking within texts. Considering the importance of semantic content in what people find funny, it is clear that the ideational metafunction is central to humor. In cartoons where text and image interact, the textual metafunction plays a role. However, the extent to which the interpersonal metafunction is involved in creating humor is less clear. This chapter will investigate the roles of the interpersonal and textual metafunctions in creating humor, and thus we need to examine the metafunctions more closely.

Language use involves all three metafunctions simultaneously; in fact, some models also include a fourth metafunction, the *poetic* (see Finch 1998). Each metafunction may be expressed through different resources; however, particular communications may emphasize one of the metafunctions over the others. Incongruity theories of humor, for instance, suggest that the ideational metafunction is stressed in jokes. The ideational metafunction, as originally conceived, involves the use of language for the expression of content (Halliday 1970) and thus concerns the propositional content of sentences (Lyons 1970). This core purpose of language was then specified and expanded to refer to language 'as the means of the expression of our experience, both of the external world and of the inner world of our own consciousness – together with what is perhaps a separate sub-component expressing certain basic logical relations' (Halliday 1973:66). Thus, some models of the ideational metafunction subdivide it into two subsidiary functions, the experiential and the logical (Thompson 2004). Other authors emphasize the centrality of using language to express experience and rename the higher-level metafunction as the *experiential* (Lock 1996), while still others reclaim the importance of semantics and term it the *conceptual* metafunction (Widdowson 1984). The linguistic features associated with the ideational metafunction are transitivity, tense, and semantic content (Halliday 1973, 1994; Thompson 2004). When viewed as semantic content, the ideational metafunction is at the root of much humor, as seen in the discussion of incongruity theory above.

The second main metafunction, the interpersonal, concerns language 'as the mediator of role, including all that may be understood by the expression of our own personalities and personal feelings on the one hand, and forms of interaction and social interplay with other participants in the communication situation on the other' (Halliday 1973: 66). Also termed the *communicative* metafunction (Widdowson 1984) and the *interactional* metafunction (Rowley-Jolivet 2002), the interpersonal, involving establishing and maintaining relations, influencing others and expressing perspectives, looks at the pragmatic meaning of utterances through speech function, modality, attitude and semantic prosody (Halliday 1973, 1994; Thompson 2004).

Within the field of humor, previous research involving interpersonal concerns has focused on the sociological effects of humor rather than how humor is created. For instance, Bonaiuto, Castellana, and Pierro (2003) examine how humor can facilitate negotiations, and Hobbs (2007) looks at the use of humor as a persuasive tool by lawyers. In such research, the main focus is on the social impact of humor, not the linguistic elements comprising humor. Within linguistics research, on the other hand, much research into the interpersonal metafunction has focused on the linguistic elements used to achieve certain goals ranging from politeness and honorification (e.g. Kashyap 2008) to hedging and stance in academic discourse (e.g. Webber 2005). Given the vast amount of valuable linguistic research into the interpersonal metafunction in areas other than humor, it is surprising that the interpersonal metafunction has been overlooked in research into how humor is created. Indeed, as far as I am aware, there have been no previous linguistic-oriented attempts to link the interpersonal metafunction with the creation of humor.

The third metafunction, the textual, 'breathes relevance into the other two' (Halliday 1994: xiii) by providing links both within a text (such as through the coherence and cohesion relations in discourse) and to the wider context (Halliday 1970; Lock 1996; Lyons 1970; Thompson 2004). Linguistic features associated with the textual metafunction include theme-rheme, deixis and conjunction (Halliday 1973). For cartoons, multimodal textual links between pictures and captions may be central in creating humor.

## 2.4 Purposes of the study

As mentioned above, most previous work in humor, especially when based on incongruity theories, has focused on the ideational metafunction of language. In a few analyses of cartoons, examinations of the relationship between text and image have concerned the textual metafunction. In linguistic-oriented humor research in general, the interpersonal metafunction has been the poor cousin, nearly always overlooked. However, the same semantic content and the same script incongruities can be expressed in different ways to serve different social purposes or to illustrate different perspectives on the content, and some of these ways may be considered funnier than others. Thus the interpersonal metafunction may play a major role in creating humor. In this study, the roles of the interpersonal metafunction in humor are investigated by comparing interpersonally different ways of expressing the same ideational content. In addition, by using cartoons as the medium of humor examined, the role of the textual metafunction in creating humor is also investigated.

## 2.5 Methodology

If humor is 'any sudden episode of joy or elation associated with a new discovery that is self-rated as funny' (Davis 2008: 547), then research into humor may rely at least in part on subjects' ratings of funniness. One relatively common approach in humor research utilizing such ratings is to present original and adapted versions of jokes which subjects then rate for funniness. Adaptations of the original can be designed to focus on specific features under investigation. For instance, Avery (1990) and Deckers and Avery (1994) compared original jokes with adapted versions where the punchline had been replaced by logical and illogical endings, and Huber and Leder (1997) compared original cartoons with adapted versions that were either more or less compact. This approach to rating funniness is used in this study. Subjects were asked to choose between an original and an adapted cartoon, where the caption in the adapted version had been altered to reduce it as far as possible to its pure semantic content by removing language which primarily served interpersonal functions.

### 2.5.1 The materials

To investigate the role of the interpersonal metafunction in humor, jokes where the original format contains elements which primarily serve interpersonal functions are needed. To investigate multimodal aspects of the textual metafunction, cartoons are required. One source which fulfills both criteria and which is widely perceived as very funny is Gary Larson cartoons (Paolillo 1998). While some Gary Larson cartoons rely solely on the picture, his typical cartoon is a single-frame cartoon with a text of one or two sentences. For the purposes of this study, ten such cartoons were chosen. The captions of each of these cartoons contained elements focused on the interpersonal metafunction.

To investigate the role of the interpersonal metafunction in creating humor, the captions of the ten cartoons were adapted in such a way as to place a greater emphasis on the ideational metafunction and to de-emphasize the interpersonal metafunction. For each of the ten cartoons, elements associated with the interpersonal metafunction were removed from the captions. These included markers of personal feeling and semantic prosody, language focused on maintaining and building relationships, and lexis expressing attitude and affect. For instance, one of the cartoons consists of a picture of two men in a lighthouse at night with Superman closely circling the lighthouse. The original caption reads, 'For God's sake, kill the lights, Murray ... He's back again!' In this text, 'For God's sake' illustrates panic, an expression of perspective associated with the interpersonal metafunction. Similarly, 'kill the lights' evokes an urgency missing from the more neutral expression of the same semantic content 'turn off the lights,' and naming 'Murray' is an attempt to influence the actions of another person. Removing or changing these elements would lead to a caption reading 'Turn off the lights. He's back again.' This caption, while still serving the interpersonal function of influencing another's behavior, is far more neutral interpersonally than the original caption, and thus emphasizes the ideational content over the interpersonal. Similar adaptations were made to the other cartoons, and an expert informant checked these adaptations for the retention of ideational content and the removal of interpersonal elements. The original and adapted cartoons are summarized in Table 2.1.

The ten cartoons in two versions (same picture but with two different texts) were presented to subjects in a questionnaire. Each cartoon in a pair was labeled A or B with letters randomly assigned to the original and adapted versions. Space was provided for writing reasons after each cartoon pair. The instructions to the subjects were: 'Below are 10 pairs of cartoons. Please look at the cartoons and read the captions. Then, choose the cartoon which you think is funnier by circling the letter. Then, please try to give a reason why you think the cartoon you have chosen is funnier.' Thus, the subjects were asked to do two things to complete the questionnaire: first, choose whether they found the original or adapted version of the cartoon funnier; second, give a reason for their choice.

### 2.5.2 The subjects

The questionnaire was distributed to 40 native speakers of English (primarily American, British or Australian) residing in Thailand. Native speakers were chosen as subjects since they are the target audience for Gary Larson cartoons. Subjects were asked to complete the questionnaires and return them at their convenience. Most completed the questionnaires within the same day, and 32 questionnaires were returned.

Table 2.1 Summary of the original and adapted Gary Larson cartoons

| Cartoon | Description of picture | Original text | Adapted text |
|---|---|---|---|
| 1 Aliens landing | A UFO on tall stilts has landed and is surrounded by several people. A set of steps from the UFO to the ground has been lowered. There are 3 aliens at the top of the steps. One alien is spread-eagled at the bottom having tripped down the steps. | Wonderful! Just wonderful! ... So much for instilling them with a sense of awe. | Now we won't be able to instill them with a sense of awe. |
| 2 Murder in a clock shop | In a clock shop, a body is slumped over a table. There are bullet holes in the walls and the clocks. All of the clocks read ten past ten. A detective holding a machine gun is talking to a policeman. | We've got the murder weapon and the motive ... now if we can just establish time-of-death. | We know the murder weapon and the motive, but we still need to establish the time of death. |
| 3 On the beach | On a beach, a large lady is sitting next to a man who looks like he has been steam-rollered. | My goodness, Harold ... Now there goes one big mosquito. | That is a very big mosquito. |
| 4 The Viking ship | On a Viking longship, two men are standing at the prow talking. The rowers on the port side behind them are all beefy and strong, while those on the starboard side are weakly wimps. | I've got it too, Omar ... a strange feeling that we've just been going in circles. | I have a strange feeling that we're going in circles. |

| | | | |
|---|---|---|---|
| 5 Superman and the lighthouse | Two men are in a lighthouse at night with Superman closely circling the lighthouse. | For God's sake, kill the lights, Murray … He's back again! | Turn off the lights. He's back again. |
| 6 The boomerang couple | A male boomerang is walking out of a house while his wife, with hands on hips, berates him. | Ho! Just like every time, you'll get about 100 yards out before you start heading back. | As you do every time, you'll go about 100 yards and then start heading back. |
| 7 The locked UFO | One alien is peering through the cockpit bubble while another looks at a group of people and a dog running towards the UFO over a hill. | Well, here they come … You locked the keys inside, you do the talkin'. | They are coming. Since you locked the keys inside, you need to do the talking. |
| 8 Scorpions | With the feet of a dead person in the background, two scorpions standing on a shoe are talking. | There I was! Asleep in this little cave here, when suddenly I was attacked by this hideous thing with five heads. | I was asleep in this little cave when suddenly I was attacked by a hideous thing with five heads. |
| 9 Target soldiers | An army of medieval soldiers is besieging a castle. They are wearing uniforms with targets on them, while arrows rain down, killing several. | What did I say, Boris? … These new uniforms are a crock! | I think these new uniforms are a crock. |
| 10 The alligator defendant | In a courthouse, an alligator is in the dock being questioned by a lawyer while a judge looks on. | Well, of COURSE I did it in cold blood, you idiot! I'm a reptile! | I killed him in cold blood because I'm a reptile. |

### 2.5.3 Data analysis

The data collected from the questionnaire are of two types. First, there are frequency data concerning which of the original and adapted versions of the cartoon was chosen for each of the ten cartoons. Second, there are the open-ended reasons for the choice.

The frequency of choosing the original and adapted versions of the cartoons was counted, and then calculated as percentages. It should be noted that several subjects did not express a preference between some pairs of cartoons, so that in total 313 choices between original and adapted versions of the cartoons were made.

In total, 261 reasons for a choice between cartoons were given (reasons were not given when no choice was made, and some choices were not justified). The total word count for these reasons was 2,253 words, giving an average word length per reason of around 8. These reasons were categorized into themes.

In order to thematically categorize the reasons, an approach based on both the purposes of the research and inductive identification of themes from a word frequency count (following Watson Todd 2006) was used. Given that the purpose of the research was to investigate the roles played by the interpersonal and textual metafunctions in creating humor, preliminary themes related to these two metafunctions were set up. Conducting a word frequency count and identifying keywords in the data allowed for these preliminary themes to be refined and for other potential themes to be identified. Keywords indicate particular focuses of content in a text (in this case, the corpus of all reasons given) through their relatively high frequency (Scott 1997). It should be emphasized that it is relative, not absolute, frequency that is crucial in identifying keywords. After all, in most written texts in English, *the* is the most frequent word but it would not usually be considered a keyword. Keywords are those words which appear in a given text with a frequency much higher than their general overall frequency in English. In this study, relative frequency was identified by comparing the frequency of words in the corpus of reasons with their general frequency in the *British National Corpus* using log likelihood. Log likelihood is used to calculate whether the frequency of a given word in a given text is noticeably different from its frequency in general English use (see Rayson and Garside 2000). Higher log likelihoods show that the words are used more frequently in the corpus than in everyday use, with a value of 30 or more generally taken as showing a relatively high frequency and thus that the word may be indicative of a theme. Semantically similar sets of words with high absolute and relative frequencies (measured through word frequency and log likelihood) therefore form the basis for identifying themes.

The most frequent word in the corpus of reasons, unsurprisingly, was *the* (frequency = 190; log likelihood = 18.70). The second most frequent word was *more* (f = 101; LL = 421.70), the word with the highest log likelihood and indicating that subjects were comparing the two cartoons in giving reasons. Several other frequent words also indicated comparisons between cartoons: *better* (f = 27; LL = 28.38), *less* (f = 12; LL = 40.41) and *funnier* (f = 8; LL = 93.21).

Some frequent words were concerned with the interpersonal metafunction. Keywords related to the expression of personal feelings included *sarcastic* (f = 9; LL = 99.34), *personal* (f = 8; LL = 32.80), *sarcasm* (f = 7; LL = 73.16) and *urgency* (f = 5; LL = 43.22). Other interpersonal keywords were related to social interaction with others including *name* (f = 10; LL = 37.94), the names *Harold* (f = 6; LL = 50.31) and *Boris* (f = 5; LL = 45.26), and, as a marginal keyword, *interaction* itself (f = 3; LL = 18.09). These findings suggested that two themes concerning the interpersonal metafunction were needed, one for the expression of personalities and personal feelings and one for social interaction.

The analysis of keywords also gave some indication of the relevance of the textual metafunction. Since the pictures in the original and adapted versions of the cartoons were identical whereas the captions differed, most of the reasons given by the subjects concerned the captions. However, several of these linked the captions with the pictures as indicated by keywords such as *cartoon* (f = 10; LL = 102.88), *situation* (f = 8; LL = 34.24) and *picture* (f = 7; LL = 33.25) indicating that a theme concerning the textual metafunction was appropriate.

Three further semantic sets of keywords indicated that three further themes were needed. First, there was a set of keywords concerning the authenticity of the dialogue: *natural* (f = 14; LL = 77.35), *conversational* (f = 6; LL = 59.73); *formal* (f = 6; LL = 32.69), *real* (f = 6; LL = 18.49) and *colloquial* (f = 3; LL = 32.09). Second, within the reasons given for choosing the adapted version of the cartoon only, the phrase *to the point* occurred five times with a log likelihood of 70.98 (*short* also occurred in this subset of the corpus three times with a log likelihood of 15.81), suggesting the need for a theme concerning brevity. Third, the word *story* occurred eight times (LL = 36.83), each time in a context suggesting that it was being used to describe a genre, suggesting a theme of genre identification.

Analyzing the keywords in the research subjects' reasons for choosing cartoons reveals six main thematic categories. These categories help identify the reasons participants gave for choosing either the original or adapted version of the cartoon:

1. interpersonal expression of feelings;
2. interpersonal social interaction;
3. textual connections;
4. caption authenticity;
5. caption brevity;
6. genre identification.

In addition to these themes, a category for unenlightening reasons is also needed for those statements that do not provide much insight into why a specific version was chosen, e.g. 'Wording is a lot better,' 'The vocab is easier to read,' 'Duh!' The reasons given by subjects for choosing a version of a cartoon were then categorized against these themes. The themes can be used to show the extent to which the interpersonal (especially themes 1 and 2) and textual metafunctions (especially theme 3) influence perceptions of funniness of the cartoons.

## 2.6 Findings

### 2.6.1 Choices between original and adapted versions

The frequency data for choices between original and adapted versions of the cartoons show whether the subjects preferred the versions incorporating the interpersonal metafunction (the original versions) or versions relying almost solely on the ideational metafunction (the adapted versions). The results are presented in Table 2.2.

Table 2.2 shows a resounding overall preference for the original versions of the cartoons which incorporate clear interpersonal elements. Unsurprisingly, Gary Larson's originals are perceived as funnier than the adaptations. These preferences suggest that the interpersonal metafunction plays a key role in creating humor. A clearer picture of how and why this happens, however, should emerge from examining the extent to which the interpersonal metafunction was mentioned in the reasons for choosing the original versions. The reasons should also allow us to see why there was no clear preference for the original version for two of the cartoons (cartoons 4 and 6).

### 2.6.2 Reasons for choices

As noted above, the reasons were categorized into six main themes. Examples of reasons for choosing each of these themes are given in Table 2.3.

*Table 2.2* Frequency of choices between original and adapted versions

| Cartoon | Number of subjects preferring original | Percentage of subjects preferring original | Number of subjects preferring adapted | Percentage of subjects preferring adapted |
|---|---|---|---|---|
| 1. Aliens landing | 26 | 81.25 | 6 | 18.75 |
| 2. Murder in a clock shop | 26 | 86.67 | 4 | 13.33 |
| 3. On the beach | 26 | 83.87 | 5 | 16.13 |
| 4. The Viking ship | 18 | 58.06 | 13 | 41.94 |
| 5. Superman and the lighthouse | 30 | 93.75 | 2 | 6.25 |
| 6. The boomerang couple | 15 | 46.88 | 17 | 53.13 |
| 7. The locked UFO | 29 | 90.63 | 3 | 9.38 |
| 8. Scorpions | 29 | 90.63 | 3 | 9.38 |
| 9. Target soldiers | 27 | 90.00 | 3 | 10.00 |
| 10. The alligator defendant | 30 | 96.77 | 1 | 3.23 |
| TOTAL | 256 | 81.79 | 57 | 18.21 |

*Table 2.3* Examples of reasons for themes

| Theme | Examples |
|---|---|
| Interpersonal expression of feelings | The alien sounds much more sarcastic (cartoon 1) |
| | "My goodness Harold" exclamation of surprise makes it sound funnier (cartoon 3) |
| | The language indicates surprise and panic (cartoon 5) |
| | Creates the urgency of them expecting to be caught - can feel the panic of not having the keys (cartoon 7) |
| | Just the opening comment gives the character a more irate feel making it funnier (cartoon 8) |
| Interpersonal social interaction | Adding a name always makes the joke more personal (cartoon 3) |
| | It seems more inclusive of others in the storyline (cartoon 4) |
| | There is an interaction between the characters that makes it more real (cartoon 9) |
| | Talking with people and having an agreement (cartoon 9) |
| Textual connections | This is at odds with the surrealism of the visual (cartoon 1) |
| | A better description of the picture (cartoon 3) |

(*continued*)

Table 2.3 Continued

| Theme | Examples |
|---|---|
| | It shows they have been trying to reason with the fact for some time. It's an ongoing discussion with an obvious reason (cartoon 4) |
| | "They are coming" draws your attention to the whole picture (cartoon 7) |
| | I imagine prior to this the aliens would have been arguing (cartoon 7) |
| | Makes you think he had complained before (cartoon 9) |
| Caption authenticity | The language is more colloquial and hence more authentic (cartoon 2) |
| | I prefer the use of "still need" to lead up to the punchline. It feels more natural (cartoon 2) |
| | More realistic conversational tone (cartoon 4) |
| | Because of its more natural rhythm – it is what you would expect to hear in the circumstance but maybe not from aliens (cartoon 7) |
| | More natural realistic speech (even though he's a lizard) (cartoon 10) |
| Caption brevity | I feel that A flows more and is sharper and simpler (cartoon 1) |
| | The blunter shorter comment makes the picture funnier (cartoon 3) |
| | B is too long to convey the same meaning (cartoon 4) |
| | In a cartoon getting your point across in less words is better (cartoon 4) |
| | Less words, same message (cartoon 8) |
| Genre identification | "There I was" made me associate the scorpion with raised claws as being the storyteller (cartoon 8) |
| | Seems he's telling his friend a story (cartoon 8) |
| | Helps set the scene as more of a story (cartoon 8) |

The themes illustrated in Table 2.3 appear with different frequencies in the reasons. As shown in Table 2.4, interpersonal expressions of feelings are both the most common theme overall and the most common theme cited in reasons for choosing the original version of the cartoons, suggesting that the interpersonal metafunction may play a key role in creating humor. The second most common theme, caption authenticity, suggests that captions containing both interpersonal and ideational

*Table 2.4* Frequencies of themes associated with original and adapted versions

| Theme | Overall frequency | Frequency in reasons for choosing the original version | Frequency in reasons for choosing the adapted version |
| --- | --- | --- | --- |
| Interpersonal expression of feelings | 67 | 61 | 6 |
| Interpersonal social interaction | 20 | 20 | 0 |
| Textual connections | 22 | 15 | 7 |
| Caption authenticity | 44 | 41 | 3 |
| Caption brevity | 31 | 8 | 23 |
| Genre identification | 8 | 7 | 1 |
| Unenlightening reasons | 69 | 60 | 9 |
| TOTAL | 261 | 212 | 49 |

elements are seen as more natural than captions primarily stressing ideational elements. Reasons reflecting the textual metafunction, while not very frequent, do not depend on the version of the cartoon chosen (since both the original and the adapted versions create textual links). From Table 2.3, we can see that the textual connections come in two forms: links between the caption and the picture, and links between the caption and an imagined preceding discourse. In the participants' reasons for choosing the adapted versions of the cartoons, caption brevity dominates (in general the captions of the adapted versions are shorter than for the original versions since elements associated with the interpersonal have been removed).

From Table 2.4, we can see that most of the themes of reasons are associated with choosing the original versions of the cartoons, with caption brevity associated with choosing the adapted version. There are also associations between the themes and individual cartoons, and these are shown in Table 2.5.

Table 2.5 shows that certain linguistic features in the captions are associated with certain themes. The original captions of the two cartoons with the highest frequencies of interpersonal expression of feelings contain phrases which are clearly linked with affect in the reasons given by the subjects. In cartoon 1, 'Wonderful! Just wonderful!' is identified as indicating sarcasm in the reasons for choosing both the original and adapted versions of the cartoons; and in cartoon 5 'For God's sake' is taken as showing urgency, panic, and distress as a reason for choosing the original version. For interpersonal social interaction, the four cartoons with

Table 2.5 Frequencies of themes associated with individual cartoons

| Cartoon with original caption (O) and adapted caption (A) | Interpersonal expression of feelings | Interpersonal social interaction | Textual connections | Caption authenticity | Caption brevity | Genre identification | Unenlightening reasons |
|---|---|---|---|---|---|---|---|
| 1 O: Wonderful! Just wonderful! ... So much for instilling them with a sense of awe.<br>A: Now we won't be able to instill them with a sense of awe. | 16 | 1 | 1 | 3 | 4 | 0 | 11 |
| 2 O: We've got the murder weapon and the motive ... now if we can just establish time-of-death.<br>A: We know the murder weapon and the motive, but we still need to establish the time of death. | 5 | 0 | 1 | 7 | 2 | 0 | 8 |
| 3 O: My goodness, Harold ... Now there goes one big mosquito.<br>A: That is a very big mosquito. | 6 | 5 | 3 | 2 | 3 | 0 | 8 |
| 4 O: I've got it too, Omar ... a strange feeling that we've just been going in circles.<br>A: I have a strange feeling that we're going in circles. | 2 | 5 | 5 | 3 | 9 | 1 | 3 |
| 5 O: For God's sake, kill the lights, Murray ... He's back again!<br>A: Turn off the lights. He's back again. | 10 | 4 | 1 | 5 | 2 | 0 | 5 |

| | Dialogue | | | | | | | |
|---|---|---|---|---|---|---|---|---|
| 6 | O: Ho! Just like every time, you'll get about 100 yards out before you start heading back.<br>A: As you do every time, you'll go about 100 yards and then start heading back. | 5 | 0 | 0 | 5 | 5 | 0 | 7 |
| 7 | O: Well, here they come … You locked the keys inside, you do the talkin'.<br>A: They are coming. Since you locked the keys inside, you need to do the talking. | 7 | 0 | 3 | 9 | 1 | 0 | 3 |
| 8 | O: There I was! Asleep in this little cave here, when suddenly I was attacked by this hideous thing with five heads.<br>A: I was asleep in this little cave when suddenly I was attacked by a hideous thing with five heads. | 7 | 0 | 0 | 3 | 3 | 7 | 6 |
| 9 | O: What did I say, Boris? … These new uniforms are a crock!<br>A: I think these new uniforms are a crock. | 3 | 3 | 8 | 4 | 1 | 0 | 7 |
| 10 | O: Well, of COURSE I did it in cold blood, you idiot! I'm a reptile!<br>A: I killed him in cold blood because I'm a reptile. | 6 | 2 | 0 | 3 | 1 | 0 | 11 |

the highest frequencies (cartoons 3, 4, 5 and 9) all contain names in the captions, which gives a more personal feeling of interaction. The role played by the textual metafunction is most apparent in cartoon 9 where many of the reasons given show connections to an imagined preceding discourse (e.g. 'Plays with the ongoing discussion,' 'referencing earlier frustration at the uniforms,' 'implies that he's already been moaning about the problem'). Genre identification comes to the fore in cartoon 8, where 'There I was' is taken as indicating the genre of story telling. Finally, caption brevity as a reason for choosing the adapted version is highlighted in cartoons 4 and possibly 6, the two cartoons where there was no clear preference for the original version.

## 2.7 Discussion

This study does not intend to downplay the importance of the ideational metafunction in creating humor. Although the subjects were asked to state why they preferred one version over the other, several responses suggested that both versions were considered funny with one perceived as funnier. For instance, one respondent added the unelicited general comment, 'Larson rules! My all-time favorite! Hilarious!' Such responses indicate that the semantic content of the cartoons was perceived as funny irrespective of differences in the interpersonal metafunction, highlighting the importance of the ideational metafunction in creating humor.

However, this study also shows that the same incongruous semantic content can be interpreted as being more or less funny depending on its interaction with other metafunctions. The interpersonal metafunction, divided in this study into expressing personal feelings and social interaction, has a clear effect on how funny a given set of incongruous semantic content is perceived to be. The findings show that expression of personal feelings is the most important determinant of funniness in the cartoons, with cartoons where feelings are manifest being perceived as funnier than ones where feelings are absent. Similarly, social interaction also plays a role in determining funniness, especially for those cartoons where characters are named. The interpersonal metafunction builds personalities (even for aliens and 'a lizard') and relationships into the discourse making the characters come alive, and it is thus related to authenticity. It also creates an affective context within which the semantic content of the humor is interpreted. Thus, while the prime source of humor in the cartoons is ideational, the interpersonal metafunction can influence the perceived funniness of this humor. In the words of the old adage: It's not (only) what you say, but how you say it.

The other three themes associated with increased perceptions of funniness – textual connections, caption authenticity, and genre identification – may all influence subjects' perceptions of funniness in ways similar to the interpersonal metafunction. The vast majority of the reasons given for choosing the original versions of the cartoons suggests that the subjects were going beyond the input of picture and caption. They were reacting to the input to create broader interpretations by imagining a wider context for the cartoon, by imagining how the cartoon would happen in real life, and by imagining themselves within the context of the cartoon. This imagination could be stimulated through interpersonal expression of feelings (e.g. 'Discussion about the feeling makes everyone who reads it part of the cartoon' – the reason given for choosing the original version of cartoon 4), through social interaction (e.g. 'Makes the event more personal' – cartoon 3), through textual connections (e.g. 'I imagine prior to this the aliens would have been arguing' – cartoon 7), through authenticity (e.g. 'Because this is similar to what I would say' – cartoon 1), and through genre identification (e.g. 'Helps set the scene as more of a story' – cartoon 8). Thus, it appears that, while humor is created through semantic or ideational incongruity in the cartoons, the extent to which this humor is perceived as funny depends, at least in part, on how the cartoons stimulate imagination. This, in turn, is related to the authenticity of language use (and most authentic language use integrates the ideational and interpersonal metafunctions), to the interpersonal metafunction especially in terms of expressing feelings, and to the textual metafunction especially in terms of creating a discourse context.

While this study has highlighted the roles of the interpersonal and textual metafunctions in supporting the ideational metafunction to create humor, it should be stressed that this may apply only to a single genre of humor: the single frame gag cartoon. It seems likely that the textual metafunction would play a diminished role in short, purely text-based jokes than in multimodal cartoons, and in some genres of humor the interpersonal metafunction may not be particularly important. For instance, riddles such as the 'ground beef' joke discussed earlier rely almost exclusively on ideational content for humor. However, in other genres such as stand-up comedy routines, the interpersonal metafunction may be crucial for humor (Rutter 1997). Such differences may allow the development of a continuum of humor genres based on the relative importance of the interpersonal metafunction in creating humor, potentially allowing new insights into both the psychology and sociology of humor. This study is based on the premise that, from an incongruity theory perspective, the semantic content of the cartoons in the form of conflicting schemata is a necessary but perhaps insufficient condition for creating humor. Cartoons which fulfill the necessary semantic conditions can be perceived as more or less funny when other factors are varied. Non-ideational linguistic elements that most contributed to cartoons being perceived as funny were elements associated with the interpersonal expression of feelings, with other elements associated with social interaction, textual connections, and caption authenticity also playing a role. It would therefore appear that the interpersonal and textual metafunctions play a major role in perceptions of how funny a moment of humor might be.

## References

Attardo, S. (2001) A primer for the linguistics of humor. In V. Raskin (ed.) *The Primer of Humor Research*, pp. 101–56. Berlin: Mouton de Gruyter.

Attardo, S. and Raskin, V. (1991) Script theory revis(it)ed: joke similarity and joke representation model. *Humor* 4 (3–4): 293–348.

Avery, P. (1990) *Testing the Semantic Theory of Humor through the Measurement of Puzzlement and Funniness Ratings to Altered Joke Endings.* Unpublished honors thesis, Ball State University. Retrieved on 9 September 2009 from http://www.bsu.edu/libraries/virtualpress/student/honorstheses/pdfs/A93_1990AveryPamela.pdf

Bonaiuto, M., Castellana, E. and Pierro, A. (2003) Arguing and laughing: the use of humor to negotiate in group discussions. *Humor* 16 (2): 183–223.

Brownell, H. H., Michel, D., Powelson, J. and Gardner, H. (1983) Surprise but not coherence: sensitivity to verbal humor in right-hemisphere patients. *Brain and Language* 18 (1): 20–27.

Cook, G. (2000) *Language Play, Language Learning.* Oxford: Oxford University Press.

Davis, D. (2008) Communication and humor. In V. Raskin (ed.) *The Primer of Humor Research*, pp. 543–68. Berlin: Mouton de Gruyter.

Deckers, L. and Avery, P. (1994) Altered joke endings and a joke structure schema. *Humor* 7 (4): 313–21.

Deckers, L. and Buttram, R. T. (1990) Humor as a response to incongruities within or between schemata. *Humor* 3 (1): 53–64.

Finch, G. (1998) *How to Study Linguistics.* Basingstoke and New York: Palgrave Macmillan.

Halliday, M. A. K. (1970) Language structure and language function. In J. Lyons (ed.) *New Horizons in Linguistics*, pp. 141–65. Harmondsworth, Middlesex: Penguin.

Halliday, M. A. K. (1973) *Explorations in the Functions of Language.* London: Edward Arnold.

Halliday, M. A. K. (1994) *An Introduction to Functional Grammar*, 2nd edition. London: Edward Arnold.
Hempelmann, C. F. and Samson, A. C. (2008) Cartoons: drawn jokes? In V. Raskin (ed.) *The Primer of Humor Research*, pp. 609–40. Berlin: Mouton de Gruyter.
Hobbs, P. (2007) Lawyers' use of humor as persuasion. *Humor* 20 (2): 123–56.
Huber, O. and Leder, H. (1997) Are more compact cartoons more humorous? *Humor* 10 (1): 91–103.
Kashyap, A. K. (2008) On honorifics. In C. Wu, C. M. I. M. Matthiessen and M. Herke (eds.) *Proceedings of ISFC 35: Voices Around the World*, pp. 201–06. Sydney: International Systemic Functional Congress.
Lock, G. (1996) *Functional English Grammar*. Cambridge: Cambridge University Press.
Lyons, J. (1970) Introduction to 'Language structure and language function.' In J. Lyons (ed.) *New Horizons in Linguistics*, pp. 140–41. Harmondsworth, Middlesex: Penguin.
Martin, R. A. and Lefcourt, H. M. (1983) Sense of humor as a moderator of the relation between stressors and moods. *Journal of Personality and Social Psychology* 45 (6): 1313–24.
Minsky, M. (1985) *The Society of Mind*. New York: Simon & Schuster.
Nassaji, H. (2002) Schema theory and knowledge-based processes in second language reading comprehension: a need for alternative perspectives. *Language Learning* 52 (2): 439–82.
Norrick, N. R. (2003) Issues in conversational joking. *Journal of Pragmatics* 35 (9): 1333–59.
Paolillo, J. C. (1998) Gary Larson's *Far Side*: nonsense? Nonsense. *Humor* 11 (3): 261–90.
Priest, R., and Swain, J. (2002) Humor and its implications for leadership effectiveness. *Humor* 15 (2): 169–89.
Raskin, V. (1985) *Semantic Mechanisms of Humor*. Dordrecht, Holland: Kluwer.
Raskin, V. (2008) Theory of humor and practice of humor research: editor's notes and thoughts. In V. Raskin (ed.) *The Primer of Humor Research*, pp. 1–16. Berlin: Mouton de Gruyter.
Rayson, P. and Garside, R. (2000) Comparing corpora using frequency profiling. In *Proceedings of the Workshop on Comparing Corpora, held in conjunction with the 38th annual meeting of the Association for Computational Linguistics (ACL 2000)*, pp. 1–6.
Ross, A. (1998) *The Language of Humour*. London: Routledge.
Rowley-Jolivet, E. (2002) Visual discourse in scientific conference papers: a genre study. *English for Specific Purposes* 21 (1): 19–40.
Rutter, J. (1997) *Stand-up as Interaction: Performance and Audience in Comedy Venues*. Unpublished PhD Thesis, University of Salford. Retrieved on 22 July 2010 from http://citeseerx.ist.psu.edu/viewdoc/download?doi=10.1.1.123.729&rep=rep1&type=pdf
Scott, M. (1997) PC analysis of key words – and key key words. *System* 25 (2): 233–45.
Scovel, T. (2001) *Learning New Languages: A Guide to Second Language Acquisition*. Boston: Heinle & Heinle.
Thompson, G. (2004) *Introducing Functional Grammar*, 2nd edition. London: Arnold.

Tsakona, V. (2009) Language and image interaction in cartoons: towards a multimodal theory of humor. *Journal of Pragmatics* 41 (6): 1171–88.
Watson Todd, R. (2006) Continuing change after the innovation. *System* 34 (1): 1–14.
Webber, P. (2005) Interactive features in medical conference monologue. *English for Specific Purposes* 24 (2): 157–81.
Widdowson, H. G. (1984) *Explorations in Applied Linguistics 2*. Oxford: Oxford University Press.

# 3
# Metaphors and Topoi of H1N1 (Swine Flu) Political Cartoons: A Cross-cultural Analysis

*Jill Hallett and Richard W. Hallett*

Beginning with the notion that the discourse of political cartoons reflects and reinforces public opinion (Edwards and Winkler 1997; Michelmore 2000; Greenberg 2008; and Dwivedi 2009), this chapter takes a multidisciplinary approach to political cartoons about the H1N1 virus (swine flu) to elucidate how fears are addressed through language and media cross culturally. Seventy three cartoons were culled from India (n=31), the United States (n=24), and other countries (n=18). The analysis, focusing on the metaphors (Lakoff and Johnson 1980) and *topoi* (Medhurst and DeSousa 1981: 200) found in these cartoons, shows how a nation's swine flu cartoons play on associations and fears relevant to that particular nation's culture(s).

## 3.1 Political cartoons

Cartoons do more than entertain. Hull (2000) examines the cartoons of Matt Groening for evidence of Foucaultian philosophy, Han (2006) examines Japanese cartoons and their reflection of Japanese-Korean relations, and O'Brien (2008) discusses the power of visual texts, specifically political cartoons, as catalysts for student writing. Other scholars of history, media, anthropology, linguistics, and other fields have researched political cartoons for visual and rhetorical presentation of stereotypes, identity promotion, and appeal to audience. In her analysis of political cartoons surrounding Indian Partition, Kamra (2003: 1) notes, 'While letters to the editor, articles, aphorisms, gossip columns are lively indeed, and obsessively political, editorial cartoons surpass these forums of opinion in their wrestling with a vexed political process by recoursing to heavily inflammatory visual rhetoric.' Likewise, Douglas, Harte, and O'Hara refer to cartoons as 'encyclopedias of popular culture' (1998: 1, cited in Michelmore 2000: 38) to which scholars may turn for 'clues to ideological forces, beliefs, assumptions, and prejudices at work in society,' or as Morris puts it, '[e]ditorial cartoons are a metalanguage for discourse about the social order' (1991: 225). Thus, analysis of political cartoons presents itself as a viable genre for research on shared ideologies and culture of a particular readership.

Political cartoons are unique in their place in a newspaper, where a fictional or at least exaggerated account is created for a current, newsworthy situation. According to Greenberg (2008: 195), 'Though they speak of the world in hyperfigurative terms, political cartoons are but one mode of opinion news discourse that enables the public to actively classify, organize and interpret what they see and experience in meaningful ways.' For Michelmore (2000: 37), '[c]artoons do not just illustrate the news. They are graphic editorials, and like all editorials they analyze and interpret a situation; they pass judgment. They tell readers what to think and how to feel about what is happening – amused, sympathetic, chagrined, angry, afraid.' Not only do political cartoons tell readers what to think and how to feel, they 'legitimate (and thus facilitate) the grounds upon which some things can be said and others impeded' (Greenberg 2008: 184). The visual image serves as a situational field, or 'politicized context,' (Edwards and Winkler 1997: 305) in which ideologies are grounded, recontextualized, and evaluated according to the cartoonist's agenda. For Lorenz (in Steffen 1995:144), a (political) cartoon has the ability to influence a reader as well as reflect some internal part of the reader.

Through a variety of rhetorical and visual strategies, the cartoonist compels polarity from readers regarding a given topic, 'motivat[ing] differing senses of community' (Edwards and Winkler 1997: 305). Individual readers will have varying degrees of shared cultural knowledge, and the understanding of a cartoon depends both on this knowledge and on the 'interpretation strategies suggested by the (near) identical circumstances under which the cartoons are accessed' (Forceville 2005: 247). Shared cultural knowledge is key to reader understanding of and connection to a particular cartoon (see Kamra 2003). For Medhurst and DeSousa (1981: 220), the 'culturally-induced message' is the source for different interpretations of cartoons by readers; as such they construct a taxonomy to uncover assumed shared cultural background by political cartoonists.

The notion of shared cultural knowledge, in addition to examination of metaphors employed and fears exploited, factors heavily into our analysis of H1N1 cartoons. There are many rhetorical devices at cartoonists' disposal by which readers may be brought into a shared discourse. According to Greenberg, these devices 'capture the essence of an issue or event graphically' (2008: 183) and are used by cartoonists to 'render normative judgments about social issues' (2008: 185). It is this normativity, or shared knowledge, that not only renders the issue and its representation salient for an audience, but also creates a cultural 'in-group' of the readers who 'get it.' Medhurst and DeSousa (1981: 220) find that this shared knowledge is manifested at several levels, allowing a variety of readers a variety of interactions with the cartoon based on which 'layers' resound with them individually.

## 3.2 Metaphor theory

One way in which cartoonists successfully exploit shared cultural knowledge among readers is the use of metaphor. In *Metaphors We Live By*, Lakoff and Johnson examine the use of metaphor in human conceptualization, often of complex concepts: *'The essence of metaphor is understanding and experiencing one kind of thing in terms of another'* (1980: 5, italics theirs). An editorial cartoon may represent a social or political complexity by employing visual and/or linguistic metaphor that is assumed to be cognitively available to its readers.

Edwards and Winkler (1997) discuss metaphor in cartoons as familiar items (shared knowledge) recontextualized. Indeed, in political cartoons metaphors may be as crudely presented as a drawing of something undesirable (the devil, for example), with the issue's name scrawled somewhere within the drawing (Bush's foreign policy, for example). In this case, Bush's foreign policy is equated with the devil, begetting the metaphor BUSH'S FOREIGN POLICY IS THE DEVIL and the subsequent entailment thereof (BUSH'S FOREIGN POLICY IS EVIL, IS IMMORAL, and so on). Other types of metaphor include CONTAINER metaphors, in which one item is conceived in terms of holding (or with the capacity to hold) another item, as in the idea of a 'full' or 'empty' life (Lakoff and Johnson 1980:51). The shared knowledge implicit in a metaphor attests that there are commonalities among a given readership in terms of ideology and cultural background.

Medhurst and DeSousa advocate using metaphor theory in understanding political cartoons (1981: 222). According to Edwards and Winkler (1997: 305), 'Not only does the parodied context of a representative form identify the specific circumstances which inspire the ideology's application, it also draws attention to key elements of the ideology at issue. Cartoonists direct the audience's attention by the addition, omission, substitution, and/or distortion of visual elements.' Metaphors boil down the issue at hand to its essence, achieved quite effectively through visual and linguistic means; thus, metaphors are ideal for the commentary limited by space, such as that of the cartoon.

Morris (1993: 231) elaborates on cartoonists' penchant for condensation and recontextualization: 'Cartoonists "domesticate" (Goffman, 1979) such abstractions as Basque nationalism, such distant personalities and events as the President of Indonesia or the war in Afghanistan, by representing them in familiar guises: as wolves, turkeys, bemused infants, bearded fanatics or scroungers.' Han (2006) further expounds on the cartoonist's skill in making a cartoon meaningful to the widest possible audience through the appropriation of 'symbols' that would be obvious to any reader, in his case discussing the tendency for other countries to be represented in cartoons by animals or clothing styles.

Michelmore (2000) presents a similar understanding of the cartoonist's task of compacting a message to fit the medium through what she terms 'symbolic simplicity':

> Because the symbols they create must be understandable to the general public, cartoonists deal in widely and instantly understood referents and allusions. Because cartoons translate complex and abstract ideas into simple and concrete form, cartoonists claim the right to exaggerate, stereotype, and distort. This necessity, along with the latitude often given to and by humor, allows cartoons considerable freedom of expression. (Michelmore 2000: 37)

Metaphor is one strategy that achieves this condensed exaggeration, stereotyping, and distortion in its use of equation and recontextualization. Medhurst and DeSousa (1981: 219–20) concur that readers are given a considerable responsibility in that they 'are unpacking one or more layers of available cultural consciousness which the cartoon has evoked from them. Cartoons "work" to the extent that readers share in the communal consciousness, the available means of cultural symbology, and are able to recognize that shared locus of meaning as expressed by the caricature.'

Various scholars have examined exaggeration and metaphor in cartoon analysis, for example Nystrom (2002), Kamra (2003), Manning (2007), Marín-Arrese (2008). Metaphors in cartoons can be textual and visual. Morris (1993: 196) is concerned with the importance of metaphor in visual semiotics, pleading for the enthusiastic inclusion of visual rhetoric alongside its better-known verbal counterpart: 'In this respect, the whole idea of visual rhetoric, similar to that of visual semiotics, assumes that art is a language and that the success of linguistic models is strong evidence that this metaphor should be applied to visual communication.'

Researchers of cartoons often manage a multimodal approach to their study. In her study of *New Yorker* cartoons over time, Steffen (1995) looks for linguistic elements such as onomatopoeia, puns, and irony, and she also addresses direct and indirect metaphor. She finds that the interplay of language and visual rhetoric is the key to the appeal of the cartoon. O'Brien (2008) advocates the analysis of visual texts through rhetorical methodologies, citing the deictic stamp of the author's name and place of publication as a deliberate element of the cartoon that demands examination of the message and context.

As cartoons combine elements both linguistic and visual, their metaphorical analysis is quite called for.

## 3.3 Methodology

O'Brien (2008: 188) asks, 'What role do such visual texts play in shaping the consciousness of readers in a specific country?' We seek the same answer, to which end we selected and analyzed 73 editorial cartoons in order to determine how metaphorical equation and fears are illustrated through language and media within and across cultures. The distribution of cartoons was as follows: India (n=31), the United States (n=24), Canada (n=5), Singapore (n=3), Thailand (n=2), and Australia, Bulgaria, France, Jordan, Mexico, Nigeria, Norway, and the UAE (n=1 for each of these countries). All of the cartoons had some written component, the language of which was always English. The cartoons were obtained from newspapers and online sources, from the summer of 2009.

Our analysis was informed by four topoi: political commonplaces and literary/cultural allusions (intertextuality) assumed common among readers, personal character traits of cartoon characters, and transient situational themes that are present in the cartoons (Medhurst and DeSousa 1981: 200). For each cartoon, we entered the country of origin, the artist, news outlet, and source into a spreadsheet. We added elaborate prose descriptions of the cartoons, and then analyzed the cartoons for Medhurst and DeSousa's (1981) four topoi, as well as for metaphors employed and fears demonstrated visually or in print.

Along with these topoi, we analyzed the cartoons for metaphors present both visually and textually, and fears addressed (outside of H1N1 itself). As Morris (1991:230) states, 'For Press (1981) the cartoonist expresses the interests and outlooks of a social group or segment of society; by close examination of the drawings one can determine what this may be.' Metaphors and fears were organized by country and analyzed to determine the fears and metaphors unique to a culture or common cross-culturally as demonstrated by the themes produced in these cartoons. We then determined (as gleaned from the data) 16 overarching categories of metaphors and 16 overarching categories of fears, and counted frequencies for each country for each overarching metaphor or fear. Finally, we plotted frequencies of fears and metaphors with respect to total metaphors/fears for India, the US, and one catch-all category for other countries.

By examining the fears presented in these cartoons, often presented as equivalent to or lesser than H1N1, and performing a cross-cultural analysis focusing on India, the US, and other countries, we purport to have some idea what the prevalent fears of a given society (defined by a readership) may be. We acknowledge, however, the subjectivity involved in this analysis and take full responsibility for our work.

## 3.4 Data and analysis

In this section, we first present the data qualitatively in terms of topoi. In subsequent sections, we analyze metaphorical analogy and fears exploited in the cartoons.

### 3.4.1 Topoi

Following Medhurst and DeSousa's (1981) taxonomy, four topoi emerge as assumed knowledge basic to the understanding and engagement of a reader with a cartoon: political commonplaces, literary/cultural allusions, personal character traits, and situational themes. This chapter does not seek to quantify anything in the realms of these categories but rather aims to demonstrate representative instances of each topos in the cartoons examined.

*Political commonplaces*

The topos of political commonplaces consists of 'those topics which are available to any cartoonist working within a modern nation-state. Such topics include the state of the economy, the defense of the nation, foreign relations, the political process, and the electoral framework [...]. Political commonplaces form the *raison d'être* of political cartoons in the sense that one cannot create a political cartoon without focusing reader attention, at least momentarily, on some agreed upon component of politics' (Medhurst and DeSousa 1981: 200).

One cartoon from by R. K. Laxman from the *Times of India* we describe as follows:

(1) A man sits at a desk, wears a Nehru hat. Another man walks into the office clutching papers, saying, 'Not a word about your visit to our city, your speeches, etc. All the paper's space is taken over by reports on swine flu and its activities, sir.'

From example (1), the following political commonplaces may be assumed: that a Nehru hat is indicative of a politician, and that politicians are necessarily worried about news coverage of their activities. A reader unfamiliar with Indian politics would probably miss out on these subtleties.

In addition to visual commonplaces such as those in (1), terminological commonplaces are apparent. Consider our description of example (2), from Ninan at the *Times of India*:

(2) Five people walk forward. Politician in the middle (in Nehru hat and vest) has judge on left wielding a gavel, military guy on right wielding a gun, bureaucrat in back wielding a 'NO' stamp, and doctor in front wielding a syringe. Onlooker comments to another onlooker, 'His upgraded security will protect him from terror, H1N1, PILs and RTI.'

To access the meaning of this cartoon, readers must be familiar with the idea that politicians require protection. Readers would also need an understanding of the acronyms and their implications: PILs are 'public interest litigations,' and RTI is the 'Right to Information Act,' giving citizens of India access to records of the Central Government and State Governments.

Political commonplaces may also reflect political fears, to be explored in more detail later. Example (3), from Ninan at *The Times of India*, brings up some such commonplaces:

(3) Two military personnel in uniform, with guns and backpacks, stand beneath a sign that reads 'AF-PAK DRONE UNIT.' They examine a large poster entitled 'WANTED DEAD' in red ink. On the poster are four headshots of what appear to be Taliban agents wearing typical headdress. A fifth picture appears at the bottom of the poster, a white circle of cloth with a tuft of hair beneath. One military agent to the other: 'The bottom

one may either be a satellite photo or he has covered himself due to swine flu.'

In example (3), it is important and expected that the reader is aware of the Indian military's concern about activity in Afghanistan and Pakistan, and that the Taliban is seen as a threat to India. This cartoon would work in few contexts outside of India, due to these assumed political commonplaces.

US cartoons similarly exploit political commonplaces particular to that location. Example (4) comes from Pat Bagley of the *Salt Lake Tribune*:

(4) Left: 'DIA DE LOS MUERTOS*', under which is a drawing of skeletons playing guitars and trumpets and wearing serapes, under which is '*MEXICAN HALLOWEEN.' Right: 'DIA DE LOS PUERCOS,**' under which is a pig skeleton, upright, holding a scythe (as the Grim Reaper), and wearing a serape, under which is '**MEXICAN SWINE FLU OUTBREAK'

A reader might not understand this cartoon if he or she did not understand the US position with respect to Mexico, replete with feelings of superiority and concern over immigration of Mexicans to the US, and the passing cultural awareness many Americans have of holidays such as *Día de los Muertos* (not, incidentally, 'Mexican Halloween'). These political commonplaces also indicate a conservative US political stance.

Example (5) comes from Mike Keefe at *The Denver Post* (US):

(5) On the left, Barack Obama juggles chainsaws, each emblazoned with a word: 'N.KOREA,' 'IRAQ,' 'AFGHAN.,' 'RECESSION,' 'DETROIT,' 'HEALTH CARE,' 'TORTURE,' 'BANKS,' 'CLIMATE.' Sound effects: RINGDINGDING, BRAPPA BRAPPA, RAPRAPRAP. Sign below the multi-limbed, tuxedoed Obama (in script): 'The First 100 DAYS.' From the right, a bubble exclaiming 'CATCH!' and a snotty pig with 'SWINE FLU' is bounced at Obama.

Readers in the US would know that the first 100 days is an unofficial trial period for first-term presidents. Obama is under immense pressure to perform, and quickly. Furthermore, the pig comes bouncing in from the right, a possible allusion to 'the (Republican) Right,' who would be keen for Obama to fail.

These examples of political commonplaces serve to demonstrate the assumed shared political knowledge of a cartoon's readership. Those readers who catch the references constitute a cultural in-group.

*Literary/cultural allusions*

Relatively self-explanatory, literary and cultural allusions are those references in cartoons that are assumed to have a base in the literature or culture of the readership; readers are expected to 'get' the references.

Indian cartoons pull from literature and culture both Western and Indian. References include *Gone with the Wind*, biblical quotations, and 'The Three Little Pigs' (present in many international cartoons as well). The examples given here are more particular to Indian allusions. Example (6) is from the Indian website 'Engage Voter,' by an anonymous author. The description of the cartoon is quite simple:

(6) Children sit in school uniform, wearing saffron, white, and green flu masks, next to armed fort. On the ground nearby are papers labeled, '15 AUG' and 'SWINE FLU.' The title of the cartoon is 'Celebrations under swine flu shadow.'

Two literary/cultural allusions present in this cartoon are the 15 August anniversary of India's 1947 independence, and that saffron, white, and green are the colors of the Indian flag. Every Indian reader would be expected to know these pervasive allusions.

In example (7), a deeper knowledge of Indian culture is assumed:

(7) Title: 'Swine flu spreads from the west.' Wife stands arms on hips next to husband, who is seated at the breakfast table reading a paper with the headline, 'SWINE FLU.' Husband to wife: 'Suddenly I've started feeling lucky that we don't have any NRI relative.'

Not only would the reader need to know what NRI was ('non-resident Indian,' or a person born in India and living abroad), but that many Indians have at least one relative living abroad. Many NRIs live in the West, and travel frequently from their countries of residence to India and back.

Don Wright of Tribune Media Services gives us a glimpse into US allusion in example (8), in which the number of cultural references is staggering:

(8) At the bottom right a large man with a small head holds a large sign that reads, '**SWINE FLU** IS A **HOAX** TO DISTRACT

FROM FIAT COMMUNISTS PUTTING POISONED CHINESE BATTERIES IN CHRYSLERS PROGRAMMED TO EXPLODE TO COVER UP INVESTIGATIONS INTO OBAMA'S FOREIGN BIRTH CERTIFICATE IN WHICH HE SWEARS TO CONFISCATE OUR GUNS!!!'

Included among the allusions are the issues of problems with foreigners, including the mid-twentieth-century communism scare and issues of the Vietnam War and Cold War that persist among older Americans, the recent discovery of Chinese-origin toxic toothpaste, the crumbling of US automakers such as Chrysler, the insistence of some conservative Americans that Barack Obama is foreign (making him ineligible for the US presidency) or Muslim (making him different and scary), and the conservative US tendency to cling to the second amendment, or the constitutional right to bear arms.

Knowledge of US law also enters into example (9), from Steve Benson of the *Arizona Republic*:

(9) At the top: 'THE OL' CROWDED THEATER ROUTINE.' In a crowded movie theater, someone in the back lets out a small, 'ACHOO!' and someone in the front stands and yells '**SWINE!!!**' People try to flee in a burst of chaos.

Even if US audiences are not aware of the particulars of the Supreme Court case in which Justice Oliver Wendell Holmes referred to the unnecessary and dangerous invocation of one's freedom of speech to cause panic, they will likely be aware of the phrase 'shouting fire in a crowded theater' as panic-inducing whether or not there is an actual threat. Fears of fire in this case have been replaced by fears of swine flu.

Some cultural references seem to permeate cartoon discourse for a number of countries. In addition to the references to 'The Three Little Pigs' previously mentioned (present in cartoons from France and India), the Grim Reaper figures in cartoons from Canada, Mexico, Singapore, and the US.

*Personal character traits*

The personal character traits in Medhurst and DeSousa's models of topoi are more about politicians, but in our analysis are used to address the traits of the people, animals, or characters featured in the cartoons. A number of the Indian cartoons portray ordinary citizens trying to live ordinary lives in a swine flu-obsessed world. In examples (10) and (11), both from Ninan at *The Times of India*, attempts at a quotidian existence are thwarted by swine flu panic:

(10) A man and a woman visit a museum. Statues are in various poses. They stop next to one statue of a couple in passionate embrace, wearing what appears to be gas masks. Man to woman: 'Don't you think they're carrying this swine flu thing a bit too far?'

(11) Two men walk up to a theater poster, which hangs under a 'NOW SHOWING' sign. A dog sleeps under the poster, encircled by flies on a garbage-strewn street. The poster is for 'GONE WITH THE WIND,' and shows Rhett Butler and Scarlett O'Hara in an embrace, wearing flu masks. The man on the right says to the man on the left, 'Somehow these remakes are never quite the same as the originals.'

Both (10) and (11) place their main characters in an evaluative position, acknowledging the desperation caused by the swine flu panic and speaking as if somehow they are outside the panic. These characters may be viewed as 'voices of reason' for the reader, who may feel justified, like the characters, in their feelings that swine-flu panic is an overreaction, or perhaps shamed or embarrassed into adopting similar attitudes.

Also from Ninan at *The Times of India* is another example of people who are quite fed up with the swine flu panic:

(12) Two men at a restaurant. The man on the right holds a menu with the name 'Vegicacy' in script. At the top of the cartoon are three drawings: a rib-eye steak, a chicken drumstick, and a pair of sausages. Each is in a circle with a line through it. Beneath, a man prepares a vegetable kebab tableside. Man on right to his dinner partner: 'After mad cow disease, bird flu and swine flu, there will be goat flu and fish flu. It's a vegetarian conspiracy.'

This cartoon features vegetarians assigned the trait of plotting conspirators and the diners as unhappy conspiracy theorists. An obviously outlandish claim, particularly in a country with a high rate of vegetarianism, this vegetarian conspiracy theory places the diners in a light of frustration and desperation for answers. While silly, the vegetarian conspiracy theory appears in consideration of the two previous pandemics of bird flu and mad cow disease, in which citizens were advised to avoid the consumption of beef and chicken, thus providing a possible answer for a bizarre set of recent animal-related scares. While these three cartoons do not incorporate themes particular only to India, this notion of trust, such as that of the media (thought to cause unnecessary panic, as in (10) and (11)), or with respect to conspiracy theory, as in (12), occurs quite often in Indian cartoons, as will be elaborated later in the chapter under 'Fears.'

Americans are portrayed in US cartoons in a much different light, generally as apathetic or conservative. Example (13) comes from Joel Pett of the Lexington, Kentucky, *Herald-Leader*:

(13) In the upper left-hand corner, a patient lies sick in a tattered bed that is labeled 'PREVENTABLE DISEASE EPIDEMIC.' In front of the bed, two people look mournfully over a water bucket, labeled 'WATER CRISIS,' that sits under a tap with one falling drop. To their left, two hands reach for an empty bowl labeled 'HUNGER EPIDEMIC.' Under the bowl, three tiny crosses and a stuffed animal lie on the ground next to a tombstone reading 'INFANT MORTALITY.' On the bottom right, a man sits in a recliner, with a beer, next to a sleeping dog, reading a newspaper with the headline, 'SWINE FLU.' Man: 'HAVE YOU HEARD?... THE WORST MAY BE YET TO COME!'

The American (and we of course know his identity, given the recliner, beer, and dog) enjoys the comforts of home while all around him are the scourges of the world. He is presented as apathetic in that only swine flu seems to have any resonance with him, perhaps because it is the only one that could potentially affect him. The typical American is then portrayed as uninterested in the problems of the rest of the world and unlikely to effect any kind of positive change with respect to those problems.

Another common US stereotype is that of the old white xenophobe. Consider example (14) from Steve Sack at the *Minneapolis Star-Tribune*:

(14) At coffeeshop. Old, bald curmudgeon on left, holding paper with headline 'SWINE FLU VACCINE.' To man on right with mustache, seed-corn cap, suspenders and flannel: 'THE GOVERNMENT BETTER NOT OFFER FLU SHOTS TO ILLEGALS! I ONLY WANT TO BE PROTECTED FROM AMERICAN GERMS!'

Through both the imagery and the text we are presented with the idea of the conservative American who is afraid of all things foreign. It should also be noted that this type of rhetoric (and fashion style) is common to rural coffee shops nationwide in the US, and the formulaic expression '... only ... American' hearkens from campaigns in the latter half of the twentieth century to buy products made the US in order to keep Americans employed.

Globally, cartoons overwhelmingly tend to employ personal character traits related to the personification of pigs. These are found in cartoons from Australia, Bulgaria, Canada, India, Nigeria, Singapore, and the US.

While political commonplaces, literary/cultural allusions, and situational themes establish shared knowledge of politics, culture, and events, personal character traits establish shared knowledge of 'types' of people. The artist may then exercise an agenda to show affiliation with or abhorrence of these types by assigning these traits to the people or animals featured. We would argue that the reader's stance places him or her within the in-group (read, conservative-minded Americans against "illegal" immigrants) or the out-group (either more progressive-minded Americans or immigrants themselves, whether documented or not) with respect to the cartoonist and other readers.

*Situational themes*

According to Medhurst and DeSousa (1981: 202), situational themes 'may have an immediate impact and contain a timely message, but they have little salience beyond their immediate context.' Swine flu itself certainly fits into this category; at the time of writing it has already made its way out of most cartoons.

However, this incessant media coverage did wear people down. We have already seen an example of this coverage in example (1), in which '[a]ll the paper's space is taken over by reports on swine flu and its activities...' Example (15) by Manjul (India) shows similar media hype:

(15) Title: 'Swine flu death toll climbs to 4, panic spreads.' Workers carrying out TVs. First worker to second, 'Strange minister! First he wanted to install a TV in each home to curb the population and now he wish (sic) to remove them to prevent swine flu panic.'

Not only does the cartoon allude to the swine flu panic-ridden media, it also alludes to the Indian effort to curb population by putting televisions in homes to begin with. These ideas are equated in their transience.

In fact, this relation between panic and population appears again in example (16), from Morparia in *Time Out Delhi*:

(16) Four frames. First frame: Stork #1, 'THAT SWINE FLU SCARE THAT HAPPENED A WHILE AGO WAS SOMETHING, EH?' Stork #2, 'OH, YEAH!' Second frame: Stork #1, 'PEOPLE GETTING INTO A REAL PANIC, SCHOOLS, COLLEGES CLOSING, THEATRES, MULTI-PLEXES SHUTTING, EXHIBITIONS, LECTURES GETTING CANCELLED, RESTAURANTS – EMPTY.' Stork #2, 'UH HUH!' Third frame: Stork #1, 'EVERYONE WAS SCARED TO VENTURE OUTSIDE – PEOPLE JUST STAYED PUT AT HOME.' Stork #2, 'SURE.' Fourth frame: The two storks fly among many other storks, carrying babies in baskets. Stork #1, 'A LITTLE OVER NINE MONTHS AGO, WASN'T IT?  WHAT HYPE! WHAT PANIC! ... AND ALL THOSE MASKS...' Stork #2, 'NOT THAT THEY WERE USELESS FOR US...'

The situational theme here is the expected population boom that would follow from people staying home out of swine flu panic. When other activities are canceled, naturally they find other ways of keeping busy.

In the US cartoons, a much more political commentary emerges. Example (17) is from Ted Rall of the Universal Press Syndicate:

(17) Upper left corner: 'Swine flu could infect half the US population this winter. The good news: Swine flu is one of the few illnesses they'll treat without medical insurance.' Woman with left arm off (leaving a bloody stump) sits in a doctor's office. She tells the doctor, 'IT'S SWINE FLU, I TELL YOU! I DONE COUGHED IT RIGHT OFF!' Her arm is in her purse, leaking blood next to her.

In this example, the central character makes desperate and ridiculous claims that swine flu is the reason for her dismemberment, in order to be entitled to free health care. Of course, to understand the situational theme of the coverage of the swine flu treatment even without health insurance, the reader would have to be aware that health care coverage is an ongoing controversy in the US. It is further notable that the woman's language is most certainly of a non-prestige variety ('I done coughed it right off'), possibly signifying that the woman is uneducated and/or rural, perhaps an indicator of her socioeconomic status and her lack of access to health care.

The situational theme of Mexico as location or source of swine flu shows up in six of the US cartoons, already seen in (4) and alluded to in (14). In example (18), from Dick Locher of *The Chicago Tribune*, anti-Mexican discourse in the US flows publicly with the freedom allowed by the transience of the situation:

(18) Two fat, sick-looking mosquitoes point their noses over a dashed line. On their side of the line is a sign reading 'MEXICO.' The mosquito on the left, whose back reads, 'SWINE FLU,' says, 'SO WE'RE CROSSING INTO THE UNITED STATES. WHO'S GOING TO STOP US?'

In this situation, business is *not* as usual on the US/Mexico border. If it were, someone indeed would stop border crossers from entering (though certainly not mosquitoes): US Immigration and Customs Enforcement, or ICE, the US government agency concerned with keeping immigrants, particularly Mexican immigrants, from entering the US illegally. Of course, this background knowledge, in addition to the portrayal of Mexican citizens as mosquitoes sick with swine flu who can think of nothing else but how to get into the US, makes the situational theme that much more apparent.

The situational theme of the novelty of swine flu is common to cartoons worldwide, presenting it as a new employee (Canada), a newly arrived tourist (Canada), a sneezing sneak attacker (India, UAE), a new member of the World Pandemic Organization (India), a pig dropping from the sky (India), a decaying pig carried by a bird skeleton (Singapore), a pig in the hand of a bigger pig (US), a monster/alien pig hovering over a world rife with problems (US), an object to juggle (US), a type of disaster drill for schoolchildren (US), a newspaper headline/television news topic (US), and a giant hog approaching the White House (US). In many of these cartoons, swine flu is somehow added to an already sizeable number of dire issues.

### 3.4.2 Metaphors

For each cartoon, metaphors emerged as analogies between visual and rhetorical discourse and more complex ideas. For example, in (18) above, we determined the following analogies: MOSQUITOES ARE MEXICAN CITIZENS/ILLEGAL IMMIGRANTS; MOSQUITOES ARE SWINE FLU CARRIERS; thus MEXICAN CITIZENS/ILLEGAL IMMIGRANTS ARE SWINE FLU CARRIERS. Following the analysis for all 72 cartoons, 16 supercategories of metaphor emerged from the collective data. These supercategories included five types of metaphors: container metaphors, metaphors in which the effect entails the cause, metaphors expressing a part/whole relationship, metaphors in which one problem is presented in terms of another problem, and metaphors in which a problem is personified. The remaining 11 categories represent metaphors in which at least one member of the category is present. In the 'death' category, for example, would be included the presentation in example (13) of the abandoned stuffed animal next to the crosses, understood as the metaphor A LONELY DOLL IS AN ABSENT/DEAD CHILD. As there are often a number of images or sentiments conveyed in each cartoon, the total number of metaphors exceeds the number of cartoons.

The supercategories of metaphors are named as follows: *Container, Death, Disease/illness (general), Effect/cause, Emptiness/absence/loss, Mask, Mexico/Mexicans/immigration, Other (non-porcine) animals, Part/whole, Pigs/hogs, Problem = another problem, Problem = person, Supernatural, Swine flu,* and *Other metaphors*. A detailed list of the metaphor supercategories and the individual metaphors included within them for India and the United States may be obtained by contacting the authors.

*Overall*

Metaphors used for each country broke down as follows, with India and the US highlighted in bold, are given in Table 3.1.

*Table 3.1* Metaphors used in swine flu cartoons by country (raw numbers)

| | Container | Death | Disease/illness (general) | Effect/cause | Emptiness/absence/loss | Mask | Mexico/Mexicans/immigration | Other animals | Part/whole | Pigs/hogs | Problem = another problem | Problem = person | Person = another person | Supernatural | Swine flu | Other metaphors |
|---|---|---|---|---|---|---|---|---|---|---|---|---|---|---|---|---|
| Australia | 0 | 0 | 0 | 0 | 0 | 0 | 0 | 0 | 0 | 0 | 0 | 0 | 0 | 0 | 3 | 0 |
| Bulgaria | 0 | 0 | 1 | 0 | 0 | 0 | 0 | 0 | 0 | 0 | 0 | 0 | 0 | 0 | 0 | 0 |
| Canada | 0 | 3 | 3 | 1 | 0 | 0 | 2 | 0 | 0 | 0 | 0 | 0 | 0 | 0 | 1 | 0 |
| France | 0 | 0 | 2 | 1 | 0 | 0 | 0 | 0 | 0 | 0 | 0 | 0 | 0 | 0 | 0 | 0 |
| **India** | **2** | **3** | **7** | **1** | **0** | **16** | **0** | **1** | **3** | **6** | **2** | **10** | **4** | **0** | **19** | **8** |
| Jordan | 0 | 0 | 0 | 0 | 0 | 0 | 0 | 0 | 0 | 0 | 1 | 0 | 0 | 0 | 1 | 0 |
| Mexico | 0 | 1 | 0 | 0 | 0 | 0 | 0 | 1 | 0 | 3 | 0 | 0 | 0 | 1 | 0 | 0 |
| Nigeria | 0 | 0 | 0 | 1 | 0 | 0 | 1 | 0 | 0 | 0 | 0 | 0 | 0 | 0 | 0 | 0 |
| Norway | 0 | 0 | 0 | 0 | 0 | 0 | 0 | 0 | 0 | 0 | 0 | 0 | 0 | 0 | 0 | 0 |
| Singapore | 0 | 0 | 0 | 3 | 0 | 0 | 0 | 0 | 1 | 0 | 0 | 5 | 0 | 0 | 0 | 0 |
| Thailand | 0 | 0 | 0 | 0 | 0 | 0 | 0 | 0 | 0 | 2 | 0 | 0 | 0 | 0 | 0 | 0 |
| UAE | 0 | 0 | 1 | 0 | 0 | 0 | 0 | 0 | 0 | 1 | 0 | 0 | 0 | 0 | 0 | 0 |
| **US** | **3** | **7** | **9** | **0** | **5** | **0** | **14** | **6** | **2** | **10** | **0** | **0** | **1** | **0** | **12** | **3** |

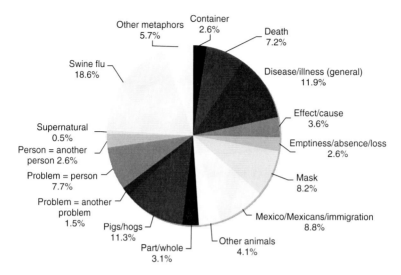

*Figure 3.1* Overarching metaphors employed in cartoons across all countries.

We found 194 total metaphors in the cartoons from all countries. Metaphors involving the direct equation of swine flu with another item accounted for 18.6% of the overall data. Metaphors involving disease and illnesses in general accounted for 11.9%, and those featuring pigs or hogs accounted for 11.3% across all countries. Figure 3.1 shows the breakdown of metaphors overall.

*India*

The most frequent metaphors for the Indian cartoons were those involving swine flu (23.2%), the problem personified (12.2%), and other metaphors (9.8%).

A metaphor involving swine flu as one of the components can be found in example (19) from India's 'Engage Voter':

(19) Title: 'Swine coming after you...' Masked man runs frightened to the north of India (seems to be somewhere in the region of Eastern Jammu and Kashmir) across crudely drawn subcontinent to escape reaching hand labeled 'SWINE FLU' with dripping bloody fingers.

One of the metaphors apparent in this example is BLOODY REACHING HAND = SWINE FLU; thus, swine flu is one of the components of the metaphor.

Other Indian metaphors with swine flu as one of the components include INACCESSIBILITY TO PUBLIC = UNSUSCEPTIBILITY TO SWINE FLU, MASK = SWINE FLU (n=5), NAXALS = H1N1, NRI (NON-RESIDENTIAL INDIAN) = CARRIER OF SWINE FLU, PIG = SWINE FLU (n=2), SWINE FLU (PANIC) = OUT-OF-CONTROL FLAMES, SWINE FLU = DEATH, SWINE FLU = FACECBOOK USER, SWINE FLU = FALLING PIG, SWINE FLU = FOREIGN (FROM ABROAD), SWINE FLU = INDIA, SWINE FLU = SHADOW, and SWINE FLU PILLS = PEARLS.

Example (20), a cartoon by Belluramki (India) demonstrates the derivation of a metaphor in which a problem or an abstract concept is presented as a person:

(20) On top, pig on right to pig on left: 'Don't reveal you are a swine on Facebook. The number of hits may decrease, and who knows, Facebook may cancel your account!' On bottom right, 'SWINE SCARE IN CITY: Swine Flu suspect cases rises (sic).'

In this strange example, PIGS = SWINE FLU and PIGS = FACEBOOK USERS, so SWINE FLU IS A FACEBOOK USER. Other metaphors in Indian cartoons in which a problem or abstract concept is presented as a person include DISEASES = POLITICIANS, DOCTOR WITH BAG = HEALTH CARE, EARTH = SICK PERSON, JUDGE = PROTECTION FROM PIL (PUBLIC INTEREST LITIGATION), MILITARY AGENT = PROTECTION FROM TERROR, NRI = CARRIER OF SWINE FLU, PIG DOCTOR WITH BAG = CURRENT STATE OF HEALTH CARE WHAT WITH THE SWINE FLU AND ALL, PERSON WEARING NO MASK = UNPROTECTED, POLITICIAN = VALUABLE COMMODITY.

The third-largest category, metaphors that did not easily fit into any other categories, included ART = REALITY, FIRE EXTINGUISHER = CONTROL, MOON = FAIR WEATHER FRIEND, MOVIE POSTER = REALITY, SMALL = INADEQUATE, TELEVISION = PANIC, TELEVISION = PROCREATION. Figure 3.2 shows the breakdown of supercategories for Indian cartoons.

## The United States

The most frequent metaphors used by US cartoonists were those that included Mexico/Mexicans/immigration (19.4%), swine flu (16.7%), pigs/hogs (13.9%), disease/illness (general) (12.5%), and death (9.7%).

We have already seen some examples of Mexican metaphors in (18); Scott Stantis' cartoon in *USA Today* is another example:

(21) Top quarter of frame: Large banner in orange, green, and yellow: 'Feliz Cinco de Mayo '09.' A lonely pig sits far underneath the sign, with a mask over its snout, a tiny 'ACHOO' written next to its head.

*Metaphors in Cross-cultural Political Cartoons* 77

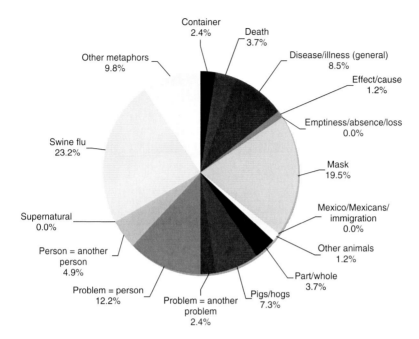

*Figure 3.2* Overarching metaphors employed in Indian cartoons

In the cartoon described in (21), the metaphor is simply MEXICANS = INFECTIOUS PIGS. Other Mexico/Mexicans/immigrants metaphors in American cartoons are MEXICO = SWINE FLU, MEXICAN = DEATH, MEXICAN CITIZENS/ILLEGAL IMMIGRANTS = SWINE FLU CARRIERS/INFECTIOUS/INFECTIOUS PIGS (n=4), MEXICO IS THREATENING TO USA, SNAKE = MEXICAN, MEXICAN ILLEGAL IMMIGRANTS WHO GET FLU SHOT = DRAIN OF RESOURCES THAT SHOULD BE FOR AMERICAN CITIZENS, MEXICANS ARE AGENTS OF TERRORISM, MOSQUITOES = MEXICAN CITIZENS/ILLEGAL IMMIGRANTS, PIG = MEXICAN, US/MEXICAN BORDER AS PROTECTION FOR AMERICANS FROM INFECTIOUS MEXICANS, and IMMIGRANTS ARE OUT OF CONTROL.

While swine flu metaphor has been shown in Indian cartoons in example (19), an example is shown in (22) below for a US cartoon by Marty Two Bulls, Sr. for the *(American) Indian Country Reporter*:

(22) Grim reaper with scythe and false pig nose held on by elastic, saying, 'Oink Oink.'

For Two Bulls' cartoon, the metaphor is SWINE FLU = DEATH IN DISGUISE. Other swine flu metaphors in American cartoons include MEXICO = SWINE

FLU, ALIEN PIG = FLU PANDEMIC, PIG = SWINE FLU (n=2), SWINE FLU = COMMUNIST/ CHINESE/OBAMA-GENERATED HOAX ON AMERICANS, SWINE FLU = DELIBERATE FEAR TACTIC, SWINE FLU = HARDER TO DEAL WITH THAN/EQUAL TO JUGGLING CHAINSAWS, SWINE FLU = PREVENTABLE, SWINE FLU = SCAPEGOAT?, SWINE FLU = SUDDEN, and SWINE FLU = TRANSIENT/RAPIDLY MOVING.

Pig and hog metaphors also abound, as the animal is equated with death, disease, bikers, and Mexicans, for example, (23) from Chan Lowe of the *South Florida Sun-Sentinel*/Tribune Media:

(23) A pig in a trench coat and hat pops out of a gangway between two buildings (that look like typical NY walk-ups), with one side of his coat open, displaying vials of something. Pig's bubble: 'PSST!' (typical scene, minus the pig, of guy selling illicit items). Man, with boy, on the right of the frame: 'LET ME GUESS... BLACK MARKET SWINE FLU VACCINE.' Overarching caption: 'AND THE FREE-FOR-ALL BEGINS'

The metaphor PIG = FLASHER/ILLICIT DEALER comes from example (23). Other pig/hog metaphors from American cartoons include MEXICAN CITIZENS/ ILLEGAL IMMIGRANTS = SWINE FLU CARRIERS/INFECTIOUS/INFECTIOUS PIGS (n=4), ALIEN PIG = FLU PANDEMIC, ALIEN PIG = WORST THING CONCEIVABLE, CONSERVATIVE AMERICANS = PIGS, MOTORCYCLE = HOG, PIG = DEATH (n=2), PIG = DISEASE (n=2), PIG = ETHNICITY, PIG = GREED, PIG = MEXICAN, PIG = SUSPECTED TERRORIST, PIG = SWINE FLU (n=2), and PIG = TOUGH BIKER.

We have seen metaphors about disease and illness in general in example (13), where a SICK PATIENT IN BED = PREVENT DISEASE, or in example (9), where A SNEEZE/'SWINE' = JUSTIFIED CAUSE OF RIOT. Other examples of metaphors about disease and illness in general in American cartoons are SYMPTOM = DANGER, DISEASE = DANGER, PIG = DISEASE (n=2), POTENTIAL VICTIMS = DESPERATE CONSUMERS, SYMPTOM = DISEASE, and SYMPTOM OF DISEASE = FIRE.

Metaphors of death in US cartoons include the one above from example (13): A LONELY DOLL IS AN ABSENT/DEAD CHILD. Other examples include MEXICAN = DEATH, DEATH = TRANSIENT, PIG = DEATH (n=2), SWINE FLU = DEATH IN DISGUISE, TINY CROSSES = DEAD CHILDREN. Figure 3.3 shows the breakdown of supercategories for US cartoons.

*Other countries*

Countries other than India and the US had as their most frequent metaphors disease/illness (general) (17.5%), pigs/hogs (15.0%), effect/cause (15.0%), swine flu (12.5%), the problem personified (12.5%), and death (10.0%). As we've already seen examples of the rest of the metaphors,

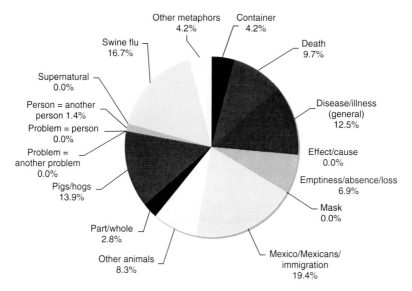

*Figure 3.3* Overarching metaphors employed in US cartoons

we will turn to examples (24) and (25) for examples of metaphors of a cause/effect relationship.

(24) Top frame: scene on a train. Man seated third from the left sneezes into a tissue. Bubble: 'AH-CHOO!' Bottom frame: man sits alone on train.

(25) Smiling pig with a sombrero and a skull for a nose wears a t-shirt that reads, 'SWINE DUDE'

Example (24) is by Theo Moudakis of the *Toronto Star* (Canada), demonstrating the metaphor SYMPTOM = DISEASE. Example (25) is by Tayo, of *This Week* in Lagos, Nigeria, and demonstrates the metaphor SKULL = DEATH. In both instances, the effect is indicative of the cause. Figure 3.4 shows the breakdown of supercategories for the cartoons of other countries.

### 3.5.3 Fears

The analysis of topoi and metaphors present in the H1N1 cartoons in a sense set the basis for exploring the fears of a culture. In these cartoons, fears are expressed as on par with or related to fear of contracting swine flu. For example, in a cartoon from Bulgaria in which a piggy bank

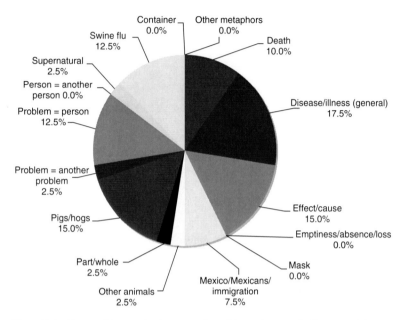

*Figure 3.4* Overarching metaphors employed in cartoons of other countries

labeled 'WORLD BANK' looks sick and sneezes, readers must have the background knowledge of children's banks being shaped like pigs, the swine flu epidemic, and the economic crisis to get the metaphor of ECONOMY = SICK PIG(GY BANK) and the exploitation of the fear of a global economic crash would fall under the supercategory of 'economy and poverty.' The individual fears for each cartoon gave rise to 12 overarching supercategories: *Other countries/foreigners, Change in quality of life, Economy and poverty, Terrorism, Health, Trust, Death, Supernatural, Environmental issues, War, Crime, and Other fears*. A detailed list of the fears supercategories and the individual metaphors included within them, and the individual fears for India and the US, may be obtained by contacting the authors.

*Overall*

Fears used for each country broke down as in Table 3.2, with India and the US highlighted in bold.

The most frequent fears exploited by the countries overall were those of other countries/foreigners (15.1%), terrorism (13.4%), other fears (13%), and economy/poverty (10.1% ). Figure 3.5 shows the breakdown of fears overall.

*Table 3.2* Fears exploited in swine flu cartoons by country (raw numbers)

| | Other countries/ foreigners | Change in quality of life | Economy and poverty | Terrorism | Health | Trust | Death | Supernatural | Environmental issues | War | Crime | Other fears |
|---|---|---|---|---|---|---|---|---|---|---|---|---|
| Australia | 0 | 0 | 0 | 0 | 0 | 0 | 1 | 0 | 0 | 0 | 0 | 0 |
| Bulgaria | 0 | 0 | 1 | 0 | 0 | 0 | 0 | 0 | 0 | 0 | 0 | 0 |
| Canada | 2 | 3 | 0 | 0 | 0 | 1 | 0 | 0 | 0 | 0 | 0 | 0 |
| France | 0 | 0 | 0 | 0 | 0 | 0 | 0 | 0 | 0 | 0 | 1 | 0 |
| **India** | **2** | **1** | **4** | **4** | **5** | **7** | **1** | **0** | **1** | **2** | **5** | **7** |
| Jordan | 0 | 0 | 0 | 0 | 0 | 0 | 0 | 0 | 0 | 1 | 0 | 0 |
| Mexico | 0 | 0 | 0 | 0 | 0 | 0 | 0 | 1 | 0 | 0 | 0 | 1 |
| Nigeria | 1 | 0 | 0 | 0 | 0 | 0 | 0 | 0 | 0 | 0 | 0 | 0 |
| Norway | 0 | 0 | 0 | 3 | 0 | 0 | 0 | 0 | 0 | 0 | 0 | 0 |
| Singapore | 1 | 0 | 2 | 0 | 1 | 0 | 0 | 0 | 0 | 0 | 0 | 1 |
| Thailand | 0 | 0 | 0 | 1 | 0 | 0 | 0 | 0 | 0 | 0 | 0 | 1 |
| UAE | 0 | 0 | 1 | 1 | 0 | 0 | 0 | 0 | 1 | 1 | 0 | 0 |
| US | 12 | 6 | 4 | 7 | 3 | 3 | 3 | 0 | 3 | 4 | 2 | 6 |

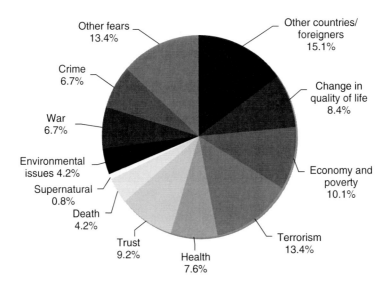

*Figure 3.5* Overarching fears employed in cartoons across all countries

## India

The most frequently exploited fears used by Indian cartoons were the catch-all category of other fears (17.9%), and fears related to trust (17.9%), crime (12.8%), health (12.8%), terrorism (10.3%), and economy and poverty (10.3%).

One example of a fear related to trust is in example (12) about the vegetarian conspiracy. Another is shown in example (26) from Indian cartoonist Manjul:

> (26) Title: 'Pandemic flu death toll passes 1000 mark: WHO.' Doctor sits at desk with red cross on front, speaks to politician in Nehru cap and vest. Text at bottom: 'The virus can't touch you. You are highly inaccessible.'

The humor in this cartoon is in the absurd idea that H1N1 would be susceptible to the same type of constraints as the typical citizen of Indian. The fear associated with inaccessibility to one's government, within the larger category of a 'trust' fear, is exploited in (26), a particularly poignant cartoon as India prides itself on its democracy.

Crime fears also abound, and include examples of robbery/getting caught by police, as in (27):

> (27) Title: 'India swine flu toll rises.' Policeman runs like mad past a mask-wearing thief holding a bag of cash.

While of course the fear of swine flu is most salient in this cartoon, the fear of theft/crime is exploited in order to offer the humorous irony of the policeman fearing the thief.

Health-related fears in Indian cartoons sometimes relate to other diseases, such as HIV/AIDS, and sometimes the state of health care in general, as in (28):

> (28) Title: '14-yr-old girl India's first swine flu victim.' Man on right holds newspaper with headline '14-YR-OLD DIES OF SWINE FLU'; looks at anthropomorphized pig in glasses, doctor's jacket, and carrying a doctor's bag reading 'HEALTH CARE.' Pig doctor looks smug, with half-smile and semi-closed eyes.

This cartoon plays on the assumed readership's fear of inadequate health care, as the smugness and/or apathy of the health care industry is conveyed to be possibly capitalizing on the swine flu epidemic.

Terrorism is a fear that is exploited in a number of ways. The cartoon in example (29) by Ninan at the *Times of India* shows the placement of fighting H1N1 as somehow on equal footing with fighting Naxals (Maoist terrorists in India):

> (29) On the left, a doctor stands before a medical vehicle with a large gun. On the right, a military agent stands in front of a CPF vehicle wielding a syringe. Both are dressed in typical professional garb, and behind them are legions of others outfitted similarly. At the bottom, a man with a clipboard yells at another man: 'War footing, yes... I ***did not*** mean paramedical forces fight Naxals and paramilitary forces fight H1N1!'

Other Indian cartoons exploiting fears of terrorism center on Pakistan, Afghanistan, and the Taliban.

The other major fear exploited in Indian cartoons is that of poverty or economic meltdown. Example (30) comes from Manjul:

> (30) Title: 'Spike in N95 mask demand due to swine flu.' Man on left, wearing flu mask, points at other man's nose. Man 2 and his whole family are wearing flu masks only around their mouths. Man 2 to Man 1: 'We need not cover our noses. Our masks are to save us from the rising prices of food items.'

Quite a sad commentary on the state of affairs, example (30) places the fear of contracting H1N1 and the fear of not being able to pay for food as roughly equivalent. Figure 3.6 shows the breakdown for fears exploited in Indian cartoons.

*The United States*

The American cartoons most often featured fears of other countries/foreigners (22.6%), terrorism (13.2%), change in the quality of life (11.3%), and other fears (11.3%).

Of the fears associated with other countries or foreigners, Mexico figured most heavily in the US cartoons. One such cartoon is (31), from Steve Breen of the *San Diego Union-Tribune*:

> (31) Color scene of blue sky over a desert. An SUV with the words 'BORDER PATROL' and the US flag on the hood drives on

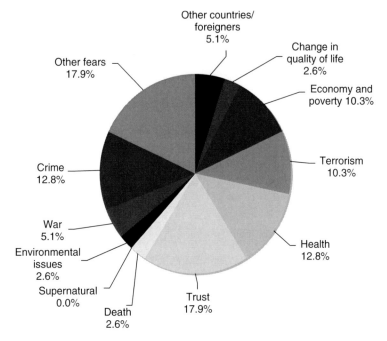

*Figure 3.6* Overarching fears exploited in Indian cartoons

the left. A giant glass wall (maybe 30 or 40 feet high) stands vertically in the desert, separating the SUV from several brown-skinned people pressed up against the glass. One man on that side sneezes, 'ACHOO!' The driver of the SUV says, 'IT'S A 2,000-MILE LONG SNEEZE GUARD.' The text at the top reads, 'COPING WITH SWINE FLU'

Example (31) achieves the portrayal of the US fear of Mexicans by situating the typical US Border Patrol scene in which 'illegal immigrants,' who are seen as wanting and taking US resources such as jobs and education, are sought and sent back, alongside the fear of Mexicans as infectious spreaders of H1N1. Anti-immigration discourse, mostly related to Mexicans, is characteristic of conservative Americans.

In fact, US cartoons addressing fears of terrorism also sometimes feature a cynical, conservative American spouting a number of problems as being on the level of H1N1, regardless of their relevance. Consider (32) from US cartoonist Nick Anderson:

(32) Top of frame: 'The SWINE/AVIAN CONTAGION...' Frame divided into four subframes. In upper left-hand frame, heavy-jawed man (who looks like several conservative American pundits) says, 'ILLEGALS ARE POURING OVER THE BORDER, INFECTING US CITIZENS....' Second frame shows the same man, continuing his speech, but now he has a pig nose: 'THEY SHOULDN'T RECEIVE TREATMENT IN THIS COUNTRY...' Third frame shows man with pig nose and ears: 'THE PANDEMIC IS DUE TO UNCONTROLLED IMMIGRATION....' In the final frame, man is fully a pig, with a cuckoo-clock-style opening in his head and a cuckoo bird flying out with a 'CUCKOO' and the man saying, 'IT'S A TERRORIST PLOT...'

The terrorist plot conspiracy theory is often used by conservative American pundits to justify their positions on immigration, so (32) is a nice example of fears of Mexican immigration along with fears of terrorism exploited in the discourse on H1N1. Readers' familiarity with the cuckoo bird as a sign of insanity enables them to grasp the cartoonist's position on this type of rhetoric. It is cases like this that demonstrate the reflection of and challenge to a mainstream discourse through the use of a political cartoon.

Change in quality of life is shown in such ways as allusion to a lack of resources (present in some cartoons about Mexican immigrants), or even, in the case of example (8) above, the loss of the constitutional second amendment right to carry guns. Another change in the quality of life demonstrated in American cartoons is that of alienation:

(33) Left foreground: Miss Piggy, alone, arms crossed, looking angry. Right background: four Muppets wearing masks, one holding a newspaper with the headline: 'SWINE FLU OUTBREAK'

Example (33) features well-loved children's television puppets, 'The Muppets.' One of the better-known of these Muppets is Miss Piggy, a porcine diva of sorts. Her quality of life has changed as her friends and colleagues have abandoned her over the fear of swine flu. The moratorium on going anywhere near anything related to swine flu, pigs, or pork has resulted in those who carry these stigmas, have H1N1, or so

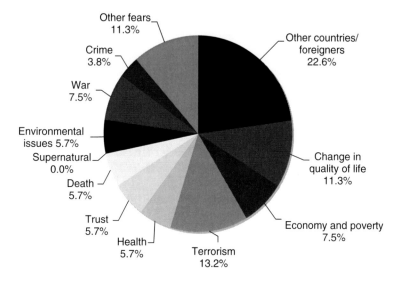

*Figure 3.7* Overarching fears exploited in US cartoons

much as sneeze to be ostracized. Figure 3.7 shows the breakdown for fears exploited in US cartoons.

It should be noted that in some cartoons, many fears are expressed at once, as in a cartoon from Indian artist Paresh Nath:

(34) Barack Obama and Hillary Clinton sit at a table with a large map. On the map, the words, 'GLOBAL WARMING,' 'N-PROLIFERATION,' 'TERRORISM,' 'RECESSION.' Obama has booklet open entitled 'STRATEGY TO FIGHT.' Obama and Clinton look terrified as they are approached from behind by a giant pig labeled 'FLU' which sneezes, 'AAACHOOO!!'

Consider a similar cartoon from US artist Monte Wolverton of Cagle Cartoons:

(35) Upper left corner: 'JUST WHEN YOU THOUGHT IT COULDN'T GET WORSE...' Monster/alien pig (whose forehead reads 'FLU PANDEMIC') hovers over a 'world'; continents are printed with 'nukes,' 'terrorists,' 'torture,' 'pirates,' 'recession,' 'environment.'

These cartoons essentially place H1N1 fears on par with fears of global warming, nuclear proliferation, terrorism, economic recession, torture, pirates, and environmental decline. This strategy, as shown above, is used both in India and the US, but not in any of the cartoons that we have found from other countries.

*Other countries*

In countries outside of India and the US, fears of terrorism are most prominent (18.5%), followed by fears relating to economy and poverty (14.8%), other countries/foreigners (14.8%), changes in quality of life (11.1%), and other fears (11.1%).

One quite sad example of a cartoon that reflects a fear of change in the quality of life comes from Gable at Toronto, Canada's *The Globe and Mail*:

> (36) A classroom. Alphabet border above chalkboard on right. 'WELCOME BACK' written on board. Buxom female teacher in a hazmat [hazardous material] suit and a bottle of hand sanitizer kneels to accept an apple from a small student in a hazmat suit and a backpack. On the bulletin board to the left hangs a small sign with 'H1N1 PRECAUTIONS.'

This cartoon demonstrates the fear of loss of childhood experiences and a changed educational environment. To our knowledge, this extreme situation has not come to pass in any classroom, but reflects the cartoonist's (and potentially readers') feelings of its possibility in light of H1N1 precautions already in place.

Other fears expressed in other countries' cartoons include economic crash (Bulgaria and Singapore), nuclear war (Jordan), the devil (Mexico), rabid animals (Mexico), Mexico (Nigeria and Singapore, of all places), terrorism (Norway and Thailand), and bird flu (Singapore). Figure 3.8 shows the breakdown for fears exploited in other countries' cartoons.

## 3.5 Conclusion

Our multimodal, multifaceted analysis of swine flu cartoons around the world yields clues about the prevalent fears in India, the United States, and other countries. While the prevailing fear(s) may differ from nation to nation, there is strong evidence of shared (socio)linguistic norms and rhetorical strategies. Of course, one possible explanation for these commonalities may be due to the fact that we only analyzed

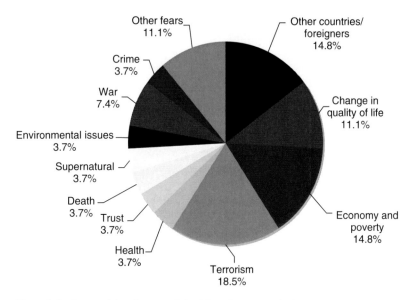

*Figure 3.8* Overarching fears exploited in other countries' cartoons

political cartoons that have texts in the English language (if they have any linguistic text at all). In her analysis of Indian editorial cartoons depicting founder of Pakistan Mohammed Ali Jinnah, Kamra (2003:3) makes the following observation:

> ... although it has proved to be near impossible to locate the cartoonists' biographies, the very choice of language – English – and political involvement they display, the texts they draw on and rhetorical sensibility they exercise suggest we are dealing with an Indian elite, the product of a colonial education. At all levels, we note a familiarity with western texts, history and ideologies, and often an internal judgement of social formations that is traceable to a 'western' value system.

Kamra's point is well taken with regard to our research: all of the cartoons we analyzed were written in English, although they were obtained from sources all over the world. That command of English is expected of the readers of these cartoons assumes not just a shared linguistic knowledge, but perhaps a set of ideologies and socio-educational background as well. To the extent that certain metaphors and fears hold within the English-speaking communities both globally and in each country, Kamra's statement encapsulates the shared values of a given readership. Indeed, many of the cartoons from outside the US and India share globally recognizable topoi, such as 'The Three Little Pigs,' terror threat levels, strategy maps, sneezing pigs, swine flu as a passing issue, and media hype. Once H1N1 was branded 'swine flu,' the pig as metaphorical component became an obvious choice worldwide. However, the common topoi of strategy maps, passing issues, and media hype are far more interesting. Types of political rule and strategies vary the world over, yet the idea of military strategists hovering over a map appears in cartoons from both India and Dubai. Swine flu is treated as a newcomer, an old trick played too many times, especially in light of animal-related scares like mad cow disease, hoof and mouth disease, and bird flu, or a replacement for some other fad-like panic in Canada, India, Jordan, Singapore, Thailand, and the US, which also speaks to the public's wariness of alarmist tendencies in the media. Those who share the sentiments of the cartoons may agree (ironically, of course, given that they are consuming media) that they are too smart to fall for the hype, to fall for another fad disease and its over-reliance on cheap masks, to give up their lifestyles to fear. Both India and Norway use depictions of Taliban agents to situate their points about swine flu. Common themes abound across nations.

All of the cartoons analyzed here were obtained from internet sources. Does this mean that globalization has caused everyone to have the same backgrounds and fears? As stated above, the commonalities among the cartoons of seemingly disparate nations are quite remarkable, although the chosen language of English carries with it certain implications for the socioeconomic and educational backgrounds of the cartoonists.

To approach a study of this nature from a cross-cultural and cross-linguistic perspective would no doubt add immense depth to the assumed shared cultural backgrounds of the readers. Future work on political cartoons should examine linguistic texts in other languages to measure the universality of themes, fears, and so on. Furthermore, metaphor and semiotic analysis has mostly been performed through Western eyes; good is up, bad is down, given is left, new is right, and so on. Comic studies as a field of inquiry could certainly benefit from application of metaphorical and semiotic analysis of cartoons that come from cultures with non-Western writing systems, such as Arabic and Chinese, which may orient the audience in a different way than those produced in areas employing left-to-right writing systems.

If, as Bakhtin explains, '[e]ach word tastes of the context and contexts in which it has lived its socially charged life; all words and forms are populated by intentions,' (1981: 293) the visual-linguistic analysis of H1N1 cartoons reveals ideologies surrounding H1N1 discourse holistically and within individual locales. These ideologies are multiply embedded in that they show up textually and visually, reflecting analogies and fears, within and across news cultures, in a Bakhtinian ventriloquation of public opinion, for public consumption.

## References

Agha, A. (2007) *Language and Social Relations*. Cambridge: Cambridge University Press.
Bakhtin, M. M. (1981) *The Dialogic Imagination*. [Trans. C. Emerson and M. Holquist.] Austin: University of Texas Press.
Bell, A. (1991) *The Language of the News Media*. Oxford: Blackwell.
Douglas, R., Harte, L., and O'Hara, J. (1998) *Drawing Conclusions: A Cartoon History of Anglo-Irish Relations, 1798–1998*. Belfast: Blackstaff.
Dwivedi, D. (2009) Of lines and letters. In S. Panja, S. Chakrabarti, and C. R. Devadawson (eds.) *Word, Image, Text: Studies in Literary and Visual Culture*, pp. 131–36. New Delhi: Orient BlackSwan.
Edwards, J. L. and Winkler, C. K. (1997) Representative form and the visual ideograph: the Iwo Jima image in editorial cartoons. *Quarterly Journal of Speech* 83 (3): 289–310.
Fairclough, N. (1999) Linguistics and intertextual analysis within discourse analysis. In A. Jaworski and N. Coupland (eds.) *The Discourse Reader*, pp. 183–211. London: Routledge.
Fairclough, N. (2001) *Language and Power*. 2nd edition. London: Longman.
Forceville, C. (2005) Addressing an audience: time, place, and genre in Peter van Straaten's calendar cartoons. *Humor* 18 (3): 247–78.
Garrett, P. and Bell, A. (1998) Media and discourse: a critical overview. In A. Bell and P. Garrett (eds.) *Approaches to Media Discourse*, pp. 1–20. Oxford: Blackwell.
Goffman, E. (1959) *The Presentation of Self in Everyday Life*. Garden City, NJ: Doubleday, Anchor Books.
Gombrich, E. H. (2000) *Art and Illusion: A Study in the Psychology of Pictorial Representation*. Princeton: Princeton University Press.
Greenberg, J. (2008) Framing and temporality in political cartoons: a critical analysis of visual news discourse. *Canadian Review of Sociology/Revue canadienne de sociologie* 39 (2): 181–98.
Hakam, J. (2009) The 'cartoons controversy': a critical discourse analysis of English-language Arab newspaper discourse. *Discourse & Society* 20 (1): 33–57.
Han, J. N. (2006) Empire of comic visions: Japanese cartoon journalism and its pictorial statements on Korea, 1876–1910. *Japanese Studies* 26 (3): 283–302.
Hull, M. B. (2000) Postmodern philosophy meets pop cartoon: Michel Foucault and Matt Groening. *Journal of Popular Culture* 34 (2): 57–67.
Kamra, S. (2003) The war of images: Mohammed Ali Jinnah and editorial cartoons in the Indian nationalist press, 1947. *ARIEL* 34 (2–3): 1–36.
Lakoff, G. and Johnson, M. (1980) *Metaphors We Live By*. Chicago: The University of Chicago Press.
Manning, P. (2007) Rose-colored glasses? Color revolutions and cartoon chaos in postsocialist Georgia. *Cultural Anthropology* 22 (2): 171–213.

Marín-Arrese, J. I. (2008) Cognition and culture in political cartoons. *Intercultural Pragmatics* 5 (1): 1–18.

Medhurst, M. J. and DeSousa, M. A. (1981) Political cartoons as rhetorical form: a taxonomy of graphic discourse. *Communication Monographs* 48: 197–236.

Michelmore, C. (2000) Old pictures in new frames: images of Islam and Muslims in post World War II American political cartoons. *Journal of American and Comparative Cultures* 23 (4): 37–50.

Morris, R. (1991) Cultural analysis through semiotics: Len Norris' cartoons on official bilingualism. *Canadian Review of Sociology and Anthropology* 28 (2): 225–54.

Morris, R. (1993) Visual rhetoric in political cartoons: a structuralist approach. *Metaphor and Symbolic Activity* 8 (3): 195–210.

Nystrom, E. A. (2002) A mixed message: African-American identity in WWII newspaper comics and cartoons. *Storytelling* 2 (1): 30–44.

O'Brien, A. (2008) Drawn to multiple sides: making arguments visible with political cartoons. In C. David and A. R. Richards (eds.) *Writing the Visual: A Practical Guide for Teachers of Composition and Communication*, pp. 183–200. West Lafayette: Parlor Press.

Press, C. (1981) *The Political Cartoon*. New Brunswick, NJ: Fairleigh Dickinson University.

Steffen, T. (1995) Linguistic features of cartoon captions in the New Yorker. *Popular Culture Review* 6 (1): 133–47.

Van Dijk, T. (1993) Principles of critical discourse analysis. *Discourse & Society* 4 (2): 249–83.

# 4
## Comics, Linguistics, and Visual Language: The Past and Future of a Field

*Neil Cohn*

### 4.1 Introduction

Many authors of comics have metaphorically compared their writing process to that of language. Jack 'King' Kirby, one of the most influential artists of mainstream American comics, once commented, 'I've been writing all along and I've been doing it in pictures' (Kirby 1999). Similarly, Japan's 'God of Comics' Osamu Tezuka stated, 'I don't consider them pictures ... In reality I'm not drawing. I'm writing a story with a unique type of symbol' (Schodt 1983). Recently, in his introduction to *McSweeny's (Issue 13)*, comic artist Chris Ware stated that 'Comics are not a genre, but a developing language.' Furthermore, several comic authors writing about their medium have described the properties of comics like a language. Will Eisner (1985) compared gestures and graphic symbols to a visual vocabulary, a sentiment echoed by Scott McCloud (1993), who also described the properties governing the sequence of panels as its 'grammar.' Meanwhile, Mort Walker (1980), the artist of *Beetle Bailey*, has catalogued the graphic emblems and symbols used in comics in his facetious dictionary, *The Lexicon of Comicana*. Truly, there seems to be an intuitive link between comics and language in the minds of their creators, a belief shared by several researchers of language who discuss properties of comics in a linguistic light. Exploring these works can provide insight into the extent to which this comparison might hold, its limitations, and how it can guide future research.

In many respects, comics do not fall within the normal scope of inquiry for contemporary linguistics, not because they are an inappropriate topic but because language is a human behavior while comics are not. Comics are a social object resulting from two human behaviors: writing and drawing. Believing 'comics' are an object of inquiry would be akin to linguists focusing on 'novels' as opposed to studying English, the language that novels are written in. Analogously, the sequential images used in comics constitute their own *visual language* (details of which will be expanded on at length further on). Thus, the behavioral domains of *writing* (written/verbal language) and *drawing* (visual language) should be the object of linguistic inquiry, stripping away the social categories like 'comics,' 'graphic novels,' 'manga,' etc.

Comics then become the predominant place in culture where this visual language is used, often joining writing (a learned importation of the verbal modality into the visual-graphic). That is, contrary to the metaphor used by their authors, *comics themselves are not a language*, but comics *are written in* visual languages the same way that novels or magazines are *written in* English. This makes comics potentially written in both a visual language and a writing system, reflecting the multimodality of human expression found in co-speech gestures (e.g., Clark 1996; McNeill 1992) which have received much attention in linguistics.

Several guiding questions of linguistic inquiry can thus be applied to the study of the visual language that comics are written in:

1. How is the *form* of the expressive system organized?
2. How is *meaning* conveyed by a form?
3. How do perceivers encode both *form* and *meaning*?
4. How do perceivers draw connections between and encode *sequential units*?
5. How do perceivers *learn* all this, given *cultural variability* across systems?

These are some fundamental motivating questions in linguistics which apply to the study of the expressive system used in comics (i.e., phonology, morphology, semantics, grammar, acquisition). While these questions have specific ways in which they are answered in linguistics proper, direct analogies between the verbal and graphic modalities (e.g., *words* = *panels*) are less important than the fact that the same *questions* are being addressed in both the verbal and graphic forms.

Finally, these questions will be situated as an examination of cognition. The approach outlined here will examine how people comprehend the structures used in comics. That is to say, the essay examines commonality between verbal and sign languages with the visual language used in comics – both with and particularly without written language – using the methodologies that linguists use to look at language. Research in other domains of cognition (such as visual attention and perception) will augment a complete theory of the comprehension and appreciation of comics. Indeed, numerous studies on comics come from other fields of cognitive science, particularly cognitive psychology, cognitive neuroscience, and developmental psychology. However, this discussion will mostly stay constrained to linguistics, keeping in mind that other important and applicable cross-disciplinary work contributes to the overall endeavor.

## 4.2 Survey of linguistic approaches to comics

Comics have been studied using many of the formalisms of linguistic inquiry, from structuralism and generative grammar to cognitive and applied linguistics. These approaches have framed their analyses, so the contributions of each will be discussed in an overview of the linguistic research that has been done using comics.

Approaches to comics from a structuralist and semiological perspective emerged as early as the 1970s in Europe, summarized in Nöth (1990) and later in D'Angelo and Cantoni (2006) and in Mey (2006). Structuralism looked at 'language' as a set of cultural codes, making comics one place that cultural codes could be found and reduced to minimal units, along with several types of art and media analyzed in this vein. Several authors aimed at describing the minimal units of comics' representations at various levels of representation. Koch (1971) and Hünig (1974) created taxonomies of unitization building from the inner parts of comic panels' graphic representations of characters, places, etc. (*logemes*) to whole panels (*syntactemes*), up through whole sequences (*texts*). Gubern (1972) also differentiates aspects of the form of representations (*morphemes*) with their color (*coloremes*). With more specific aims, several works in particular have attempted to identify morpho-graphemic minimal units using Charles Schulz's *Peanuts*, highlighting how individual graphic elements of hands, eyes, noses, etc. combine to create differences in full-blown representations of various characters (Gauthier 1976; Kloepfer 1977; Oomen 1975). Other approaches have focused on comics' sequences by identifying elementary units of narrative forms, consistent trends in plots and stories, and comparing them with other narrative genres (Fresnault-Dervelle 1972; Hünig 1974). More recent works from this tradition include Groensteen (1999), who has couched his approach within the equation that 'comics are a language,' yet has eschewed the search for minimal units, while semiological approaches appear in dissertations by Dean (2000) and Miller (2001).

Aside from looking at the structure of codes, some works have also used the philosopher C.S. Peirce's (1931) 'semiotics,' which focuses on the expression of meaning through various types of reference. Peirce's formulation of semiotic types – particularly his distinction between icons, indexes, and symbols – has been applied to comics in broad strokes by Magnussen (2000) and specifically to the differences between cartoony, realistic, and abstract representational styles by Manning (1998). Peirce's philosophies also provide the framework for Cohn (2007, 2010b), who describes the semantics and other systematic patterns in the morphology of graphic expression.

While European work drew from structuralist traditions, contemporary works on comics in a linguistic light, both in the US and abroad, have exploded since the publication of Scott McCloud's (1993) graphic book *Understanding Comics*, in which McCloud posited cognitive principles to explain how people understand both individual images and sequential ones. For example, McCloud's principle of 'closure' describes where the mind 'fills in the gaps' between panels in order to comprehend the sequence of images. McCloud characterizes the linear relationships between juxtaposed panels with six types of 'panel transitions' based on changes in actions, characters, or the environment, among others. He also quite overtly compares the medium to language, claiming that the 'iconography' of graphic meanings constitutes the vocabulary and closure is the grammar.

McCloud's approach has permeated nearly all linguistically driven studies since its publication. Both Saraceni (2000) and Stainbrook (2003) focused their dissertations on adapting McCloud's panel transitions to theories of verbal discourse, while Narayan (2001) compared them to cognitive theories about event structure. Similarly, Saraceni (2001), Bridgeman (2005), and Lim (2006) all invoke McCloud's ideas in their discussions of multimodal texts integrating images and words.

One highly productive framework for analyzing comics is cognitive linguistics, which attempts to relate the comprehension of meaning in linguistic structure to aspects of general cognition. Cognitive linguistic research on conceptual metaphor (Lakoff and Johnson 1980), which involves the mapping of one conceptual domain onto another, has been of particular interest to comic theorists. Forceville (2005) has examined the expressions of anger in *Astérix* comics, especially related to the metaphor ANGER IS HOT FLUID IN A CONTAINER – which in verbal form appears in sentences like 'He was *steamed*' or 'His rage *erupted*,' where concepts about anger map onto aspects of boiling water coming out of a container. In the graphic form, this metaphor appears in the common trope of steam coming out an angry character's ears, as if the emotion were bubbling over in the container of the head. A recent volume edited by Forceville looks at multimodal metaphor, including further examination of emotion in *Astérix* (Eerden 2009) and similar metaphors in Japanese manga (Shinohara and Matsunaka 2009), as well as metaphor in other contexts such as political cartoons (El Refaie 2009) and editorial cartoons (Schilperoord and Maes 2009; Teng 2009). Additional articles have examined metaphor in Neil Gaiman's *Sandman* comics (Narayan 2000), in McCloud's *Understanding Comics* (Horrocks 2001; Narayan 1999), in post-September 11 political cartoons (Bergen 2004), superhero comics (Potsch and Williams, this volume), and in comic advertisements for the *Chicago Tribune* newspaper (Cohn 2010a).

A wide variety of other linguistic methods have been drawn upon for diverse purposes as well. Cohn (2005) has argued for sociological distinctions for comics' underlying structure similar to the split between speech acts and the linguistic system made by de Saussure (1972) and the distinction between the internal cognitive understanding of a language and its external sociocultural understanding made by Chomksy (1986). Chomsky's (e.g. 1965) approach to generative grammar, which hypothesizes that sentences are generated by hierarchical rules guided by a system of constraints, has been drawn upon by Cohn (2003) to describe sequential images in a way different from panel-to-panel transitions. Meanwhile, an approach to a 'visual lexicon' (Cohn 2007) has drawn upon theories of construction grammar (e.g. Goldberg 1995) with sensitivity to concatenation of form–meaning patterns stored in memory ranging from small-scale graphic components to full multimodal sequences. Finally, Laraudogoitia (2008, 2009) has used computational linguistics methods of using computer algorithms to explore the patterns and structures of language to examine the sequential structures of comics.

Despite this growing trend of associating linguistics and comics, few of these disparate works are motivated by a central theory of language, graphic expression, or comics (acknowledged in D'Angelo and Cantoni 2006). They all share an intuition that linguistics is the proper discipline with which to study these phenomena; however, despite all tapping into the questions posed at the outset, nothing binds these works together. Moreover, few of them present any unifying vision to accomplish this, much less provide a gateway for future linguistic inquiry and/or connect graphic expression to the verbal or manual domains. At their best, these studies use comics as further support for particular linguistic theories (Forceville 2005), and at their worst, they make equations that 'comics are a language' without grounding such claims in adequate contemporary linguistics knowledge (e.g. Groensteen 1999). However, a 'unified theory' is possible, though it requires a deeper consideration of the object of inquiry.

## 4.3 What is 'visual language'?

All of these cases address the graphic form used in comics with methods used to analyze a linguistic system. Ultimately, this research contributes towards filling a gap in the cultural category regarding the channel of graphic expression. While verbal communication ('speaking') is readily acknowledged as using a *system* of expression ('spoken *language*'), graphic communication ('drawing') has no equivalent system recognized. While language is viewed as a rule-governed system acquired through a developmental period, drawing is viewed as a 'skill' subject only to the expressive aims of the artist and their abilities, which often are assumed to be developed through explicit instruction or practice. While sentences can be grammatical or ungrammatical, the predominant conception holds that there is no unacceptable way to structure images. However, there is no principled reason why these beliefs about the visual-graphic domain are held, and the growing literature discussing 'comics' in linguistics points towards systematization in the same way as language.

Indeed, humans only use three modalities to express concepts: creating sounds, moving bodies, and creating graphic representations. A theoretical extension can then be proposed: when any one of these modalities takes on a structured sequence governed by rules that constrain the output, i.e., a grammar, that form becomes a type of language. This leads structured sequential sounds to be spoken languages of the world, structured sequential body motions to become sign languages, and structured sequential images literally to become *visual languages*. An analogy can then be made: individual manual expressions (which have no grammar) are to sign languages (that use a grammar) what individual drawn images (no grammar) are to visual languages (grammar).

This notion of a 'visual language' fills the gap in categorization for describing the cognitive system at work in graphic expression. When individuals acquire or develop systematic patterns of graphic representation, along with the structures necessary to string them into sequences, they effectively use a visual language. Just as spoken language does not have a universal manifestation, visual language varies by culture.

We would expect diverse cultural manifestations of visual languages throughout the world, perhaps not even resembling comics at all, such as the sand narratives used by native communities in Central Australia.[1] This context explains why, for example, Japanese and American comics feature different graphic styles and sequential patterns (Cohn 2010b): they are written in different visual languages, used by differing populations. However, while 'Japanese Visual Language' and 'American English Visual Language' feature patterns that are unique to their 'speakers,' they still feature patterned sequential images expressing concepts that contribute toward their inclusion in the broader 'visual language' the same way that English and Japanese are both types of 'verbal language.'

## 4.4 Future research: 'visual linguistics'

If visual language is to be studied as a language, research can easily involve all the traditional areas of linguistic inquiry framed by the questions posed at the outset of this chapter, though re-imagined in the graphic modality. The various areas of study can thus follow the major branches of linguistics that address these questions: *graphemics*, *photology*[2] (visual-graphic analogues to *phonetics* and *phonology*), *morphology*, *semantics*, *grammar* as well as *multimodality* and *acquisition*. Note that these fields address the questions posed at the outset of this discussion. For each field described, I will frame the overall inquiry, followed by some pertinent existing research that fits into its endeavor. Following this, suggested future research endeavors will be outlined.

As a caveat, the intent of this chapter is *not* to force equivalencies between the structures used in verbal and visual forms (i.e., not *words = panels*). The goal is not to make direct mappings between spoken and visual language, but rather to understand how the two systems use analogous functions and units of organization. In some cases, correspondences should flow naturally (semantics could apply in the same way to aspects of meaning no matter the modality) while in others, a rigid analogy is wholly inappropriate (the graphic form likely has no 'phonology,' but has a 'photology' with its own properties that suits the visual-graphic modality). That said, the labels from linguistics are ultimately less important than the functional roles that they describe within the system as a whole.

### 4.4.1 Graphetics

The greatest disparity between visual language and verbal or sign languages is the modality itself: the visual-graphic channel is quite different from both the verbal-auditory and manual-visual channels. For example, while the verbal form is conveyed temporally, the graphic form is (in most cases) static. The auditory form mandates 'linearity' to the expression, while the analogue nature of vision leaves linearity to be guided by specific layouts of panels. However, both systems have physical form through which the information is conveyed. As such, both spoken and visual languages have constraints on form, creating the topic for this level of analysis. There are two major issues concerning the investigation of the form of visual language: (1) examination of the physical manifestation involved in graphic comprehension and (2) study of the organization of the units of form. In verbal language, these goals comprise the fields of phonetics and phonology, while corresponding fields of 'graphetics' and 'photology' can apply to the visual-graphic form.

The study of graphetics should first seek to describe what information is encoded such that drawings can be produced and perceived. Verbal language uses 'phonemes' of sound, e.g., /k/, /m/, and /a/, that are created by physical articulations in the human vocal tract. Similarly, basic 'graphemes' are combined to form larger representations. Such basic shapes like dots, lines, and spirals combine to form angles, squares, circles, etc., and create the basic shapes of various graphic iconographies, which are deployed in drawn representations (Liungman 1991). However, articulation of the hands (or feet, or elbows, etc.) allows huge variability to create graphic forms – there are many ways to draw a line with a finger or to hold a pen. In most visual languages, drawers have unlimited time and a vast array of media and techniques at their disposal. This contrasts with verbal and manual languages, which are produced in a rapid manner and are a direct trace of the articulatory gestures used to create the signal via the hands or mouth. Thus, while the physical production of verbal and manual language has great importance for encoding, for visual language, articulation appears to have little bearing on the final graphical structure.

Nevertheless, important constraints guide the perception of visual language. Viewers must be able to recognize the objects depicted in scenes, meaning that the elements of visual language must be drawn in a way that facilitates visual object and scene perception. Drawers must thus be constrained by the general principles of object recognition: panels must depict scenes that follow Gestalt principles of organization (Palmer 1992; Palmer and Rock 1994; Wertheimer 1923), principles of figure and ground, and others.

Some work has described rules for grapheme combinations that resemble phonological constraints. For example, all languages of the world restrict illegal combinations of sounds, e.g., 'tlk' or 'mp' cannot appear at the beginning of words in English. Similarly, Willats (2005) draws upon the work of Huffman (1971) and Reith (1988) to formulate acceptable versus ill-formed combinations of lines in drawings based on rules allowing types of configurations of line junctions. For example, Figure 4.1 illustrates that the circled portions near the shoulder and leg of the kangaroo have awkward line junctions that do not accurately show the occlusion of objects. Thus, graphetics is concerned with making sure that the graphical units combine to form recognizable objects and scenes.

The limitations on graphics shown here involve only the graphic form; they have little effect on the meaning of either the parts or whole representation. While the comprehension of graphic elements must rely on general aspects of perception, these restrictions appear based on purely graphical components. Thus, research must identify the restrictions on structure in the graphic form, as well as their departures from, and similarities to, general aspects of vision. To this end, as with phonetics, graphetics is the part of the cognitive architecture that interfaces most seriously with processes from other domains (here, visual perception; in phonetics, motor programming, acoustic perception).

Finally, the graphic patterns used in visual language can be explored within individual drawers and broader cultures. Given that a 'language' at the societal level comprises the grand average of similar patterns

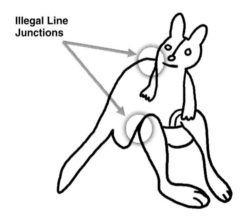

*Figure 4.1* Ungrammatical line junctions in a child's drawing of a kangaroo. Adapted from Reith (1988)

in individuals' heads, identifying such patterns can lend towards the documentation of graphic dialects. For example, consistent patterns are recognizable between the authors of most Japanese manga such that it can be called a 'Japanese style' (big eyes, pointy chins, big hair), though distinguishable subgenres vary from this 'standard' (Cohn 2010b). These distinctions can be understood as accents or dialects of the broader Japanese visual language; similar research can be undertaken for all visual languages.

### 4.4.2 Photology

Beyond identifying basic visual units, research must also describe how these units combine together and the cognitive principles that motivate such combinations. For these qualities in sound, this would be the endeavor of phonology. By analogy, 'photology' studies the organization of the graphic modality. Early structuralist approaches looked at how basic graphemes combine to influence meaning. Particularly, studies investigating the morpho-graphemic structures in *Peanuts* (Gauthier 1976; Kloepfer 1977; Oomen 1975) frame the search for consistent patterns of graphic representations similar to looking for phonemes, a practice that goes all the way back to Töpffer's (1965[1845]) analysis of regularities in his own drawings.

This level of organization, however, should not relate to the basic needs of object recognition, but should be more arbitrary. This notion may best be demonstrated with an example. Note the two faces in Figure 4.2.

*Figure 4.2* Variation in photology

Here, the simple representation uses simple lines and negligible detail. The complex face, however, appears to use three-dimensional depth and his eyes, nose, hair, and ears all are depicted with detail. Crucially, even though the complex face is drawn with nostrils and the simple one is not, we are not supposed to infer that the simple drawing has no nostrils. Rather, the *photology* of the simple face allows a single graphical element such as a line to depict a lip or a nose, whereas the photology at work for the complex face requires more complicated configurations. One way to describe this difference would be to postulate a 'minimality constraint' that dictates the minimum number of graphical elements capable of depicting an object. While the simple photology allows single graphical elements to be a unit of representation (e.g., a circle for an eye, a line for a nose), the complex face's photology requires more than one, i.e., requiring the nose and eyes to be more than a single line. This restriction would be similar to prosodic minimality constraints observed in natural languages that dictate the minimal size that words can take (e.g., English and Dutch morphemes must consist of at least a heavy syllable, a syllable with a long vowel or a short vowel followed by a consonant; see Booij 1995; Hammond 1999). This constraint has nothing to do with object perception; both faces are easily recognized as such. Rather, these are arbitrary rules for the way graphical elements may combine.

The study of photology must account not only for the combination of individual graphemes to form basic shapes, but also individual characters all the way up through full scenes. Just because combinations might build to form objects and full scenes, it does not necessarily mean they involve morphology: comprehension must still account for the purely visual-graphic aspects of the modality at various levels of processing, as shown in Figure 4.3. This is similar to sound in verbal language, which features phonological combinations within words, sentences, and even discourse with aspects of intonation, prosody, and metrical structure.

As in the verbal and manual domains, we might expect that building larger levels of graphic structure requires the mind to encode units of schematic information (both of simple graphemic shapes and combined chunks) that can be combined together using systematic rules. Research must characterize this schematic information, its combinatorial system, and how this purely graphic information maps to conceptual structures (be it through iconic or symbolic reference). One approach to this issue emerged in investigating children's drawings. Along with subsequent research supporting that most children imitate drawings from other sources (Lamme and Thompson 1994; Smith 1985; Wilson 1999), Wilson and Wilson (1977) have hypothesized that drawing makes use of schematic encodings of these imitated graphemic structures, which then become 'averaged' to produce novel forms. That is, a drawer either

Visual Language 103

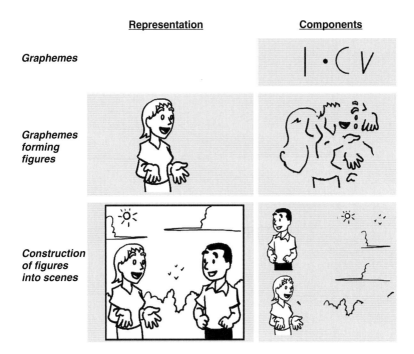

Figure 4.3 Various levels of photological constructions

acquires (from imitating external sources) or creates schematic patterns of graphic representations and then produces them in a generative fashion to create novel drawings.

Working out the mechanisms guiding these combinations is the primary challenge to a field of photology. As with graphetics, photology must interface with the concerns of general studies of vision and perception. Exploring the relations between these fields as well as their disconnections will provide an explication of the cognition guiding the form of visual language.

### 4.4.3 Morphology

In all modalities, morphology is the study of how explicit forms encapsulate meaning. Unitized meaning in visual language can be divided similarly to standard linguistic understandings, with productive versus non-productive morphemes, open and closed classes, and varying levels of lexical items. Following the construction grammar view that a lexicon is the cognitive encoding of stored units of form-meaning pairings in the linguistic system (Goldberg 1995), Cohn (2007) described the overview of a 'visual lexicon' that ranges from individual meaningful emblems[3] (such as symbolic word bubbles or lightbulbs over heads) to fully productive and patterned representations at the level of full panels, even to constructional patterns of sequential images.

Among the work necessary in visual morphology, one topic might include cataloguing the graphic emblems used in various cultures' visual languages. Walker (1980) has provided perhaps the most extensive collection, though humorous in intent, while several works have noted differences in morphology between cultures, particularly disparities between Japanese and American emblems (Cohn 2010b; McCloud 1993). Shipman (2006) has also contrasted the signs used by American and French authors. Still further contrast comes from visual languages found outside of the cultural objects of 'comics,' such as Aboriginal Australian sand narratives which carry their own morphological systems (Green forthcoming; Munn 1986; Wilkins 1997).

Like verbal languages, visual languages make distinctions between the attachment and binding to other signs that follow bound forms versus free forms. For example, words might affix a morpheme that cannot stand on its own like -*ness* to a root like *happy* or *sad* to make *happiness* or *sadness*. Similarly, research can study the combinatoriality of various morphological signs in the graphic form (Cohn 2007). For example, motion lines and speech balloons are 'bound morphemes' that cannot appear without affixing to a root object like someone running or speaking. They usually do not appear on their own and even imply a root if one is not shown. Other bound signs might have more restrictive areas of binding, such as those appearing above individuals' heads (lightbulbs, hearts, rainclouds, etc.), but not to the side of the head or body, as in Figure 4.4. These above-the-head signs also require a degree of 'agreement' with the facial expression: a happy face with rainclouds or an angry face with a lightbulb would be an ungrammatical combination. Again, it is important to emphasize that these graphic components are not implied as equivalent to roots and affixes in verbal language, but that both types engage similar relationships in their concatenation of bound and free forms.

Other morphological processes are similar to language as well. Some signs use 'suppletion,' the substitution of all or part of an entire morpheme as a variant for another, such as *people* or *men* for the plural of *person* or *man*. Graphically for example, hearts, stars, dollar signs, etc. can be substituted for a character's eyes to add meaning (see Figure 4.4), or full form suppletion of dotted lines for the lines of a character's body stand for invisibility. Reduplication is also a common process where morphemes repeat to create expanded meaning, as in the repetition of the second morpheme in the Hebrew *klavlav* for 'puppy' expanded from *kelev* for 'dog' (Moravcsik 1978). A similar process appears graphically when the lines of a character or object are repeated and layered on top of each other to show shaking, as in Figure 4.4.

Again, the processes involved in the combinatorial strategies of graphic morphology are not necessarily viewed as direct unit-to-unit analogues to the verbal form (i.e. there is no search here for the graphic equivalent of a 'word'). However, the same general strategies appear in both the graphic and verbal forms. Both modalities attach meaningful elements together (affixing), substitute them within each other (suppletion), and repeat them (reduplication) to varying effects on the meaning of an expression. Exploring the constraints mediating these usages in the graphic modality echoes morphological research on other types of languages.

Further, work should explore the complex relationships between elements within an individual image. Studies in semiotics have sought to do this by describing the vectors between semantic roles of characters in an image (Kress and van Leeuwen 1996). However, others have argued that generative approaches can also work for individual images such as maps and diagrams, meshed with morphological cues (Engelhardt 2002; Sonesson 2005). Further study of this domain must explain how both semantic and formal aspects of individual images facilitate understanding, as well as how the comprehension of individual images might differ from general visual perception.

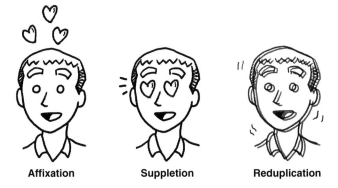

*Figure 4.4* Morphological processes in graphic form

### 4.4.4 Semantics

Semantics may be the field of linguistics that is shared the most clearly between all modalities. Beyond the categorization and combination of meaning in stored morphological signs, meanings in visual language can take on far more complex representations. One route can involve directly exploring the features of emblems and signs, such as Cohn's (under review a) treatment of the semantic features underlying word balloons and thought bubbles. Other explorations of semantics take on greater complexity, such as the cognitive linguistics approaches to conceptual metaphor and blending in both individual images (e.g. Bergen 2004; Forceville 2005) and sequential images (Cohn 2010a) as reviewed above.

Common semantic phenomena like metonymy and synecdoche also appear in the graphic form (Cohn 2010a; Kennedy 1982; Kukkonen 2008). Metonymy is expressed in a variety of ways, especially the substitution for one thing to mean a related thing. For example, in the sentence 'The White House issued a statement,' we know that 'The White House' stands for the administration that works in that building, not the building itself. Similarly, this same metonymy is commonly used in the comic *Doonesbury*, which shows word balloons coming from the White House to represent things said by people inside the building. Synecdoche, referring to the whole through the parts, also appears graphically. For example, panels using 'extreme close-ups' of a person or object use a sliver of information to refer to the whole. Further exploration of these types of conceptual correspondences will inform existing research on metonymy in the graphic form.

Finally, a primary focus of semantic research on visual language has focused on *inference*, the drawing of non-provided meaning from the existing forms. Discourse studies have particularly looked at inference, as how, for example, in the sentences 'The fireman sprayed the water on the house. Smoke rose from the building.' the reader derives the meaning that the house was on fire and went out, though such concepts are never mentioned overtly. Essentially, McCloud's (1993) invocation of 'closure' as the process of 'mentally filling in the gaps' is a view-setting inference between every pair of panels. However, inference may happen *within* a particular panel. For example, note the sequence in Figure 4.5. The fifth panel in the Mafoud cartoon uses the morpheme of an 'action star' to take up the entire panel (Cohn 2009). By showing only the action star, the actual event of the security guard being hit by the backpack is never shown and only implied. Studying the comprehension of

*Figure 4.5* Inference generated by an action star. Grrl Scouts art © 2002 Jim Mahfood. Used with permission

predictable inferences (as here with the information of the backpack being thrown) versus cases that are less predictable seems a formidable task for any study of visual language semantics.

### 4.4.5 Grammar

Grammar functions as a central object of linguistic research and thus must also factor into research of visual languages. For the graphic form, one question to ask is 'what are the constraints placed on sequential images that allow some sequences to be acceptable and others not (at least between 'normal' narrative sequences and purely random panels)?'

Semiological European approaches addressed this sequential structure by analyzing elementary narrative functions, often related to types of plotlines and their components (Fresnault-Dervelle 1972; Hünig 1974). The first theory with any sort of cognitive aim in mind was in McCloud's (1993) panel transitions, which characterized various ways in which panels linearly connect to their juxtaposed neighbors, as in Figure 4.6, which uses action-to-action transitions to describe a progression of an action between panels.

Cohn (2003, 2010c) has criticized transitional approaches that rely on individual panel relationships and has presented an alternative perspective drawing upon tools of generative grammar. This hierarchical approach has shown that juxtaposed panel transitions are not enough to account for distance dependencies and structural ambiguities (Cohn 2003, 2010c, under review b). In verbal language, distance dependencies arise when one unit must connect to another much further unit, such as in 'My roommate, who is a total bore, watches TV all day' where an embedded clause separates the subject 'My roommate' from its predication 'watches....' Similarly, Figure 4.7 shows an embedded clause with a hierarchical approach in contrast to the analysis of panel transitions in Figure 4.6.

Additional ideas related to grammar have come from construction grammar, which describes schematic patterns at the various levels of

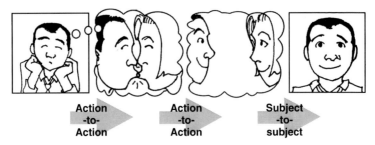

*Figure 4.6* Panel transitions analyzing a sequence of images

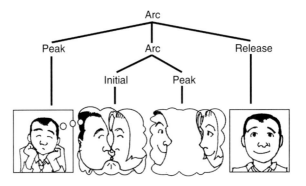

*Figure 4.7* Sequential images with center-embedded clause analyzed with a tree diagram highlighting the narrative structure. Within the maximal node of an Arc, categories describe different roles played by panels in the overall architecture. The culmination of a predication occurs in Peaks, the initiation of an interaction in the Initial, and an aftermath of the predicate in a Release.

language grammar (Goldberg 1995) including syntax. For example, the pattern 'Verb-ing the TIME Away' construction (Jackendoff 1997) appears in 'twistin' the night away' or 'lounging the afternoon away.' This same approach has framed schematic patterns in visual language grammar (Cohn 2007, 2010a). For example, sometimes a scene might flip back and forth between panels showing each character individually before coming together at a panel that shows both together, as in Figure 4.8.

The pattern in the Stan Sakai cartoon alternates between panels depicting only the ninja and those depicting only the samurai rabbit before both characters appear together in the final panel. This pattern

Visual Language 109

*Figure 4.8* 'Convergence' construction from a sequence in *Usagi Yojimbo* Usagi Yojimbo art © 1987 Stan Sakai. Used with permission

of 'convergence' prevails in sequential images and could possibly be explained using the following schema:

*Convergence*:
[[(A) (B)]$^{Xn}$ [(AB)]]

*Convergence in Figure 4.8*:
[[(Ninja) (Rabbit)] [(Ninja) (Rabbit)] [(Ninja Rabbit)]]

   Here, square brackets represent clauses, parentheses represent panels, while A and B stand for different characters. The construction begins with a clause X with panels showing A and B, that can repeat *n* number of times before coming together in a panel with both A and B. In the case of Figure 4.8, two clauses of individual panels precede the final joined panel. Further research of the visual grammar can identify and describe additional constructions like this using the formal tools of linguistics.
   Future study of visual language grammar must explain the rules and restrictions guiding graphic sequences, while also addressing the elements from semantics and morphology that may motivate grammatical categorization. For example, in the verbal modality nouns can be nearly any semantic class (objects, events, etc.), though adjectives are most often properties (Jackendoff 1990). Comparatively, the visual-graphic form may have correspondences to meaning that limits the roles that panels play in a sequence. For example, the action star in Figure 4.5 can substitute for a full event, but would seem awkward starting a sequence; what are the semantic properties that motivate it to take the sequential role that it does? Additionally, study of this grammar must involve identifying and codifying types of grammatical patterns (such

as convergence), as well as diagnostic tests for their constraints (such as in Figure 4.8, where switching the order of the third and fourth panels would result in a less felicitous reading: why?). Finally, like other fields, beyond understanding the cognitive system itself, research will benefit from cross-cultural comparisons, such as McCloud's (1993) contrasting of panel transitions in American, European, and Japanese comics. Like syntactic categories in the verbal form, it may be the case that the same grammatical elements are used cross-culturally in diverse ways.

### 4.4.6 Multimodality

Most speech in discourse contexts appears with gesture, just as most visual language does not appear on its own, occurring in conjunction with written language. Again, Australian sand narratives can provide a good contrast, since their production not only accompanies speech, but also occurs with an auxiliary sign language in interactive multimodal exchanges (Green forthcoming; Wilkins 1997). Since the graphic form does not involve temporal co-occurrence the way that speech and gesture do, various interfaces are required to achieve unified expressions in the spatial form. Cohn (2003, under review a) has described these various ways in which the verbal and visual modalities interface, focusing especially on how text and image can combine to form singular units of expression. More attuned to the expression of meaning in multiple channels, McCloud (1993) has outlined a general taxonomy for the interactions between modalities, while Cohn (in preparation) has expanded this to investigate the contributions of cognitive structures in multimodal expressions of all types.

The study of multimodality must account for not only how different modalities interact in expressions and their regularities but also how cognitive structures contribute to such interactions. Do modalities share a common conceptual structure? Are there constraints on how much each modality contributes to the whole of meaning? What is the architecture of a grammatical model that can distribute semantics into various modalities at the same time? A great challenge is to develop semantic theories that can be applied to a range of modalities at the same time.

### 4.4.7 Acquisition

Research on how children learn to draw and create sequential images has largely been addressed in art education, with minimal focus from linguistics and developmental psychology. However, a focus centered on advising education carries different assumptions from those aiming simply to describe the processes of cognitive development. Nevertheless, several insightful works can inform the visual language paradigm, especially since drawing and language appear to have similarities in development. For example, like language acquisition, drawing begins with a period of 'babbling' before progressing to increasingly complex forms (Kindler and Darras 1997; Willats 2005). Additionally, drawing does appear subject to a critical learning period. It has been well established that children in the US (and those of most cultures) show a 'drop off' in the development of drawing skills at puberty (Kindler and Darras 1997). However, this drop off does not occur in Japan (Toku 1998, 2001), where children imitate and draw Japanese comics throughout childhood (Wilson 1999; Wilson and Wilson 1987). In this light, such a 'drop off' appears to be the apex of a critical developmental period, which American children do not receive adequate stimulus to overcome, but Japanese children do.

Some theories of child drawing directly compare the structure and acquisition of drawing to language. For example, Willats (2005) outlines a trajectory of development for children's drawings tied to the perceptual theories of Marr (1982) and inspired by ideas in generative grammar. Also key are the insights of Brent and Marjorie Wilson, who have compared the acquisition of graphic schemas to language (Wilson and Wilson 1977), and have described numerous instances in which cultural knowledge and imitation factor into drawing ability both for individual images (Wilson 1997, 1999) and sequential images (Wilson and Wilson 1987). Additional work has looked directly at how age and expertise influence comprehension of comics (e.g. Nakazawa 2005; Pallenik 1986). While many other studies are worth exploring, such research can lay a foundation towards understanding how children acquire graphic abilities, which thereby feeds back into understanding the visual language itself.

### 4.4.8 Further inquiries

The fields described above are merely broad strokes describing how 'visual linguistics' might proceed, though any subfield of linguistics should apply to the visual form. These might include (though are not limited to):

- *language variation and change.* How might the structure of a language change over time and what are the properties of languages no longer in use? (See Meesters, this volume.)
- *comparative linguistics/language contact.* How does the structure of a language change with exposure to another, and might existing languages share historical roots?

- *perceptual dialectology* and *interactional sociolinguistics*. How is a language used in sociocultural settings, and how does it frame a person's identity? What biases does that engender towards the perceptions of other dialects? (See Walshe, this volume.) How do speakers enact or perform their identities through code choice? (See Beers Fägersten, Bramlett, and Breidenbach, this volume.)
- *anthropology*. What are the characteristics of the cultures that arise around languages of the world?
- *linguistic typology*. What is the range of variation in the languages of the world and are there consistencies that lead to underlying universals?
- *computational linguistics*. How might statistical modeling be used to study the properties of a language? How might the construction of large-scale corpora help explain socially-situated language use (i.e., register)? (See Meesters, this volume.)
- *neurolinguistics*. What brain areas are associated with the processing of structures of language and how might they be similar or different from those used in other linguistic and non-linguistic domains?
- *cognitive deficits*. What can cognitive impairments (aphasias, genetic disorders, etc.) teach us about the biological and neural structuring of language?

The questions that motivate all of these fields apply equally to verbal, sign, and visual languages. Insofar as visual language is a real and actual linguistic system, research in nearly any domain related to language should both be conceivable and possible.

In addition to applying to the subfields of linguistics, visual language can be analyzed using nearly all extant theoretical approaches: generative, cognitive, applied, computational, etc. In some sense, visual language can be seen as an equalizing force between such schools of linguistics, since all are applicable given the proper goals of the research. Truly, the theory of visual language not only applies to linguistics, but can serve to unify such disparate research as comics with that of sand narratives, and in fields as diverse as art history, linguistics, cognitive science, and art education. This broad accessibility is a testament to the linguistic status of visual language and the potential for a future 'visual linguistics' across the entire field of the study of language.

### 4.5 Visual language versus comics

It is worth reiterating here the relationship between comics and linguistics, particularly the important difference between 'visual language' and 'comics.' While 'visual language' is the biological and cognitive capacity that humans have for conveying concepts in the visual-graphic modality, 'comics' are a sociocultural context in which this visual language appears (frequently in conjunction with writing). By dividing the mode of conceptual expression from the sociocultural artifact, the definition of comics is not founded on structural properties. This split is in direct contrast to approaches that define comics by features of images and/or text, such as the requirement of sequential images by McCloud (1993), the dominance of images in Groensteen (1999), or the need for multimodal text–image interactions by Harvey (1994).

By recognizing 'visual language' as a system divorced from its predominant sociocultural context, a 'comic' can use any combination of writing and images: single images, sequential images, some writing, no writing, dominated by writing, etc. All these permutations do in fact appear in objects identified as 'comics' (Cohn 2003, 2005). Such a division allows visual language to be used outside of the cultural institution of comics as well, such as the appearance of sequential images in instruction manuals, illustrated books, and various other sociocultural contexts not typically labeled as 'comics.' Indeed, both illustrated books and comics use sequential images and/or writing to (most often) tell stories, but they play far different roles in culture, not to mention carrying different stereotypes. Both may use visual languages, but both are not called 'comics' because they belong to different cultural contexts.

Furthermore, making this separation between 'comics' and the 'visual language they are written in' will halt the recasting of the modern label of 'comics' onto historical instances of sequential images like cave paintings, medieval carvings, or tapestries (Kunzle 1973; McCloud 1993). Rather than call these artifacts historical 'comics' (or 'protocomics'), these cases can be viewed simply as visual languages tied to their own unique and specific cultural and historical contexts (Horrocks 2001).

This split between the sociocultural object/context ('comics') and the structural/cognitive system ('visual language') is the key to future research of the graphic form in the linguistic sciences. It also changes the spotlight of inquiry: the focus is not just on 'comics,' but on the system they are written in and how the mind works to create meaning through various modalities, particularly graphic expression and its relation to other systems.

In the context of visual language, the ultimate object of inquiry in linguistics is not physical or social phenomena 'out in the world' at all. Rather, the units of investigation are the abstract representations and principles in the human mind that motivate comprehension of various domains, from the understanding of the form and meaning to its use in social settings. Arguably, these principles are not tied to any sociocultural context like comics, and indeed may be cognitive artifacts abstract enough to engage both the verbal and visual domains. Thus, while glossed over as the study of comics, really the linguistic study of this visual language illuminates the links between domains that can paint a broader picture of the nature of human expression.

## Acknowledgements

Ray Jackendoff, Eva Wittenberg, Stephanie Gottwald, Kelly Cooper, and Ariel Goldberg helped tremendously with early drafts of this paper. Dark Horse Comics is thanked for their contribution to my research corpus.

## Notes

1. Native communities in Central Australia, particularly the Walbiri (Munn 1986) and the Arrernte (Green, forthcoming; Wilkins 1997), use language-like systems of sand narratives in combination with spoken and signed languages. While the most elaborate sand drawings are used in specific storytelling contexts, this system is not conceived of as an auxiliary system and is used in daily conversation. Wilkins (1997) explains that this system is integrated into the Arrernte's notion of everyday 'speaking.' These graphic systems use highly systematized signs that maintain an aerial viewpoint. For example, the prevalent sign for 'person' looks like the letter 'U,' representing the imprint that people make in the sand (Munn 1986; Wilkins 1997). Both Cox (1998) and Wilkins (1997) have noted children must reconcile the differences between sand drawings and Western representations. Sequentially, sand narratives unfurl temporally in a single area, as opposed to the discrete sequence of images in comics. As Wilkins (1997) notes about this difference, the Arrernte seem to struggle somewhat to make meaningful connections between static images as opposed to an image that evolves over time. It seems that this system would fall under the scope of a 'visual language,' though its properties are vastly different from those in the visual languages used in comics around the world.
2. Or, alternatively, 'graphology.' My personal preference is asymmetric, with 'photology' for the field and 'graphemes' for the minimal graphic units (and thus 'graphetics' instead of 'photetics'). Future researchers can make this labeling decision. Although sign language researchers have opted to maintain 'phonology' for the structure of the manual modality and its articulation, for the graphic modality this seems inappropriate. If the study of a language's modality used a 'domain-neutral' term instead of 'phon-' referencing sound, differing names for fields would likely be unnecessary.
3. 'Emblem' comes from gesture research (McNeill 1992) to mean a conventional expression of meaning that cannot act as a syntactic unit. In gesture this includes the 'thumbs up,' 'Okay,' or 'middle finger' hand positions,

which are used individually but not as novel productive gestures. In graphic form, emblems include word balloons, smoke emerging from someone's ears, hearts for eyes, and other conventionalized signs.

## Graphic references

Mahfood, J. (2002) Grrl Scouts in 'Just Another Day.' In D. Schutz (ed.) *Dark Horse Maverick: Happy Endings*. Milwaukie: Dark Horse Comics.
Sakai, Stan (1987) *Usagi Yojimbo: Book One*. Seattle: Fantagraphics Books.

## References

Bergen, B. (2004) To awaken a sleeping giant. In M. Achard and S. Kemmer (eds.) *Language, Culture, and Mind*, pp. 23–36. Palo Alto: CSLI.
Booij, G. E. (1995) *The Phonology of Dutch*. Oxford: Clarendon Press.
Bridgeman, T. (2005) Figuration and configuration. In C. Forsdick, L. Grove, and L. McQuillan (eds.) *The Francophone Bande Dessinée*, pp. 115–36. Amsterdam: Rodopi.
Chomsky, N. (1965) *Aspects of the Theory of Syntax*. Cambridge, MA: MIT Press.
Chomsky, N. (1986) *Knowledge of Language*. New York: Praeger.
Clark, H. H. (1996) *Using Language*. Cambridge, UK: Cambridge University Press.
Cohn, N. (2003) *Early Writings on Visual Language*. Carlsbad, CA: Emaki Productions.
Cohn, N. (2005) Un-defining 'comics.' *International Journal of Comic Art* 7 (2): 236–48.
Cohn, N. (2007) A visual lexicon. *Public Journal of Semiotics* 1 (1): 53–84.
Cohn, N. (2009) Action starring narrative and events. Paper presented at the Comic Arts Conference, July. San Diego, CA.
Cohn, N. (2010a) Extra! extra! semantics in comics! *Journal of Pragmatics* 42 (11): 3138–46.
Cohn, N. (2010b) Japanese visual language. In T. Johnson-Woods (ed.) *Manga*, pp. 187–203. New York: Continuum Books.
Cohn, N. (2010c) The limits of time and transitions. *Studies in Comics* 1 (1): 127–47.
Cohn, N. (in preparation) Meaning in Multiple Modalities.
Cohn, N. (under review a) Beyond speech balloons and thought bubbles. *Semiotica*.
Cohn, N. (under review b) Visual narrative structure. *Cognitive Science*.
Cox, M. V. (1998). Drawings of people by Australian aboriginal children. *Journal of Art and Design Education (JADE)* 17 (1): 71–80.
D'Angelo, M. and Cantoni, L. (2006) Comics: semiotics approaches. In K. Brown (ed.) *Encyclopedia of Language and Linguistics*, pp. 627–35. Oxford: Elsevier.
de Saussure, F. (1972) *Course in General Linguistics*. [Trans. R. Harris.] Chicago: Open Court Classics.
Dean, M. (2000) *The Ninth Art*. Doctoral dissertation, University of Wisconsin-Milwaukee.
Eerden, B. (2009) Anger in Asterix. In C. Forceville and E. Urios-Aparisi (eds.) *Multimodal Metaphor*, pp. 243–64. New York: Mouton De Gruyter.

Eisner, W. (1985) *Comics & Sequential Art*. Florida: Poorhouse Press.
El Refaie, E. (2009) Metaphor in political cartoons. In C. Forceville and E. Urios-Aparisi (eds.) *Multimodal Metaphor*, pp. 173–96. New York: Mouton De Gruyter.
Engelhardt, Y. (2002) *The Language of Graphics*. Doctoral dissertation, University of Amsterdam.
Forceville, C. (2005) Visual representations of the idealized cognitive model of anger in the Asterix album *La Zizanie*. *Journal of Pragmatics* 37 (1): 69–88.
Fresnault-Dervelle, P. (1972) *La Bande Dessinée*. Paris: Hachette.
Gauthier, G. (1976) Les Peanuts. *Communications* 24: 108–39.
Goldberg, A. (1995) *Constructions*. Chicago: University of Chicago Press.
Green, J. (forthcoming) Multimodal complexity in Arandic sand story narratives. In L. Stirling, T. Strahan, and S. Douglas (eds.) *Narrative in Intimate Societies*. Amsterdam: John Benjamins.
Groensteen, T. (1999) *Systeme De La Bande Dessinée*: Paris: Presses Universitaires de France.
Gubern, R. (1972) *El Lenguaje De Los Comics*. Barcelona: Peninsula.
Hammond, M. (1999) *The Phonology of English*. Oxford: Oxford University Press.
Harvey, R. C. (1994) *The Art of the Funnies*. Jackson: University of Mississippi Press.
Horrocks, D. (2001) Inventing comics. *The Comics Journal* 234: 29–39.
Huffman, D. A. (1971) Impossible objects as nonsense sentences. In B. Meltzer and D. Mitchie (eds.) *Machine Intelligence, Vol. 6*, pp. 295–323. Edinburgh: Edinburgh University Press.
Hünig, W. K. (1974) *Strukturen Des Comic Strip*. Hildensheim: Olms.
Jackendoff, R. (1990) *Semantic Structures*. Cambridge, MA: MIT Press.
Jackendoff, R. (1997) Twistin' the night away. *Language* 73 (3): 534–59.
Kennedy, J. M. (1982) Metaphor in pictures. *Perception* 11 (5): 589–605.
Kindler, A. M., and Darras, B. (1997) Map of artistic development. In A. M. Kindler (ed.) *Child Development in Art*, pp. 17–44. Reston, VA: National Art Education Association.
Kirby, J. (1999) Interview with Ben Schwartz. *The Jack Kirby Collector* 23 (February): 19–23.
Kloepfer, R. (1977) Komplentarität von Sprache und Bild, Karikatur und Reklame. In R. Zeichenprosesse Posner and H.-P. Reinecke (eds.) *Zeichenprozesse. Semiotische Foxschung in den Einzelwissenschaften*, pp. 129–45. Wiesbaden: Athenaion.
Koch, W. A. (1971) *Varia Semiotica*. Hildensheim: Olms.
Kress, G. and van Leeuwen, T. (1996) *Reading Images*. London: Routledge.
Kukkonen, K. (2008) Beyond language. *English Language Notes* 46 (2): 89–98.
Kunzle, D. (1973) *The History of the Comic Strip, Vol. 1*. Berkeley: University of California Press.
Lakoff, G. and Johnson, M. (1980) *Metaphors We Live By*. Chicago: University of Chicago Press.
Lamme, L. L. and Thompson, S. (1994) Copy?...real artists don't copy! *Art Education* 47 (6): 46–51.
Laraudogoitia, J. P. (2008) The comic as a binary language. *Journal of Quantitative Linguistics* 15 (2): 111–35.
Laraudogoitia, J. P. (2009) The composition and structure of the comic. *Journal of Quantitative Linguistics* 16 (4): 327–53.

Lim, V. F. (2006) The visual semantics stratum. In T. Royce and W. Bowcher (eds.) *New Directions in the Analysis of Multimodal Discourse*, pp. 195–213. Mahwah, NJ: Lawrence Erlbaum Associates.
Liungman, C. G. (1991) *Dictionary of Symbols*. Santa Barbara: ABC-CLIO Inc.
Magnussen, A. (2000) The semiotics of C.S. Peirce as a theoretical framework for the understanding of comics. In A. Magnussen and H.-C. Christiansen (eds.) *Comics and Culture*, pp. 193–207. Copenhagen: Museum of Tusculanum Press.
Manning, A. D. (1998) Scott McCloud *Understanding Comics: The Invisible Art*. *IEEE Transactions on Professional Communication*, 41 (1): 66–9.
Marr, D. (1982) *Vision*. San Francisco: Freeman.
McCloud, S. (1993) *Understanding Comics*. New York: HarperCollins.
McNeill, D. (1992) *Hand and Mind*. Chicago: University of Chicago Press.
Mey, K.-A. L. (2006) Comics: pragmatics. In K. Brown (ed.) *Encyclopedia of Language and Linguistics*, 2nd edition. Vol. 2, pp. 623–27. Oxford: Elsevier.
Miller, J. (2001) *Critical Analysis of Comic Strips*. Doctoral dissertation, State University of New York, Buffalo.
Moravcsik, E. (1978) Reduplicative constructions. In J. H. Greenberg (ed.) *Universals of Human Language, III* , pp. 297–334. Stanford: Stanford University Press.
Munn, N. D. (1986) *Walbiri Iconography*. Chicago: University of Chicago Press.
Nakazawa, J. (2005) Development of manga (comic book) literacy in children. In D. W. Shwalb, J. Nakazawa, and B. J. Shwalb (eds.) *Applied Developmental Psychology*, pp. 23–42. Greenwich,CT: Information Age Publishing.
Narayan, S. (1999) *The Language of Comics*. Unpublished manuscript, University of California, Berkeley.
Narayan, S. (2000) *Mappings in Art and Language*. Senior Honors thesis, University of California, Berkeley.
Narayan, S. (2001) A window on a trail. Paper presented at the Comic Arts Conference, July, San Diego, CA.
Nöth, W. (1990) *Handbook of Semiotics*. Indianapolis: University of Indiana Press.
Oomen, U. (1975) Wort-Bild-Nachricht. *Linguistik und Didaktik* 24: 247–59.
Pallenik, M. J. (1986) A gunman in town! Children interpret a comic book. *Studies in the Anthropology of Visual Communication* 3 (1): 38–51.
Palmer, S. (1992) Common region. *Cognitive Psychology* 24 (3): 436–47.
Palmer, S. and Rock, I. (1994) Rethinking perceptual organization. *Psychonomic Bulletin and Review* 1: 29–55.
Peirce, C. S. (1931) Division of signs. In C. Hartshorne and P. Weiss (eds.) *Collected Papers of Charles Sanders Peirce: Vol. 2*, pp. 134–73. Cambridge, MA: Harvard University Press.
Reith, E. (1988) The development of use of contour lines in children's drawings of figurative and non-figurative three-dimensional models. *Archives de Psychologie*, 56 (217): 83–103.
Saraceni, M. (2000) *Language Beyond Language*. Doctoral dissertation, University of Nottingham.
Saraceni, M. (2001) Seeing beyond language: when words are not alone. *CAUCE* 24: 433–55.
Schilperoord, J. and Maes, A. (2009) Visual metaphoric conceptualization in editorial cartoons. In C. Forceville and E. Urios-Aparisi (eds.) *Multimodal Metaphor*, pp. 213–40. New York: Mouton De Gruyter.

Schodt, F. L. (1983) *Manga! Manga! The World of Japanese Comics.* New York: Kodansha America Inc.

Shinohara, K., and Matsunaka, Y. (2009) Pictorial metaphors of emotion in Japanese comics. In C. Forceville and E. Urios-Aparisi (eds.) *Multimodal Metaphor*, pp. 265–93. New York: Mouton De Gruyter.

Shipman, H. (2006) *Hergé's Tintin and Milton Caniff's Terry and the Pirates.* Paper presented at the Comic Arts Conference. San Diego, CA.

Smith, N. R. (1985) Copying and artistic behaviors. *Studies in Art Education* 26 (3): 147–56.

Sonesson, G. (2005) From the linguistic model to semiotic ecology. *Semiotics Institute Online, Lecture 4.* Retrieved on 25 August 2008 from http://www.chass.utoronto.ca/epc/srb /cyber/Sonesson4.pdf

Stainbrook, E. J. (2003) *Reading Comics.* Doctoral dissertation, Indiana University of Pennsylvania.

Teng, N. Y. (2009) Image alignment in multimodal metaphor. In C. Forceville and E. Urios-Aparisi (eds.) *Multimodal Metaphor*, pp. 197–211. New York: Mouton De Gruyter.

Toku, M. (1998) *Why Do Japanese Children Draw in Their Own Ways?* Doctoral dissertation, University of Illinois at Urbana-Champaign.

Toku, M. (2001) What is manga? *Journal of National Art Education* 54 (2): 11–17.

Töpffer, R. (1965[1845]) *Enter: The Comics* [Trans. E. Wiese.] Lincoln: University of Nebraska Press.

Walker, M. (1980) *The Lexicon of Comicana.* Port Chester, NY: Comicana, Inc.

Wertheimer, M. (1923) Untersuchungen zur Lehre von der Gestalt. *Psychol. Forsch.* 4: 301–50.

Wilkins, D. P. (1997) Alternative representations of space. In M. Biemans and J. van de Weijer (eds.) *Proceedings of the CLS Opening Academic Year '97 '98*, pp. 133–64. Nijmegen: Nijmegen/Tilburg Center for Language Studies.

Willats, J. (2005) *Making Sense of Children's Drawings.* Mahwah, NJ: Lawrence Erlbaum.

Wilson, B. (1997) Child art, multiple interpretations, and conflicts of interest. In A. M. Kindler (ed.) *Child Development in Art*, pp. 81–94. Reston: National Art Education Association.

Wilson, B. (1999) Becoming Japanese. *Visual Arts Research* 25 (2): 48–60.

Wilson, B. and Wilson, M. (1977) An iconoclastic view of the imagery sources in the drawings of young people. *Art Education* 30 (1): 4–12.

Wilson, B. and Wilson, M. (1987) Pictorial composition and narrative structure. *Visual Arts Research* 13 (2): 10–21.

# 5
# Constructing Meaning: Verbalizing the Unspeakable in Turkish Political Cartoons

*Veronika Tzankova and Thecla Schiphorst*

This study focuses on the relationship between culture, use of language, and the changing sociopolitical situation in the Republic of Turkey as reflected in the textual content of political cartoons. It is based on the presumption that the increasing and pervasive support of fundamentalist Islam, mediated by current government enforcement of dogmatic religious beliefs and social conduct, has affected and shaped language use in Turkish political cartoons.

If we rely on a semiotic approach to analyze language as a translative system of symbols and codes, we can observe the interplay between the interpretive dimensions of language in the process of subjective self-positioning and the tendency to 'play it safe' in a hybrid and tension-filled political situation. This particular verbal encoding serves as a polycentric mapping where direct textual content is projected and translated onto a metalevel of experiential knowledge in accordance with cultural models of textual imagery. The evidence presented here is based on the articulation of meaning between two orders of translation: the first being at the linguistic level, where the translation of meaning is based on relatively stable cultural entities (grammar and vocabulary); and the second being formed by a semiotic system, making use of cross-culturally fertilized entities of meaning and points of reference which rely heavily on Western cultured scenarios and paradigms.

This chapter examines the language used in Turkish political cartoons beyond the cultural linguistics point of view which treats grammar and vocabulary as entities 'relative to cultural models and culturally defined models' (Palmer 1998: 15). It explores the use of cultural models and linguistic cognitive entities as a constructive framework for the introduction of cross-cultural world views and as a way of critically problematizing the process of Islamization of the country.

The use of a Westernized semiotic order within the traditional linguistic framework of Turkey addresses the need for a certain audience. The translation of meaning that appears in the juxtaposition of form and content shapes the characteristics of the audience. In most cases the intended message reception requires an interpretation beyond the limits of the traditional Turkish culture. The consumption of the verbal concepts, moderated by a highly Westernized culture in the form of cartoons, requires a contextual understanding of Western culture itself. This contextual appropriation of Western knowledge is a trademark of modern Turks, who frequently seize the opportunity to address the problematic co-existence of religion and secularism under the authority of one government. From this perspective, the use of Westernized semiotic content within a traditional linguistic model emphasizes the interconnectedness between culture, use of language, and the changing sociopolitical situation.

## 5.1 Introduction

The critical analysis of language used in Turkish political cartoons relies heavily on the notion that language use is defined by its social purpose. Such an analysis requires a close examination of the relationship between language, identity, and politics in the framing of contemporary Turkish sociopolitical reality. To investigate the processes which inform the articulation of meaning between orders of linguistic and cultural translation, we use recently published cartoons in Turkey's major political cartoon magazine *Penguen*.[1] Our analysis focuses on the use of language strategies that first tend to avoid any entity of specific political reference and second tend to reveal a broader spectrum of cultural and political discourse that is represented by a system of symbols between words and image. Our examination of selected cartoons concludes that Turkish political cartoonists externalize linguistic references from their conventional social use to defend against the government's increasing media censorship, control, and authoritarianism. The analysis reveals the rendering of verbal content directed by an inability to express oppositional political views explicitly as well as various 'safe' models for delivery of political messages. In other words, we argue that Turkish political cartoonists who hold opposing political views locate words and images within a structural context that allows for an interchangeability of linguistic references. Such a play of linguistic elusiveness provides a successful connection with a supporting audience and avoids a conflict with the government authorities.

Understanding Turkey's present is based on understanding its positioning both historically and geographically. From a historical perspective, the roots of national identity are encoded in the acceptance of Islam as an official religion during the period of the Seljuk Turks, from the eleventh to the fourteenth centuries. The religious transition from previous polytheism to Islam was accompanied by a commitment to the unchangeable sources of the 'ultimate truth' that were available in Arabic, as well as the structural adoption of various aspects of public bureaucratic documents written in Farsi along with various Arabic dialects. The complex embeddedness of Turkish historical sources in different languages – Arabic, Farsi, and Ottoman Turkish – poses problems of interpretation and contextualization of national roots.

From a geographical perspective, Turkey stands at the crossroads between the East and the West, forming a territory of great economic and political foreign interests. Based on its unique geographical position, Turkey's cultural exchanges with the West began with and continued along the roads of commerce, sometimes having seismic ramifications for the whole society, sometimes having an effect more akin to rippling patterns on a surface. The importance of past commercial and cultural imperialism of the West and its continuing inductive influence on non-Occidental territories today is underlined by Said:

> To a very great degree the era of high nineteenth-century imperialism is over. France and Britain gave up their most splendid possessions after World War Two, and lesser powers also divested themselves of their far-flung dominions. Yet, once again recalling the words of T. S. Eliot, although that era clearly had an identity all its own, the meaning of the imperial past is not totally contained within it, but has entered the reality of hundreds of millions of people, where its existence as shared memory and as a highly conflictual texture of culture, ideology, and policy still exercises tremendous force. (1994: 12)

Although Turkey is considered a Middle Eastern country, it has never behaved as a distinctly 'Eastern' or distinctly 'Western' cultural entity, operating more like a pendulum between Eastern and Western influences. This is why the models for national identity that have been copied, adapted, institutionalized, legalized, and in all other ways activated are a product of a complex syncretism of Eastern and Western cultural influences at different times. These tendencies enable the understanding of the current pro-Islamic political situation and pro-Islamic national positioning as an indicator of deeper identification with Islamic values and a gradual lessening of Occidental influences. Within these layers of national identity, pro-Western orientation has a distinctive demographic lobby that cannot be ignored as it evidently opposes and even balances the current drifting toward Islam-centered dogmas (Dai 2005: 33).

The discussion around the process of Islamization in Turkey must address the role of the community that unites the oppositional members. According to Zygmunt Bauman, in a community 'we all understand each other well, we may trust what we hear, we are safe most of the time and hardly ever puzzled or taken aback' (2001: 2). This statement about communities is closely linked to language as mutual 'understanding' and is bound in the first place to 'hearing' and not to an ultimate form of sensorium. In this respect, the focus of this study is on the language determinants specific for a community that counteracts political Islamization and uses cartoons as a medium of expression. It should be emphasized that the analysis of the language used in Turkish political cartoons is encoded and decoded in the medium and the inseparability of word and image: '[...] political cartoons provide a means of expressing usually critical political and social commentary through a visual format that may include images, words, or both' (Bergen 2004: 24).

It is specifically the cartoon medium that allows their use as a significant political counterforce in the Republic of Turkey. The combination of words, image, and humor softens the sharpness of political messages, which allows the oppositional views to pass through the censorship and reach their supporting audience: 'With censorship, political cartoons started to transform from a site of representation of ideas and values to one of resistance against state control; new meanings and symbol systems emerged to criticize the authorities' (Göçek 1997: 5).

## 5.2 Language culture in Turkish political cartoons

In the context of Turkey, cartoon artists have generally been on the opposite side of dominant streams in the political spectrum (*Today's Zaman* 2008). The famous Turkish cartoonist Bahadir Boysal is not afraid to confess this fact in an interview for a Turkish newspaper: 'Political cartoons have always been drawn from the opposition point of the most popular points of view. We have always been a bit socialist. If it were the Republican People's Party (CHP) – which is known to be leftist – in power, there would be cartoons satirizing them as well' (*Today's Zaman* 2008). However, the tension in the current Turkish political atmosphere brings about the need for a 'play it safe' strategy performed by political counteractants like cartoon artists. According to Boysal, the current Prime Minister of Turkey, Recep Tayyip Erdoğan, is not open to criticism and 'has filed several lawsuits at various times against comics [artists]' (*Today's Zaman* 2008).

The evidence of the Prime Minister's 'tender' response toward political criticism (*The Economist* 2005: 37) triggers first a reductionist approach toward explicit verbal information in political cartoons and second the development of narratives by means of vague, elusive, and polysemous semiotic systems in order to (culturally) meet the requirements of the changing and slippery political situation (Erdoğan 1998: 118–22; Tunc 2002: 54). In this tendency, the minimal use of text plays a complementary role to the semiotics of the image, thus producing non-conventional symbols itself. In an engagement with the field of conceptualization more than practical experience, the question of gradations of verbal abstraction in Turkish political cartoons is achieved by layering the meanings between the borders of the signifier and the signified. The use of the signifier, in the current pro-Islamic political situation, has become the liquid mirroring of the unspeakables. Changing the meaning and the social concept of the signifier is efficient and 'safe' because what is considered semantic becomes semiotic. In this way, relative social objectivity is avoided, with little conceptual space remaining for classic notions of language codes, since language codes presuppose an agreement among the recipients about the meaning.

Generally speaking, gradation of meaning in verbal representation relies heavily on the unity between the semantic and semiotic systems of language. Drawing in part on Saussure's work, we propose a circular visual model representing the development of socially coherent verbal meaning as illustrated in Figure 5.1, which employs a figure-ground perceptual archetype as the basis for the articulation of the inseparability of semantics and semiotics within a language system.

*Figure 5.1* Inseparability between semantics and semiotics within social coherence of verbal meaning

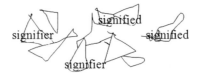

*Figure 5.2* Shattered plate: Saussurian fragmented language system attributed to Turkish political cartoons

This model agrees in principle with what McCloud observes about the appearance of non-pictorial icons in comics: 'In the non-pictorial icons, meaning is fixed and absolute' (1994 [1993]: 28). McCloud's definition of the non-pictorial includes the written word.

Our analysis, however, reveals that the language used in Turkish political cartoons offers something that may be called a sort of *shattered plate* image of language systems and knots (see Figure 5.2). In other words, the analysis reveals the postmodern character in the Turkish language, one that is fragmented, different from itself, severed in its parts, fractured, split, fissioned, divided in pieces: in short, the obvious disconnection between the semantic and semiotic systems. This phenomenon can be described as a Saussurian semiology expounded upon by Derrida:

> [Semiology of the Saussurian type] has marked against the tradition, that the signified is inseparable from the signifier, that the signified and signifier are the two sides of one and the same production. Saussure even purposely refused to have this opposition or this 'two sided unity' conform to the relationship between soul and body, as had always been done. (2004: 18)

The conceptual implausibility and the difficulty in apprehension of a fragmented semiotic language system becomes an isolated yet connected shard or part of a whole – the socially accepted structure of the Turkish language – while keeping a very specific profile of informal and uncommon meaning. To some extent, this strategy resembles and can be confused with the use of metaphor that can be described either as the effect of resemblance between referents (referentialist view) or the descriptive information associated with them (descriptivist view) (Leezenberg 2001: 69–79). In both cases, metaphorical meaning is socially defined in two ways. First, perception of resemblance and/or descriptive information is based on previously learned experience which is to a great extent a form of sociocultural experience; and second, generally speaking, metaphors are entities with socially fixed meaning that are widely used and thus attain the status of recognizable clichés, e.g., as in the case of 'a sea of troubles' (Lakoff and Johnson 2006: 103).

From this perspective, in the context of Turkish political cartoons, the strategy of detaching the signifier from the signified is not in the form of metaphor as it does not rely on resemblance or descriptive reference but rather on an intuitional, elusive, and liquid assignment of meaning. By extension, this construction of meaning allows for some flexibility in interpretation as its central intention is to propel a readership toward political awareness.

The call for a holistic and integrated political consciousness is very much addressed to the 'supporting community' within which the linguistic codes are recognized and shared, and the signifier acquires a liquid encoded sign. This process of meaning recognition within a community establishes the 'pragmatic coherence' (Kramsch 1998: 28) of the message. Pragmatic coherence can be described as a form of semiosis which 'relates speaker to speaker within the larger cultural context of communication' (Kramsch 1998: 28). What becomes observable in Turkish political cartoons is the detachment of the signified from the social entity of the signifier in a post-structural manner, which directs the readers to isolate themselves self-consciously from standardized and understood linguistic forms and seek meaning in cognitive resources not specific to common linguistic experiences. Linguistic entities, in combination with humoristic images, become structures in flux with varying meaning depending on readers' ability to process and interpret the input. This fluctuation in meaning assures a partial loss of intensity of oppositional political views.

## 5.3 Language as a work of art or a political message

Identifying the context in which the verbal information in Turkish political cartoons is decoded is an inevitable result of seeking pragmatic coherence. Is the political cartoon considered a piece of art, a political message, or an entity in between? Is it a private discourse, emerging solely in the mind of the reader, or is it instead a public discourse?

There is a certain stable logic in considering political cartoons a 'public discourse' appearing in 'public spaces,' because of the way this medium is produced and published. Here, the term 'public space' is used in the sense developed by critical theorists, namely the space in which consensus is arrived at through discourse in a free and open realm, where voices can be articulated without the threat of sanction. However, the current Turkish political climate has eliminated this 'free' expression, thereby limiting the notion of public space to a void that is filled with the government's pro-Islamic conservatism. Maureen Freely laconically reflects on this situation: 'there is civilised discourse and there is uncivilised discourse, but to ensure the former by banning the latter is not the way forward' (Freely 2006: 146).

Freely's notion concisely addresses the situation in which Turkish political cartoons come into existence and explains the 'minimalist' approach toward the use of language as a form of escape from a possible government counteraction. If this approach can be called 'simplicity of language,' then simplicity becomes an incremental articulation of the 'essential' by avoiding semantically referential structures. From this perspective, verbal information becomes and remains a 'symbolic image' (Saussure 2007: 15) that co-exists in a multimodal combination with the pictorial forms of representation and is no more or less a part of the artistic choices made to complete the visual. Blending verbal items within pictorial forms, or hiding the message behind art, induces the idea of 'neutralization' (Lomheim 1999: 190–207) of possible negative interpretation. The neutralization of negative interpretation reveals the socially (in the case of the current study, politically) constructed and induced notion of fear that has to be both confronted and obeyed. What becomes obvious in the Turkish political cartoons is an opposition to the 'naturally' appearing and commonly positioned acceptance of sociopolitical reality. 'Naturally' is not used here in terms of the binary opposition of nature-nurture, but in relation to the social collusion and politically agreed upon blindness subsuming the act of addressing an ambiguous political reality.

## 5.4 Communication of political ideas in a tense political situation

Linking language to the process of 'cultivation' within a cultural context embeds it in a specific frame of normative dimensions and behavioral patterns. In other words, language can be considered a heuristic that specifies the regulative limits between 'self,' 'society' as social relations, and 'culture' as the mediation of those relations. Language provides a means for social survival given that cultural polemics are properly addressed. A hypothetical separation of verbal expression from its cultural setting may suggest a fundamental error in the communication codes, leading to social exclusion. Within this framework, we suggest that social communication and development of meaning is constructed by linguistically coded expression of cultural nodes and culturally coded nodes of language.

Linguistic components, such as grammar rules and vocabulary for example, represent instances of what may be called 'cultural traditions,' or 'negotiating frames' (Kramsch 1998: 46). These serve as forms of mediation that in some way, whether in terms of aesthetics, style, or subject matter, resonate with the social production of hegemonic cultural codes of behavior.

If the relationship between language and culture is a commonly accepted fact, then linguistic nodes of 'neutralization' will become explicit in situations of disagreement. From a cultural perspective, conflicts are often hidden beneath the surface of normative consensual social reality. In terms of political conflicts, it has been suggested that

> while language is hardly ever the cause of such conflicts, nevertheless it is always implicated in them, whether functionally as a medium of communication or symbolically as a site of mobilisation and counter-mobilisation in games of power relations between contending parties. (Suleiman 1999: 10)

Suleiman goes on to connect the symbolic relationship between language and conflict to 'power relations' (1999:10). In the current political situation in the territory of the Republic of Turkey, the power of the political counteractants is being systematically repressed, to the degree that certain linguistic discourses have gained a status in between functionality and symbolism.

## 5.5 Exemplifications of language shifts between functionality and symbolism

Please note that we provide the web address for all cartoons analyzed in this section so that readers can see the images under discussion.

### 5.5.1 Lack of a linguistic subject

An example of this phenomenon is the observable dominant use of null-subject linguistic structures and semi-clauses in Turkish political cartoons when addressing members of the government, especially the Prime Minister. Although Turkish is a null-subject language (Kornfilt 1997: 192), which grammatically allows the subject of the sentence to be 'phonetically unrealized or unpronounced' (Shlonsky 1997: 112), the use of null-subject clauses can be considered a cultural neutralization of a conflict through avoidance of direct reference.

In this section, we analyze a political cartoon by Turkish artist Selçuk Erdem, published in *Penguen Dergisi*, volume 387. Table 5.1 contains extracts of the language used in the cartoon.

*Table 5.1* A seal's protest

| Original content | English translation | Additional details |
|---|---|---|
| Erdoğan "Fok balıkları için ayağa kalkan ınsanlık, Gazze'deki zulmü görmüyor" diyerek tepki gösterdi. | Erdoğan expressed a protest with the following words: 'The people who strive for the rights of the seals fail to notice the terror in Gaza.' | Contextualizing explanation located below Penguen's logo. |
| ABi SÜREKLi BANA LAF SOKTU, SiZDEN HiÇ BAHSETMEDi... | Big Brother, He/she/ it constantly insulted me, but never talked about you... | Located in a speech balloon above the seal's head. |
| ÖLMEK VAR, DÖNMEK YOK | There is death, but there is no way back | Located on a small banner hanging above the seal's head in the background. This Turkish expression is used to highlight the irreversibility of a situation. |

*Penguen Dergisi*, Vol. 387. http://www.penguen.com/kapak.asp?gun=20100216

This cartoon is a reaction to the words of Recep Tayyip Erdoğan, the Prime Minister of Turkey: 'The people who strive for the rights of the seals fail to notice the terror in Gaza.' This quotation from the Prime Minister's speech is provided in small font below the *Penguen* logo of the magazine and serves as a detailed conceptualization of the depicted situation. The cartoon itself contains several sources of verbal information, the most important ones being the caption and the slogan, both situated above the seal's head. The slogan says: 'ÖLMEK VAR, DÖNMEK YOK' which literally translated into English means: 'There is death, but there is no turning back / way back.' From a sociolinguistic perspective, this Turkish expression implies the irreversible cul-de-sac of a situation.

Of particular importance to our analysis is the linguistic content found in the speech balloon above the seal's head and its corresponding literal English translation. An analysis of each line allows us to locate the lack of transitivity and definiteness, which would help to define a state of linguistic neutrality and intended subversion:

ABi (short form of *ağabey*) Big Brother – refers to a male individual and shows real or feigned respect. In this case, it is likely that the seal is addressing someone in the crowd of people. Because of lack of punctuation after this reference, 'ABi' can be considered also the subject of the sentence, which is possible but unlikely if acknowledging the provided context. In this situation, the neutralization appears by the lack of personalization of the 'ABi,' as 'ABi' is a very frequently used form of addressing male individuals when the personal name is not known. In other words, the question of who the 'ABi' is remains unanswered.

SÜREKLi (constantly) BANA (to me; first person singular dative declension from *ben* meaning 'I')

LAF (chat) SOKTU (put into / in simple past tense); *laf soktu* as a phrase is equivalent to 'insulted.' This clause lacks a subject. The active insulting party can be anybody; however, it can be inferred that it references the current Prime Minister of Turkey, Recep Tayyip Erdoğan. This inference is based on the context and the quoted words of Erdoğan provided for the reader.

SiZDEN (from you; declension from *siz* meaning 'you (pl.)' in ablative) HiÇ (at all)
BAHSETMEDi ... (didn't talk [about].)
Literally, 'SiZDEN HiÇ BAHSETMEDi' corresponds to 'He/she/ it didn't talk about you at all.'

Despite being short, the text is open to interpretations in many different ways. This provides a ground for fusion within a politically neutral context, while simultaneously delivering the intended message. As a performance of counteraction against the current political situation, this text leaves no stone unturned. Every note of the political game is struck within the limits of cultural and linguistic rules: effects replace essences, instabilities replace stabilities, surfaces replace interiorities, null or neutral subjects replace identities, linguistic codes replace cores of meaning, subversion attacks hegemony, opposition compromises obedience, ambiguous signifiers replace expression, verbal discontinuity becomes better than continuity. In other words, the fluidity of linguistic signs and protean qualities of meaning are celebrated with no attempt at avoiding any possible scope of interpretational absurdity. The meaning of the message is constructed within the fluctuating framework of the text, and this 'textual flux' implies precisely that any identifications (regarding the grammatical subject of the message) are temporarily devoid of presence. Construction of social meaning implies that linguistic decisions are based on repetitive acts, generally unconscious stances which, like any stance, if the direction of the vector is changed, the polarities, charges and valences could just flip to the opposite side, or perhaps just a pastiched parody of a side.

### 5.5.2 The role of the 'I' in the construction of verbal allusion

[...] we can say that today's writing has freed itself from the dimensions of expression. Referring only to itself, but without being restricted to the confines of its interiority, writing is identified with its own unfolded exteriority. (Foucault 2007: 904)

The analysis of the 'unfolded exteriority' of writing, text, and language in Turkish political cartoons cannot be presented coherently without a discussion of the ongoing conversation between words, image, creator (author or artist) and recipient (reader, viewer) in the context of narrative development (see Figure 5.3).

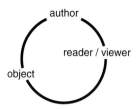

*Figure 5.3* Language as a circular connection between object, author, and reader forming an intertextual narrative

Language is the circular connection between object (pictorial protagonist), author, and reader which forms the intertextuality of the narrative. In this relationship, it is interesting to analyze the liquid role of the 'I' as a subject, as self-presence, and self-identification. The engagement of the 'I' with multiple 'beings' resembles Heidegger's precedence of the 'I.' The 'I' 'deconstructs its own voice – revealing the provisional nature of all vocabulary, discourse and understanding – in favor of the voice of the other [...]' (Ward 1995: 139). In other words, is the 'I' the visual protagonist, or the cartoonist, who presents a political perspective, or the reader, who is challenged to decode the chain between the signifier and the signified? This interchangeability of the roles assigned to the linguistic subject of the 'I' acts as a method of building towards the allusiveness of verbal expression and thus as avoidance of direct political confrontation.

The language in Turkish political cartoons reveals certain levels of abstraction and detachment from specific meaning. This intersubjectivity allows the protagonist to have no fixed identity, only a 'style,' to have a 'self' which is an 'effect' of mirroring and reflections of significations. This chain of self-projection (or, from a linguistic perspective, in the role of a subject, the 'I') is similar to the Lacanian idea of 'the infinity of reflection' (Lacan 2004: 126), which in the case of political cartoons can be considered an 'infinity of significations.' The possibility for a metaphoric insertion of the 'I' as a linguistic substitution of the 'self' provides the reader with the opportunity for personal inscriptions into the situation presented in the artifact.

In his analysis of shifts of identities reflected in Turkish political cartoons, Ayhan Akman clearly depicts the phenomenon of interchangeability of the role of the 'I':

> modernist identities in cartoons also have a symbolic character. In their abstraction, they always stand for something other (and larger) than themselves. Modernist cartoons have no individualized characters. The figures are always the prototypical representations of some larger group or social class. As such, the individual figures (when they exist) have no individuality; they are interchangeable units of a larger whole, members of a group or class [...]. The decontextual character of these cartoons also works to create the same effect; stripped of any reference to the concrete, present social context, the abstract characters of these cartoons 'drift' in homogeneous space. (1998: 123)

The next political cartoon, by Serkan Yılmaz, is a clear example of this phenomenon (see Table 5.2).

This cartoon is inspired by the Turkish Prime Minister's words that the employer providing the salary can 'kick anybody out of the shop,' implying that the owner of a business can fire any employee with no further explanation. This contextualizing information is found below the *Penguen* logo. In this cartoon image, the interplay between the different 'I's' is well represented. For the sake of clarity, we first provide an explanation of the image and translation of the language.

Table 5.2  A worker has a conversation with his boss

| Original content | English translation | Additional details |
| --- | --- | --- |
| Başbakan 'Maaşını sen veriyorsun. Sana bu dükkanda yer yok diyeceksin' sözleriyle medya patronlarının köşe yazarlarını kovabileceğini söyledi. | The Prime Minister informed the media bosses of their power to fire column writers with the following words: 'If you (the media bosses) pay the salary, you can kick anybody out (of the shop).' | This contextualizing information is located below *Penguen*'s logo. |
| BENi SEN KOVMUYORSUN, BEN iSTiFA EDiYORUM! | You are not firing me, I am the one who is quitting! | Located in a speech balloon above the figure of the employee (the first figure from left to right.) |
| OHH...OHH... TAZMiNATI DA YAKTI... | Ohh...Ohh...(a sound of pleasure), he lost his employment insurance rights... | Located in a balloon above the head of the Prime Minister – a figure hiding behind the chair of the employer (third from left to right.) |

*Penguen Dergisi*, Vol 389. http://www.penguen.com/kapak.asp?gun=20100302

Let us start from the representation of the Prime Minister of Turkey who, in this particular cartoon, is the figure hiding behind the chair (third figure from the left). The other two figures represent an employer and an employee. From the situation of the objects within the visual frame, their postures and the conversational turn, it can be inferred that the 'speaking' figure is the employee and the one sitting on the chair, the employer. As can be observed, the symbolic figure and the identity of the employee can be assigned to or adopted by, generally speaking, any working Turk. The role of the 'I' (or Turkish *ben*) is produced by the collective nature of the victimized citizen. The choice of separation between the private and the social, between solidarity with the political situation and opposition is left to the reader. If the reader decides to choose a distanced position of self-exclusion, then the 'I' will be assigned to one of two roles: first, the protagonist as a linguistic-narrative figure; or, second, the cartoonist, who uses the medium of single-panel drawings to 'represent the cartoonist's view of political issues, current events, public personalities, and contemporary social mores' (Kenney and Colgan 2003: 226). Thus, the transmission of meaning and concepts emerges from the interplay of roles and 'selves' within the framework of an interpretative system that transforms itself into a mirror of political ideology.

The characteristics of the relationship between the 'I' as a linguistic subject and the personal orientation of the more philosophic notion of the 'self' are inscribed in the direct attitude of the reader toward the protagonist, or in other words, the graphic object and its speech.[2] This choice of attitude, in the form of the choice between distantiation or self-inclusion/immersion, can be considered a structural approach toward latent freedom of expression since no dimension of potential social threat or fear of suffering is present.

Another interesting aspect of the 'I' interchangeability reflected in Turkish political cartoons is the concept of 'self' in the context of social reality defined by 'here and now' (Berger and Luckmann 1966: 37). According to Berger and Luckmann, the reality of social life can be organized around the axis of the physical location of one's body, defined by the term *here*, and the current state of one's presence, defined by the term *now*. The unity between *here* and *now* forms the scope and limits of individuals' perception of positioning in social life (Berger and Luckmann 1966: 36–37). Each element presented in the *here* and *now* relationship constructs the verisimilitude of existence. Components existing outside the *here and now* unity initiate different levels of distance, distantiation, and/or immediacy. This relationship implies that the closer a fragment is to *here and now*, the closer it is to one's perception of reality.

Regarding table 5.2, we have already considered the interchangeability of the 'I' of the employee. It may be said that the spatial notion of *here* is supported by the pictorial graphics and the visual frame. The flavor of *now* is primarily achieved by the speech of the protagonist, and especially by the tense of the verbs:

> Beni sen kov-mu-yor-sun, ben istifa edi—yor -um!
> You are not firing me, I am the one who is quitting!

The suffix *-yor* used in *kovmu-yor-sun* ('firing') and *edi-yor-um* (auxiliary verb) signifies a prolonged action in the present (similar to present progressive tense in English.) The use of this specific tense restricts the moment presented in the cartoon to *now*, which locates the illustrated situation in the center of one's perception of social reality. The achieved unity between *here* and *now* conceptualizes the validity of the situation for anybody who is willing to participate in the play of 'I' exchangeability. From this perspective, the 'I' exchangeability defines the unfixed status of social order where social roles depend merely on actions of self-identification.

### 5.5.3 Code switching and audience

From earlier and more recent research we know (a) that code-switching is related to and indicative of group membership in particular types of bilingual speech communities, such that the regularities of the alternating use of two or more languages within one conversation may vary to a considerable degree between speech communities, and (b) that intrasentential code-switching, where it occurs, is constrained by syntactic and morphosyntactic considerations which may or may not be of a universal kind. Accordingly, the dominant perspectives on code-switching taken in research have been either sociolinguistic (in the narrow sense of the term, i.e. as referring to relationships between social and linguistic structure), or grammatical (referring to constraints on intrasentential code-switching). (Auer 1998: 3)

Within this definition of code switching, we focus on three particular points: first, the use of two or more languages in one conversation; second, the appearance of code switching in specific speech communities; and third, the presence of [syntactic] considerations that may not be of a universal kind.

To speak about code switching is to speak about the practical meaning and operating ideology of cultural and linguistic codes that appear in communication processes and within the interaction framework of two or more languages. Placing this linguistic phenomenon in the general context of the Republic of Turkey, it may be suggested that code switching is most likely to appear between Turkish (as the official language of Turkey) and either languages used by minorities, such as Kurdish, Azeri, Bulgarian, and Greek, or languages of cultural and historical influence, such as Persian and Arabic. At the official level (including publishing), however, the use of Turkish is enforced by law. These regulations are enforced by the government and the Türk Dil Kurumu (TDK, meaning 'Turkish Language Society'). The role of the Türk Dil Kurumu includes ensuring the 'purity' of the Turkish language:

> The time when TDK's principal business was seeking Öztürkçe [core Turkish] replacements for Arabic and Persian words has long passed; much of the post-1983 TDK's effort goes into devising and disseminating Turkish equivalents for English words in common use. (Lewis 2002: 133)

The practice of seeking Turkish replacement words for borrowings that are generally based on English, primarily as a result of the effects of the internet culture, inevitably uncovers a tendency to language and cultural protectionism. In this stream of 'keeping with the phonetics, aesthetics, and grammar' (Lewis 2002:155) of the Turkish language, it is interesting to observe the use of pure English words as a recent trend in Turkish political cartoons.

The practice of code switching is clearly observable in a cartoon by artists Selçuk Erdem and Bahadır Baruter. This cartoon is inspired by the Turkish Prime Minister's approval of a presidential system, which, in the case of the Republic of Turkey's legal order would provide the Prime Minister's party with total authority. The assertion of English words as part of a virtual dialogue between the two characters in the cartoon creates a form which may be associated with code switching that occurs either in the cartoonist's speech creation or with code switching appearing in the virtual dialogue between the two protagonists (see Table 5.3).

Table 5.3  Prime Minister and President exchange pleasantries

| Original content | English translation | Additional details |
|---|---|---|
| Başbakan Erdoğan, başkanlık sistemine olumlu baktığını açıkladı. | The Prime Minister Erdoğan announced his approval of a presidential system. | This contextualizing information is located below Penguen's logo. |
| 23 NiSAN! | April 23rd! | Located in a caption above the Prime Minister's head (the first figure from left to right.) 23rd April is Turkey's National Sovereignty and Children's Day. |
| What? | | Located in a balloon below the head of second figure from left to right (explicitly from the graphics, the figure represents Barack Obama.) |
| NEŞE DOLUYOR İNSAAAN! | It is filling people with joy! | Located in a caption below the Prime Minister's head. |

*Penguen Dergisi*, Vol 396. http://www.penguen.com/kapak.asp?gun=20100420

The structural integration of English and Turkish within Turkish political cartoons may also be analyzed from the perspective of extralinguistic factors such as community norms, or societal, political and ideological developments (Li Wei 1998: 156–78), all of which serve as cultural affirmation addressing a particular part of the population. This structural integration functions as an encoding method of politically and ideologically charged messages, which can be decoded only by individuals that are acquainted with both languages. From this perspective, the structure of political meaning depends on how the cartoonist has set the code switching to operate. It is the transformation of linguistic codes that addresses a population turned more toward the values of the West. This process is secured by the ability of the audience to decode a message mediated by a Western language. Code switching is intended to generate new configurations by displacing concepts and then re-inscribing them into another language field where 'the differences between codes, rather than the directions of the change, seems to be important' (Sebba and Wootton 1998: 274).

### 5.5.4 Switching of linguocultural codes

In relation to the identification of code switching as a means of addressing a particular audience group, we should notice the practice of 'switching of linguocultural codes' which appears within the frames of the Turkish language. In other words, by 'switching of linguocultural codes' we mean a mechanism of 'displacing a certain word or phrase from its usual or conventional signification to an unexpected one' (Ermida 2008: 68). In this scenario, the development of meaning emerges from the contrast between (a) a socially recognized usage of a word and (b) the shift in the context of this usage. In short, we may describe this scenario as 'recontextualization of the context.'

This linguistic tactic can be considered a game of representations, concerning the *how-what* relationship between language reference, contextual use and social reality (Ward 1995: 64–78). This strategy closely relies on the re-appropriation of signification. The use of a shift in the contextual meaning of linguistic codes can be considered to a degree a trace of the meaning where the 'trace is not a presence but is rather the simulacrum of a presence that dislocates, displaces and refers beyond itself' (Derrida 1973: 156). In large part, the use of contextual differences is both (a) a justification of translating meaning within the limits of signification, and (b) a means to trigger social understandings, perception, and awareness, which are not a part of the linguistic metaphysics of the masses.

*Table 5.4* Valentine's Day in politics

| Original content | English translation | Additional details |
|---|---|---|
| Deniz Baykal, Cübbeli Ahmet Hoca'yı telefonla aradı... | Deniz Baykal phoned the 'dressed in a robe' Ahmet Hoca | This contextualizing information is provided below *Penguen*'s logo. |
| ŞU AN ÜZERINDE NE VAR? | What are you wearing right now? | Located in a speech balloon visually leading to the figure of Deniz Baykal (a Turkish politician). |
| CÜBBE! | A robe! | Located in a balloon appearing from the telephone. |
| Sevgililer Gününüz Kutlu Olsun! | Happy Valentine's Day! | Located in a red ribbon at the bottom of the image. |

*Penguen Dergisi*, Vol 386. http://www.penguen.com/kapak.asp?gun=20100209

The linguocultural codes shift explicitly in this 'Valentine's Day issue' political cartoon by Bahadır Baruter (see Table 5.4). In this image, Deniz Baykal, a prominent male figure in the political life of Turkey, is talking to Ahmet Hoca, another male figure frequently appearing in the media and political life of the Republic of Turkey. He is known for discussing and giving public advice in the field of Islam, Islamic norms, and Islamic lifestyle. This short explanation is provided to the reader under the *Penguen* logo.

The pose of the semi-dressed protagonist, the illustration of hearts depicted at the right bottom area, and the general reddish-pink color scheme of the cartoon imply the articulation of a romantically erotic conversation. The gradation from romance to sexual intensity is provoked by the question 'What are you wearing right now?', which strengthens the sexual connotations of the dialogue, especially in the sociocultural context of the Republic of Turkey.

The humor in this cartoon is constructed around shock caused by the displacement of the use of the word *Cübbe* (associated with male individuals). When this cartoon is considered from the perspective of the programmatic code of lexical choice isolated from the pictorial representation, then the humor of the situation leads the reader to the depths of the sexual orientation of the protagonist (a political figure) and the status of his 'masculinity.' Here, it should be indicated that in the context of Turkey, there has lately been a higher level of intolerance and negative reaction to men behaving in gender-nonconforming ways.

In this specific context, the word *Cübbe* has been inscribed in a chain of simultaneously conflicting orders of meaning associated with Islam as a religious system and also with homosexuality, two mutually exclusive terms from a conceptual perspective. This displacement of significations addresses a transcendental public group recognizing the grasp of the political system. In this case, the word *Cübbe* signifies not only a 'robe,' but the sarcasm of the author triggered by the current flaws in the political separation of religion and state. The double signification here comes from the nonconformity of the contextual use of the signifier and its inscription into an elaborate chain of social and political positioning. It can be concluded that equivocality is initiated by the shift appearing in the process of cognitive denotation and connotation in which the 'text exceeds its meaning, permits itself to be turned away from, to return to, and to repeat itself outside its self-identity' (Derrida 2004: 65).

The strategy of 'switching of linguocultural codes' undertakes a deconstruction of the unspeakable by a reinscription of significations into a play of oppositional designation. That is why it is correct to say that 'switching of linguocultural codes' is possible: the nature of language is such that meaning cannot be fixed to 'a system of fundamental constraints' (Derrida 2004:5) and social inertia. The reappropriation of the 'context–meaning' relationship can be considered a metaphoric attempt to take back the authority over the value of social truth.

## 5.6 Summary

This chapter has focused on the relationship between culture, use of language, and the changing sociopolitical situation in the Republic of Turkey as reflected in political cartoons. It illustrated through examples the construction of meaning through various techniques of verbalizing the unspeakable. Analyses were made on the basis of the methodology of close reading of the language and language variables used in Turkish political cartoons. Examining the interpretive dimensions of linguistic choices within the systematic framework of Turkish has revealed the idiosyncratic characteristics of language both as a determinative self-positioning and as a translative system of symbols and codes associated with groups of political counteractants.

Understanding linguistic variation in Turkish political cartoons depends in large part on understanding the externalization of an interior political identity. The current tense political situation in Turkey has limited the platforms of expression, forcing oppositionists to add new variables to the verbal communication frames. These linguistic choices suggest the appearance of not just a simple expression, but rather a complex system of symbols encoded and decoded between words and image. These linguistic tendencies introduce a new structuration in linguistic codes shaped by the social context. These linguistic codes or entities are built along the lines of political stagnation and oppositional identity and are based on a semiotic field of circulating extralinguistic factors.

## Notes

1. *Penguen* is a weekly cartoon magazine that addresses current issues in the political life of the Republic of Turkey. It is well known among university students in the fields of political science and civil law. *Penguen* is available both online (http://www.penguen.com/) and in print.
2. Here, in the notion of reader, the cartoonist or the author is also included as a distanced observer of his own work.

## References

Akman, A. (1998) From cultural schizophrenia to modernist binarism: cartoons and identities in Turkey (1930–1975). In F. M. Göçek (ed.) *Political Cartoons in the Middle East*, pp. 83–132. Princeton, NJ: Markus Wiener Publishers.

Auer, P. (1998) Introduction: Bilingual conversation revisited. In P. Auer (ed.) *Code Switching in Conversation: Language, Interaction and Identity*. London/ New York: Routledge, pp. 1–25.

Bauman, Z. (2001) *Community: Seeking Safety in an Insecure World*. Cambridge/ Malden: Polity/Blackwell.

Bergen, B. (2004) To awaken a sleeping giant: cognition and culture in September 11 political cartoons. In M. Archard and S. Kemmer (eds.) *Language, Culture and Mind*. pp. 23–35. Stanford, CA: CSLI Publications.

Berger, P. and Luckmann, T. (1966) *The Social Construction of Reality: A Treatise in the Sociology of Knowledge*. Garden City, NY: Doubleday & Company.

Dai, H. D. (2005) Transformation of Islamic political identity in Turkey: rethinking the west and westernization. *Turkish Studies* 6 (1): 21–37. Retrieved on 27 August 2010 from http://dx.doi.org/10.1080/1468384042000339302

Derrida, J. (1973) *Speech and Phenomena, and Other Essays on Husserl's Theory of Signs*. Evanston, IN: Northwestern University Press.

Derrida, J. (2004) *Positions*. London and New York: Continuum.

Erdoğan, N. (1998) Popüler anlatılar ve Kemalist Pedagoji (Popular stories and Kemalist pedagogy), *Birikim* 105: 117–25.

Ermida, I. (2008) *The Language of Comic Narratives: Humor Construction in Short Stories*. Berlin: Mouton de Gruyter.

Freely, M. (2006) Cultural translation. In G. MacLean (ed.) *Writing Turkey: Explorations in Turkish History, Politics, and Cultural Identity*, pp. 145–53. London: Middlesex University Press.

Foucalt, M. (2007) What is an author? In D. Richter (ed.) *The Critical Tradition: Classic Texts and Contemporary Trends*, 3rd edition. Boston: Bedford/St. Martin's Press.
Göçek, F.M. (1997) Political cartoons as a site of representation and resistance in the middle east. *Interdisciplinary Journal of Middle Eastern Studies* 6: 1–13.
Kenney, K. and Colgan, M. (2003) Drawing blood: images, stereotypes, and the political cartoon. In P. M. Lester and S. D. Ross (eds.) *Images that Injure*, pp. 223–32.Westport, CT: Praeger Publishers.
Kornfilt, J. (1997) *Turkish*. London, New York: Routledge.
Kramsch, C. (1998) *Language and Culture*. Oxford: Oxford University Press.
Lacan, J. (2004) *Ecrits: a Selection*. New York: W.W. Norton & Co.
Lakoff, G. and Johnson, M. (2006) Metaphors we live by. In J. O'Brien (ed.) *The Production of Reality: Essays and Readings on Social Interaction*, 4th edition, pp. 103–14.Thousand Oaks, CA: Sage.
Leezenberg, M. (2001) *Contexts of Metaphor*. Amsterdam and New York: Elsevier.
Lewis, G. (2002) *The Turkish Language Reform: A Catastrophic Success*. Oxford and New York: Oxford University Press.
Li Wei. (1998) The 'why' and 'how' questions in the analysis of conversational code-switching. In P. Auer (ed.) *Code-Switching in Conversation: Language, Interaction and Identity*, pp. 156–80. London, New York: Routledge.
Lomheim, S. (1999) The writing on the screen. Subtitling: a case study from Norwegian broadcasting (NRK), Oslo. In G. Anderman (ed.) *Word, Text, Translation: Liber Amicorum for Peter Newmark*, pp. 190–208. Clevedon and Buffalo: Multilingual Matters.
McCloud, S. (1994 [1993]) *Understanding Comics: the Invisible Art*. New York: HarperPerennial.
Palmer, G. (1998) When does cognitive linguistics become cultural? In J. Luchjenbroers (ed.) *Cognitive Linguistics Investigations: Across Languages, Fields and Philosophical Boundaries*, pp. 13–47. Amsterdam and Philadelphia: John Benjamins Publishing Company.
Said, E. (1994) *Culture and Imperialism*. New York: Vintage Books.
Saussure, F. de (2007) *Course in General Linguistics*. 17th edition. Chicago: Open Court.
Sebba, M. and Wootton, T. (1998) We, they and identity: sequential versus identity-related explanation in code-switching. In P. Auer (ed.) *Code-Switching in Conversation: Language, Interaction and Identity*, pp. 262–90. London, New York: Routledge.
Shlonsky, U. (1997). *Clause Structure and Word Order in Hebrew and Arabic: An Essay in Comparative Semitic Syntax*. New York: Oxford University Press.
Suleiman, Y. (1999) Language and political conflict in the Middle East: a study in symbolic sociolinguistics. In Y. Suleiman (ed.) *Language and Society in the Middle East and North Africa: Studies in Variation and Identity*, pp. 10–38. London: Curzon Press.
*The Economist* 375, no. 8420. (2005) Censored: 37. Retrieved on May 13 2010 from Academic Search Premier, EBSCOhost.
*Today's Zaman*. (2008) Cartoons tell political journey of Turkey. Retrieved on May 13 2010 from http://www.todayszaman.com/tz-web/detaylar.do?load=detay&link=155666

Tunc, A. (2002) Pushing the limits of tolerance: functions of political cartoonists in the democratization process: the case of Turkey. *International Communication Gazette* 64 (1): 47–62. Retrieved on August 31 2010 from http://gaz.sagepub.com/content/64/1/47

Ward, G. (1995). *Barth, Derrida, and the Language of Theology*. Cambridge and New York: Cambridge University Press.

# 6
# Plurilingualism in Francophone Comics

*Miriam Ben-Rafael and Eliezer Ben-Rafael*

## 6.1 Introduction: the challenge

While interest in comics has now become more prevalent in the academy, they are particularly appreciated in French-speaking locales, like France, Belgium or Quebec, where they are named *bandes dessinées* or BDs. Designated in these countries as the Ninth Art, the BD has a status far surpassing that of equivalent English-language comic strips (Forsdick 2005). In some cases, they are the object of theoretical discussions, such as when the stories of Bécassine, the Breton female peasant working in the city, are deconstructed in the light of postcolonial approaches. BD urban landscapes are also often viewed as influenced by Le Corbusier's architecture, while many forms of language can be understood as linguistic contributions dating from the student revolutions of the 1960s. References are discernible in BD to ongoing debates about feminism and other social issues. Drawing on psychoanalytic and Marxist interpretations, Ann Miller elaborates on different possible levels of BD reading (Miller 2007). Moreover, the importance of comics in present-day literature is reflected in national and international events such as the annual International Festival of Comics in Angoulême, first held in 1974.

Some BD figures have become popular cultural symbols. Lucky Luke, the Schtroumpfs, Astérix and, of course, Tintin have become markers of the juvenile culture. As Joel E. Vessels (2010) reminds us, the status of BD is also a topic of debate: for years the dominant opinion in the academy was that BD foments rebellion, is a medium suitable only for semiliterates, and constitutes an impediment to education. It is only in recent decades that it has become widely accepted as a medium pertaining to the mainstream culture (see also McKinney 2008).

BD is now also well established as a legitimate component of youth culture and of what some French linguists call *parler jeune* (Bulot 2007). Indeed, BDs belong to those language practices which identify social categories that include individuals from different milieux and generations, but still carry common identity markers. This *parler jeune* points to models of language interactions that convey cultural orientations as well as occasionally conflicting attitudes towards social realities.

In spite of its importance for the area of written literature, the linguistic aspect of BDs has thus far been relatively neglected. Among others, one may cite Forsdick's (2005) work about different styles: realistic, comic, and schematic. Others focus on BD's vocabulary as reflecting the language of present-day youth (Bulot 2007). However, most works on comics address narratives and aesthetics, while the question of the languages and variety of registers appearing in BDs has yet to be explored. In particular, studies of the uses and roles of the different languages that meet in BDs are lacking. In francophone countries, and especially in France, the absence of studies focusing on language issues is overshadowed by the polemics over the role of English. Some commentators go as far as denouncing the presence of English as a genuine threat to the status of French in BDs (Etiemble 1964; Lederer 1988; Voirol 1980).

Against this backdrop, this chapter ameliorates the shortfall of research on plurilingualism in French BDs by discussing the forms and models that plurilingualism illustrates. English indeed plays a key role in BDs, but we would like to define here the importance of other languages which are also present. In essence, this analysis challenges the assumption that equates BD plurilingualism with the weakening, even the loss, of French.

A previous study (M. Ben-Rafael 2008b) which focused solely on English in BDs serves as our starting point. In that study, the special status of English was related to the contemporary development of globalization (Appadurai 2002) with its unprecedented flows of resources, people, and symbols across the world. Innovations, political news, and trends reach all corners of the globe, and English has become the lingua franca of our epoch (E. Ben-Rafael *et al.* 2006). Moreover, through its contact with almost every language of the world, English influences their registers in many ways. It seems that no area – sports, business, entertainment, technology or economics – is free of English borrowings.

The impact of English, however, does not yet imply that people are necessarily disengaging from their own linguistic and cultural heritage. On this point, the literature discusses tendencies for 'hybridization,' i.e., the merging of one set of symbols with patterns stemming from different sources (Pieterse 2000; Glick Schiller 1999). Researchers have documented an abundance of hybrid forms deviating from the 'normative' language among given sociocultural audiences (Rosenhouse and Kowner 2008; Crystal 2003; Görlach 2001; 2002; Maurais and Morris 2004). This hybridization is not confined to French-speaking countries, though evidence from numerous sources confirms that the French in francophone countries is now, as in other countries, strongly susceptible to the influence of English (Hagège 1987; Höfler 1982; Humbley 2002; Pergnier 1989; Truchot 1990; Walter 2001).

It is this reality that as early as the 1960s sparked the polemics mentioned above, which continue unabated. Etiemble's famous *Parlez-vous Franglais?* (1964) was followed by many others (Lenoble-Pinson 1991; Deniau 1983; Le Cornec 1981; Boly 1979; Doppagne 1979), and the question of the 'Anglicization' of French remains a much debated topic (Pivot 2004; Laroche-Claire 2004). The French Academy and other institutions charged with the preservation of *un bon français* agree with with the Etiemble tradition. That approach is challenged by linguists who assert that they discern no harm for French from the presence of English. French, they argue, is a living language which, like any other language, undergoes innovations and developments (Cholewka 2000; Yaguello 2000). Walter (2001) calls on contenders to stop worrying obsessively about the *méchant loup*, i.e. the influence of English. In the same vein, scholars believe that the use of English in the *parler jeune* simply reveals how far the young francophone aspires to be connected to that 'world' where English functions as a lingua franca. English is a means of conveying messages and 'young ideas' (Hagège 1996) while, as scholars point out, the contemporary influence of English is also spreading among adults (Walter 1998; Hagège 1987).

On the other hand, even a superficial look at francophone BD albums reveals that English is not the only language to appear in addition to French. The question which then arises concerns the roles of those other languages. Examining this aspect is the purpose of this chapter. We want to learn whether those roles are equivalent to those of English, and the extent to which their implications for BDs' French balances out the influence of English. Thus, we intend to assess the scope of the combined incidence of English and other languages in French BD. Does BD's plurilingualism represent an opening up of French to other languages and cultures or rather a defensive strategy against the threat of invasion by a single 'enemy,' English? We aspire to offer here, within the limits of our corpus, some elements of answers to this challenging issue.

## 6.2 The Study[1]

We have addressed in this study the popular French comics *Tintin*, *Spirou*, *Astérix*, *Lucky Luke*, *Le Dernier Round*, *Le Chat du Rabbin*, and *Les Schtroumpfs*, as well as the *Spirou Hebdomadaire* which is a weekly journal that offers a variety of comic strips (among others, *Tamara*, *Les Femmes Blanches*, *Les Zappeurs*, and *Comiques Strips*), the weekly magazine *Mickey*, and the youth magazines *Picsou*, *Julie*, and *Witch* (issues from 2005 to 2006). Our study begins with an early case, *Les Aventures de Tintin au Pays des Soviets* (published in the 1930s), where plurilingualism was already present, and ends with recent volumes like *Le Chat du Rabbin*: *Jerusalem d'Afrique* (2006), *Spirou le Journal d'un Ingénu* (2008a), and *Spirou 300e Album* (2008b). The characters in these comics travel the world and meet people from highly varied origins and environments. In these narratives, the use of diverse languages is frequent and in this work, we chart a (non-exhaustive) account of the uses of these languages.

Before turning to the data, these comics' protagonists merit a few words of introduction. Tintin is a courageous young Belgian reporter who leads investigations all over the world. His first adventures were published in the form of strips in *Le Petit Vingtième*, a weekly children's supplement of the Belgian newspaper, *Le Vingtième Siècle*. The strips were later anthologized in a black and white album. The story is about Tintin's visit to the Soviet Union, accompanied by his dog, Milou. He sets out from Brussels, and travels by train through Germany until he reaches the USSR. During his dangerous and turbulent trip, he meets individuals who speak German, Russian, Chinese, and other languages. In other volumes, Tintin visits numerous countries ranging from the United States to China. In his own surroundings, he meets Gypsies and people of other origins, and each story is an opportunity for new language contacts.

Lucky Luke is an American cowboy who tracks down outlaws and rescues innocent people. He goes from one adventure to another, all over the 'Wild West'. More particularly, he pursues the four Dalton Brothers who rob banks and regularly escape from prison. Many other characters are immigrants from several different places and cultures: Mexicans, American Indians, French, and Swiss.

Astérix lives in Gaul under the Roman occupation, yet thanks to a secret magic potion, his small village is still unconquered. Together with his inseparable friend Obelix and his dog Idéfix, Astérix is dispatched on special missions by the leader of the village and its Druid.

They visit numerous places, from Rome, Britain, Belgium and Spain to Switzerland, Palestine, Egypt and Greece. In each location, they encounter different languages.

Le chat du rabbin (the rabbi's cat) lives in a Jewish community in Algeria. The main human characters are a rabbi and his daughter, who is the cat's owner. Numerous events take place such as the visit of a Parisian rabbi, the wedding of the young lady, and the arrival of a Jewish Russian painter in search of Africa's Jerusalem. The old rabbi and his daughter travel to France, return to Algeria, go to Ethiopia. The French language thus comes into contact with Russian, Arabic, Hebrew, and Amharic.

Spirou is the young hero of another series. His career begins on the eve of World War II, as an elevator operator in a luxury hotel. These beginnings are narrated in a 2008 volume (*Spirou* 2008a). In some ways, the story follows in the footprints of Tintin. We analyze this volume which revolves around Spirou's encounters with German officers, Soviet spies, Spanish refugees, persecuted Jews, Italians and, of course, Belgians.

We also include in our investigation several BD sequences featuring various characters that appeared in the weekly *Spirou Hebdomadaire* (*Spirou Hebdo* in the following). We analysed 22 issues of the magazine published from 2005 to 2006. In addition, we included in our research a special collection, *Spirou 300e Album* (2008b; referred to in the following as *Spirou* 2008b), published on the occasion of the *Journal Spirou*'s seventieth birthday and which comprises the 3633–3640 issues. Among other features, these BDs include *Bone Steak, Parker & Badger, Kid Paddle, Zapping Generation, Marzi, Zappa et Tika, Pic et Zou, Yorhopia, Cucaracha,* and *Tamara*.[2]

Of these BDs, the *Schtroumpfs* (translated in English versions as 'Smurfs') merit special attention. They are little blue creatures who live in the middle of a forest and endlessly fight against a terrifying sorcerer. They speak a pseudo-language of their own in which nearly every sentence contains the term *schtroumpf*.[3]

Finally, we considered a relatively recent BD album, *Le Dernier Round* (The Last Round), written in a new style in which the emphasis is more on action than on heroes. The book's interest for this study lies in the fact that the narrative, a complex spy story, unfolds in Washington, DC, and Mexico, thus bringing French into contact with both English and Spanish.

In brief, the wide range of languages found in these BDs allows us to elaborate on the specific roles they play in French comics. Because of its overall importance in the field investigated, we start by discussing the case of English – widening the scope of previous analyses – before continuing with other languages.

## 6.3 English: framing stories and describing landscapes

A previous study of the roles of English in French comics (M. Ben-Rafael 2008b) already underlined the idea that English assumes major importance in a popular Francophone BD like *Lucky Luke* and in volumes like *Tintin en Amérique* and *Astérix chez les Bretons* which take place in English-speaking settings: the Far West, Chicago and Britain, respectively. The use of English appears there as essentially attached to the framing of the stories and descriptions of physical-geographical and human landscapes. In *Lucky Luke* (*LL*) and *Tintin* (*TT*), for example, regions, towns and road signs are given real or fictive English names (examples 1 and 2). The same applies to shops, banks, agencies, and companies (ex. 3 and 4):

(1) Carson City, Middle of Nowhere, Crazy Town (LL)
(2) Silvermount, Redskin City (TT)
(3) Barber, Sheriff's office, General Store (LL)
(4) Slift & Co, Refreshment, Petroleum and Cactus Bank (TT)

Most characters have English names; the name Lucky Luke combines the English adjective *lucky* with the English version of the French name *Luc*. In LL stories, one meets *Coffin*, *Bones*, *Big Nose Kate*, and *Crazy Dan*. In *Tintin*, one encounters *Mike*, *McAdam*, *Bill*, and *Bobby Smiles*. These characters speak French and also use English terms (ex. 5), sequences (ex. 6), interjections and onomatopoeia (ex.7):

(5) boss, building, score, kidnapping (LL); policeman, old boy, appointments (TT)
(6) home sweet home (LL); how do you do, Mister Tintin ? (TT)
(7) boy! gosh! help! ouch! youpee! (LL); hello, old chap, ow! well! old fellow! (TT)

Characters often sing in English: Lucky Luke concludes each adventure by singing '*I am a lonely cowboy*'; saloon girls sing '*The marshal packed his forty-five to get his man dead or alive.*' Moreover, passers-by read English newspapers like *The Morning News* or *The Moral Virtue*.

In *Astérix chez les Bretons*, British English is present in the form of lexical or syntactic calques: *un morceau de chance* (like: a bit of luck); *c'était*

*grand de vous avoir ici! C'était!* ('it was great to have you here! It was!').
The word order follows English models, and adjectives are set before nouns (ex. 8 and 9):

(8) une romaine patrouille (a Roman patrol) instead of une patrouille romaine
(9) voici la chaude eau (here is hot water) instead of voici l'eau chaude

Such English elements – borrowings, code switching, syntactic and lexical calques – frame the setting in which narratives take place.

### 6.4 Linguistic changes in French

Other kinds of Anglicization in *Journal Spirou* consist of the use of English markers like *cool, fan, chat, hot, light* (ex. 10) or of innovations based on English terms like *customiser, chatter, chiller, booster, scotché,* or *coolitude, testeur,* and *testeuse* (ex. 11):

(10) Les vacances se passent *cool?* Ils sont pas *cool* les vieux ils s'inquiètent (*Zack et Willie* in *Spirou Hebdo* 3536)
(11) Elle *surfe* sur la vague… je *freeride* (*Spirou* in *Spirou Hebdo* 3521)

Phatics and English interjections abound: *yeah man! okay! ok! woah! oups! crash! splash! what? yes! no it's not too difficult!* un peu *too much!* In a similar vein, *Picsou, Julie* or *Witch* (M. Ben Rafael 2007a, 2007b) illustrate the adoption of numerous English elements in areas like fashion, music, video games, internet and sport (ex. 12):

(12) Fashion: la *French touch,* coiffeur trop *fashion,* un *sweat,* des fringues *top* tendance
Music and dance: *hip hop, rock, music-bag*
Video games: une *gameboy advance*
Sport: le faucon du *skate, challenges* en hausse, *tricks* inédits

As these examples show, English borrowings are preferred even when French substitutes exist. This applies, among other patterns, to the names of consoles like *PlayStation* or *Gameboy* which are now common in French. Similarly, one says *sweat* rather than 'chandail,' *boots* rather than 'bottes,' and *e-mail* rather than 'courriel.' English terms receive semantic nuances that do not exist in their French equivalents: *new* is not always what is meant by 'nouveau,' *Miss* by 'mademoiselle,' *princess* by 'princesse.' English idioms, songs and quotes are frequent (ex. 13); some strips have English titles (ex. 14), and characters receive English names (ex. 15):

(13) All you need is love… and cash
(14) Potatoes, Manager mode d'emploi, Black le jaune, les zappeurs, Game over, Wondertown, Billy the cat, Screen Shot, Zapping Generation
(15) Parker and Badge, Gone, Roboboy, Man, Has been, Kid Paddle

The characters in these BDs easily mix French and English codes (ex. 16–18):

(16) *No comment*! Vise moi ce *string*! On n'a pas besoin d'*air bag* (*Tamara* in *Spirou Hebdo* 3536)
(17) Il a trouvé que c'était un peu *too much*. Ce style a été réalisé par Monsieur Bertschy *himself*! (*Spirou* in *Spirou Hebdo* 3536)
(18) Tout ce que vous avez toujours voulu savoir sur la vie des *depressed housewives* (*Screenshot* in *Spirou Hebdo* 3525)

English often adds a ludic dimension to the French discourse: 'Vive l'année deux mille *sex!*' for 'deux mille six!' and *beast records* on the model of *best records* (Cover slogan of *Spirou Hebdo* 3533).

In these instances, the presence of English differs from the framing function seen in the previous section. It is indicative of the language widely spoken by young people and of the role English plays in their developing a code of their own (M. Ben-Rafael 2007a). It expresses a positive disposition toward this language, most probably due to the influence of globalization and the worldwide diffusion of English-speaking cultural consumption (M. Ben-Rafael 2008a; Rosenhouse and Kowner 2008). This youth register (Cholewka 2000; Yaguello 2000), what we call *parler jeune*, comes up in BDs in a variety of contexts and forms such as intentional grammatical errors (ex. 19), ludic lexical combinations (ex. 20), idioms and slang (ex. 21, 22, and 23) or abbreviations and elisions (ex. 24):

(19) *Le saviez-tu?* for 'le savais-tu?' (did you know that?)
(20) *l' écri vain* for 'écrivain' (meaning here: bad writer)
(21) *Eclate-toi en voiture!* for 'amuse-toi bien en voiture!' (have fun with the car)

(22) Hé *man* kess tu fais?...c'est c'que j'lui *explain* depuis t'à l'heure
(23) C'est *super* bon; pas très *glamour* ton métier; trop *cool!*
(24) *Je suis accro à l'ordi* for 'je suis accroché à l'ordinateur' (I am glued to the computer); *m'enfin* for 'mais enfin' (but still); *le v'la* for 'le voilà' (here he is)

Our corpus demonstrates that more than a few English elements insert themselves in this kind of young people's French speech.

## 6.5 The roles of plurilingualism in framing narratives

Despite the importance of English, it is by no means the only foreign language in francophone comics. One encounters Hebrew, Yiddish, Spanish, Gallic, and Latin, which frame the narratives. This framing by foreign languages operates in a variety of dimensions.

### 6.5.1 Framing the physical and geographic landscapes

The frames of the stories are always indicated by linguistic features that specify time and place. When Tintin arrives in Shanghai (*Le Lotus Bleu*), Chinese appears on street signs, window panes and hotel boards. Announcements to the public are made in Chinese. A similar rule applies to Marzi's stories in Poland where Polish terms are common: *pewex* (shop) or *kolory telewizor* (color television) (*Marzi* in *Spirou Hebdo* 3251). In Russia, billboards and shop signs are in Russian (*Tintin au Pays des Soviets)*; *Stolbsty* stands for lakes of petrol; Russian signs forbid swimming in pools. In turn, posters and public signs in Berlin are in German (*Ausgang, Halt, Achtung*), and so are the street names *(Schwein Strasse)* and the hotel boards (*Zum Gasthaus*). On the train to Brussels, signs become bilingual combining German and French (*Nach Brussel / vers Bruxelles*). The use of languages sets the boundaries of geographical, cultural, and national spaces, and crossing those spaces entails language switching.

Newspapers also carry characteristic markers in different contexts: French and Chinese in Shanghai (*Le Lotus Bleu*), and Polish in Poland where characters read a journal called *Detektyw* (detective) *(Marzi* in *Spirou Hebdo* 3530). The contents of these newspapers are always in French, but the titles of the journals appear in local languages. The titles of the BDs themselves often include such markers: *Tintin et les Picaros* includes the Spanish term *picaros* (adventurers); *Le Chat du Rabbin: La Bar-Mitzvah* introduces the Hebrew term *bar-mitzvah* (rite of passage for boys) as the subtitle of the volume; *Le Lotus Bleu* is doubled by a Chinese translation. Other BD titles use English wording like *Le Dernier Round* and *Le Comeback du Cowboy*.

### 6.5.2 Naming people and identities

In many cases, the characters' names convey linguistic markers. Following the English example of *Lucky Luke*, we find legions of others in *Tintin en Amérique*, *Astérix chez les Bretons* and other BDs. We encounter Gypsy names like *Miarka*; Italian names like *Castafiore* or *Gino*; Russian names like *Nitchevo* or *Dimitrieff*; Hebrew names like *Malka* or *Sasson Nahoum*; Arabic names like *Mohammed* and *Zlabya*; Sephardic Jewish names like *Rebibo* or *Fitoussi*; Ashkenazic Jewish names like *Rosenblumenthalovitch*; American names like *Floyd* or *Lenny*; Spanish ones like *Francisco* or *Roberto*. Romans have Latin or pseudo-Latin names such as *Malosinus, Diplodocus,* or *Decubitus*. The Gauls have Gallic or pseudo-Gallic denominations like *Obelix, Astérix, Panoramix*. Similarly, in India we find *Rahazade* and *Kiça*; Spirou welcomes German visitors called *Reinhard* or *Karl* to the *Moustic* hotel; and on Arab soil (*Au Pays de L'Or Noir*), Tintin meets *Abdul* and *Abdallah Mohamed*.[4]

These characters speak in French but use linguistic elements indicative of their cultural or national identity. For instance, in *Tintin et les Picaros* which is set in South America, numerous Spanish words and expressions appear which include, among others, *buenas noches, buenas tardes, guerilleros, amigo* and *hombre*. In the same story, the indigenous people also speak their own language (ex. 25):

(25) *Nagoum wazenh! yommo! nagoum ennrgang!* (*Tintin et les Picaros*).

In some cases, translation is supplied (ex. 26):

(26) First South American Indian: *wa paisde douvan?*
Second South American Indian: *il demande si vous aimez?* ('he asks if you like')
Tintin: *je trouve ça délicieux* ('I find it delicious') (*Tintin et les Picaros*)

## 6.6 Specific uses of foreign languages: interjections and singing

All in all, foreign languages emerge in BDs in numerous ways. Below, we consider interjections and singing before turning to borrowings and code switching, which are the major patterns.

### 6.6.1 Interjections

In *Astérix*, many interjections are drawn directly from a broad diversity of languages, from Yiddish and Hebrew (*hoyoyoye* and *mazel-tov*) to pseudo-Gallic (*holala houla* and *ouie*), Spanish (*olé*), or Latin (*avé Caesar*). In *Tintin*, interjections may be in German (*mein Gott*; *prosit*) or Russian (*Hoptb!*). In volumes of *Le Chat du Rabbin* and *Spirou*, plurilingual expressions also abound (ex. 27):

(27) Arabic: yala al moussi ba; salam aleikoum (Le Chat du Rabbin)
Hebrew: shalom aleichem; has ve shalom; baruch hachem (Le Chat du Rabbin)
German: wunderbar! schnell! (Spirou 2008a)

These interjections are sometimes written either in their original writing systems like Cyrillic, Chinese, Egyptian hieroglyphs, Arabic, or Hebrew or in a pseudo writing system. Other interjections are transliterated in Latin characters.

### 6.6.2 Singing

Music and song play an undeniable role in BDs' plurilingualism. In the volumes of *Le Chat du Rabbin*, comics, prayers, and songs are mostly in Hebrew: the rabbi walks the streets humming *mazal tov vé siman tov ... yéhé laanouyéhé lanou* (good luck and good sign ... should be for us) (*Le Malka des Lions*). At the synagogue on Friday evening, the people break into *lecha dodi likrat kala péné chabaat* (to you my beloved towards Shabbat we welcome you) (*l'Exode*).

When Astérix and Obélix happen to be in Switzerland, they hear Swiss onomatopeia: *boooooooooooo, yodléiiiiii, oléeleéiiii, yodléééiiiiiiiiii* (*Astérix chez les Helvètes*). When the Druid Panoramix gets drunk, he sings in a French tainted with pseudo-Gallic intonations: *boire un petit coup ch' est une aubaiiiine, boire un petit coup ch' est doux ... mais il ne faut pas rouler dechous le dolmen* (drinking a little is a godsend, drinking a little is pleasant ... but we should not roll under the dolmen). In Spain, Gypsies use musical markers as well: *ayayayayyyy ... clapaclapa* (*Astérix en Hispanie*). The giants in Spirou (*Les Géants Pétrifiés* in *Spirou Hebdo* 3546) break into English songs: *death is death, ... it's life it's life*. Dissenters in Poland sing in Polish (*Marzi* in *Spirou Hebdo* 3533).

### 6.7 Borrowings and code switching

As mentioned, plurilingual borrowings and code switching are the most common patterns of occurrence of foreign languages in French BDs.

In *Tintin et les Picaros*, one finds, besides the Spanish elements noted previously (ex. 28):

(28) Italian borrowings: *madonna*; *signora*
Gypsy borrowings: *gadgo* (non-Gypsy); *gadgé* (non-Gypsies)

In *Tintin au Pays des Soviets*, British tourists react in their language to the 'wonders' they discover in the Soviet Union: *beautiful, very nice*; Tintin himself uses English terms that have become common in French: *penalty! goal!* quel *shot!* The Chinese men in charge of a torture chamber talk to each other in Chinese, but no translation is provided. Borrowings may also be concepts that have been widely publicized outside their specific contexts, like *guepeou* (for GPU i.e. the State Political Directorate or secret police) or *koulak* (peasant or kulak) (ex. 29 and 30):

(29) vous faites partie du *guepeou*, n'est-ce-pas? (you belong to the GPU, isn't it?)
(30) qu'on mette ce *koulak* à la torture (let us torture this kulak)

In the volumes of *Le Chat du Rabbin*, Hebrew markers are numerous and often remain untranslated. In *Le Chat du Rabbin: La Bar-Mitzvah*, one finds *tora* (Bible), *talmud mishna* (text of Talmudic Law), *gmara* (commentary on Talmudic Law). In *Le Chat du Rabbin: Le Malka des Lions*, one finds *Hakadosh Barouh Hou* (the Blessed Holy One) and *shofar* (horn).

In some cases, Hebrew terms are clarified (ex. 31):

(31) Le *Lachone Hara*, la mauvaise langue, c'est aussi grave qu'un meurtre (the *lashon hara*, slandering, is as grave as murder) (*Le Chat du Rabbin: La Bar-Mitsva*)

Additional forms of language contact appear in *Le Chat du Rabbin: Jerusalem d'Afrique*, which describes, as mentioned in the above, the arrival in Algeria of a Russian Jewish painter. He knows only Russian and finds himself in a small Jewish community where no one speaks his language. First he encounters Hebrew and French, and later Amharic. Even the rabbi who uses the Hebrew of prayers fails to communicate with him. It is only when he meets non-Jewish Russians that he is able to converse at all. Their exchanges are transcribed in Cyrillic characters and rendered in French translation. On the other hand, the French of those Russian characters is riddled with grammatical and lexical errors typical of nonnative speakers (ex. 32 and 33):

(32) *toi tu faire rien. Toi tu faire rien du tout* (you nothing to do. You to do nothing at all)
(33) *ils ne nous connaître pas encore* (they know we not yet)

The painter's dream is to discover the 'Jerusalem of Africa' which, according to legend, is in Ethiopia. During the expedition, he meets Ethiopian Jews who speak Amharic and again experiences communication difficulties. Only the rabbi's plurilingual cat understands Amharic and is able to translate into French what is said and presented in pseudo-Amharic graphics. Hence, ironically, the cat becomes the mediator between the Russian, the Algerians, and the Ethiopian Jews (*Le Chat du Rabbin: Jerusalem d'Afrique*).

In several volumes of *Le Chat du Rabbin*, one encounters Arabic terms like *djellaba, casbah, arrouah* (see *l'Exode*). In the presence of Arab individuals, figures greet each other with typical Arabic expressions like *Salam aleikoum* and *aleikoum salam* (*Le Malka des Lions*). The discussions between the rabbi and Sheikh Mohammed Sfar are in French, Arabic, and Hebrew. In the same vein, the rabbi's cat and the sheikh's donkey debate in French about the origin of the word *sfar* in Arabic which is close to the Hebrew *sofer* (writer).

Arabic terms may be written in Arabic script (*Tintin au Pays de L'Or Noir*; *Porté Disparu* in *Spirou Hebdo* 3637), but in most cases they are transcribed in Latin characters (*Tamara* in *Spirou Hebdo* 3544; *Tintin au Pays de L'Or Noir*) (ex. 34):

(34) une tempête de sable: le *khamsin!* (a sandstorm: the *khamsin!*)

In *Spirou* (2008a), we already saw German borrowings and code switchings like *Jawohl, mein Herr; wunderbar!* We may add to this list *schnell! Das Fenster offnet sich nicht mehr.* Asian or pseudo-Asian languages function similarly (*Qu'as-tu Kim?* in *Spirou* 2008b: 3240). On the other hand, Chen, Cedric's girlfriend, frequently speaks in Chinese; her words are expressed in the Chinese writing system without translation (*Cédric* in *Spirou Hebdo* 3527) while Marzi inserts Polish into her French (*tato* for papa, *mamus* for maman, *zomo* or *zomos* for gréviste/s) (*Marzi* in *Spirou Hebdo* 3536).

Similarly, numerous Spanish or Latin American characters express themselves in Spanish in *Spirou Hebdo: vamos, cantina, tequila, silencio, bandidos, el banco, gracias amigo, qué pasa?* (*Lucky Luke* in *Spirou Hebdo* 3549), or *Hasta la vista* and *un, dos* (*Cucaracha* in *Spirou Hebdo* 3530). Russian appears in *Pic et Zou* (in *Spirou Hebdo* 5335), and Chinese in *Zack et Willie* (in *Spirou Hebdo* 35423). In *Astérix et Cléopatre*, hieroglyphs or pseudo-hieroglyphs are frequent and supplemented by French translation.

## 6.8 Interlanguages and artificial languages

In tandem with the wide use of foreign languages, there are two other phenomena of particular interest in BDs: interlanguages and artificial languages.

### 6.8.1 Interlanguages

We find several examples of interlanguages[5] where the French language is presented as a character's foreign language. This is the case, for instance, of the North African immigrants (*Le Chat du Rabbin: L'Exode*), Russian immigrants in Algeria (*Le Chat du Rabbin: Jerusalem d'Afrique*), and Gypsies in Tintin's country. Hence, for instance, in *Tintin et les Picaros*, an old Gypsy woman says to him: '*toi mordu, écoute monsieur moi te dire bonne aventure*' (you bitten, listen Mister, me to tell you future). In *Tintin au Pays des Soviets*, policemen tell Tintin: '*vous … aller tout de suite comissaire…*' (you … immediately to go police station). In *Spirou* (2008a), a Spanish immigrant says to Spirou, with distorted phonology: '*Espirou! moi, yé soui espagnol*' (Spirou! Me, I am Spanish) and '*yé soui vonou ici avec ma mama*' (I have come here with my mommy); '*ma si, yé soui catholique!*' (but yes, I am Catholic) (*Spirou 2008a* ). The same feature marks the French of Poles: '*ovoi* c'est mon premier mot de français' (*au revoir* this is my first word of French) (*Marzi* in *Spirou Hebdo* 3536).

One also encounters in some BDs negative stereotypes. A well-known case is found in *Tintin au Congo* (1946) where Tintin urges the inhabitants of a village to get to work and is answered: '*Moi y en a fatigué!*' (Me have tired) and '*Mais … mais moi va salir moi!*' (But … but me to make dirty me!). Such forms of speech would be unacceptable today, and this is articulated by the Algerian rabbi when, on his way to Ethiopia, he 'meets' Tintin in Congo and upbraids him for his racist prejudices (*Le Chat du Rabbin: Jerusalem d'Afrique*).

### 6.8.2 Artificial languages and games

We also find in French BDs artificial languages and language games. The authors like to play with words, create new terms, and use humorous connotations. In *Lucky Luke*, for example, names always convey humorous associations: *Mr. Coffin* and *Mr. Bones* are undertakers, and *Mr. Gamble* is a gambler. Nicknames speak for themselves: *Big Nose Kate* or *Crazy Dan*. In *Astérix*, a Roman camp is given a pseudo-Latin name: *Babaorum*, a pun drawing on the French cake baba au rhum; in *Astérix chez Rahazade*, the Asian flu becomes the neologism *gravero asiatica*.

Another expression of linguistic creativity consists of fictitious languages. One example appears in *Zappa et Tika* where a professor teaches his students the language of an imaginary country (*Zappa et Tika* in *Spirou Hebdo* 3547). One recalls the *Poulpe* (octopus) language in *Wondertown* (in *Spirou Hebdo* 3543) and also *Yorthopia*, a pseudo-Northern language that remains unexplained, with expressions like *han her dod* or *tilbake ga tilbake* (*Yorthopia* in *Spirou Hebdo* 3541).

Above all, there is the case of the *Schtroumpfs* series which devises a whole new language, the *Schtroumpf* language. The narratives take place in the *Schtroumpf* country, an imaginary place where tiny blue creatures live and speak *Schtroumpf*. The word *schtroumpf* generates numerous nouns, verbs, adjectives and adverbs which follow the grammatical rules of French. Table 6.1 presents a list of examples and approximate translations.

As shown in Table 6.1, the word *schtroumpf* and its variants receive different meanings according to contexts. *Schtroumpfer* may signify *do*,

*Table 6.1* Examples of *schtroumpf* expressions and translation

| | |
|---|---|
| un *schtroumpf* | one smurf |
| des *schtroumpfs* | smurfs |
| une *schtroumpfette* | a female smurf |
| *schtroumpfer* | to be at something |
| j'ai *schtroumpfé* | I succeeded |
| je tombe de *schtroumpf* | I am very tired |
| se *schtroumpfer* | to get it wrong |
| *schtroumpfement* sale | very dirty |
| pourvu que ça *schtroumpfe* | let's hope it succeeds |
| une *schtroumpferie* | a trick |
| ça sent la *schroumpf* | it smells good |
| tiens *schtroumpfe* ça | taste this |
| se faire *schtroumpfer* | to be cheated |
| une *schtroumpf* souris | a nice girl (lit. 'a schtroumpf mouse') |
| un baba au *schtroumpf* | a cake with cream |
| une bête *schtroumpfée* | a crazy animal |

*mix up, remain in place, lose time* or *stand fast*. The pictures and situations described clarify the meanings of the words. The French elements used conjunctively make explicit what the *schtroumpfian* innovations stand for. While the syntax of the *Schtroumpf* language remains French, its semantics constantly adjusts to the narratives (ex. 35, 36):

(35) Mais qu'est-ce qu'ils *schtroumpfent* tout est prêt! (But what do they do everything is ready!) (*L'apprenti Schtroumpf*)
(36) Quel est le *schtroumpf* de *schtroumpf* qui a *schtroumpfé* un trou ici?...qu'on *schtroumpfe* en prison celui qui *a schtroumpfé ce trou!* (Who is the awfully crazy guy who dug this hole here?... let us jail this crazy one who dug this hole!) (*Le Schtroumpfissme et Schtroumpfonie en Ut*)

## 6.9 Conclusion

We have explored the plurilingual dimension of French BDs. English, we have seen, is the main feature of this plurilingualism and our analyses have elaborated on the openness of these BDs to its influence, an openness that primarily indicates an acceptance of English's major worldwide status in communication and its importance among French-speaking youth. English contributes to the framing of narratives and the depicting of landscapes. It adds terms and forms that penetrate into French, and, as such, constitutes an undeniable factor of linguistic development. However, this status of English does not prevent other foreign languages from appearing as well. Their roles partly converge with those of English, though they do not achieve a similar impact as agents of linguistic change. Still, the Spanish, Gallic, Latin, Hebrew or Yiddish interjections, idioms, tokens, songs, and liturgy reflect the open-mindedness of BDs toward the diversity of the world.

In brief, BDs represent spaces of rich language contact which articulate a transnational calling. This calling was already there decades ago, as illustrated in *Tintin au Pays des Soviets* but over the years it gradually became explicitly formulated as an ideology that runs through the BD literature. This tendency is discernible in many instances. In *Le Chat du Rabbin: Jérusalem d'Afrique*, the rabbi's plurilingual cat enables communication between human characters. In *Les Femmes en Blanc* (in *Spirou Hebdo* 3531), a patient is in distress because he speaks a language unknown to the nurses where he is hospitalized, leading the characters to praise the importance of knowing languages. Cédric teaches Chen, a Chinese schoolgirl, how to pronounce the French /r/ (*Cédric* in *Spirou Hebdo* 3527).

The other children react with sarcasm to her difficulties, but at the end of the story, it is Chen who teaches them the rudiments of Chinese. This positive attitude toward plurilingualism reaches a peak when it comes to interlanguages, word games, and artificial languages. These phenomena, which play major roles in BDs, underline the crucial communicative value of languages. The respect that purists request for the correctness of French forms and styles withdraws in face of direct and efficient face-to-face exchanges cutting across cultures and codes. In this perspective, languages may even be straightforwardly invented in order to safeguard the creativity of BDs. Hence, BDs are no less than agents of globalization. They hyphenate areas and countries, and they set in relation the most diverse cultures and languages. In this diversity, they find a horizon appropriate to the medium they represent. This link to global realities, we suggest, is also an important reason for BDs' contemporary popularity. As an ingredient of twenty-first century youth culture, BDs both express and articulate the curiosity for 'what's new in the world.' For example, the *Spirou Hebdomadaire* of December 25, 2005, wished its readers a 'Happy New Year' in no less than eleven different languages. Similarly, the magazine *Mickey* recently launched a bilingual strip in English and French called *Donald Speaks English*.

What our investigation elicits, in striking opposition to the approach of many defenders of 'pure' French, is a perspective on languages in BDs that can be summarized as follows:

(a) flexible attitudes towards the practice of French
(b) recognition of the global importance of English and openness to its influence
(c) recognition of the importance of other languages in general
(d) readiness to coexist with other languages in a plurilingual perspective
(e) eagerness to accept and initiate linguistic innovations, borrowings, and games.

Considered in this light, the *bandes dessinées* do not belong to canonical French literature and effectively constitute an art of their own.

## Notes

1. In the following, we use italics to indicate: (1) elements in examples which are in English or other languages besides French; (2) French calques from English; (3) sources of quotations; (4) French utterances typical of *parler jeune*; (5) all

names of comics books and strips. We use single quote marks to designate speech elements quoted in text but not in numbered examples.
2. In the text, we refer to *Spirou Journal d'un Ingénu* by using the full name or only *Spirou* 2008a. We refer to the strips of *Spirou 300e Album* as Spirou 2008b. We refer to the diverse strips of individual issues of *Spirou Hebdomadaire* by the name of the strips in addition to mentioning *Spirou Hebdo* and the numbers of the specific issue.
3. The Schtroumpfs appear in *Spirou Hebdo* and the *Spirou Album* (2008b) as well as in volumes of their own.
4. Among such foreign names, some authors play with French or Belgian-French slang expressions. The Russian Potferdeksky (in *Tintin au Pays des Soviets*) derives from the Belgian-French derogatory slang exclamation 'potferdek,' and the quasi Arabic Bab El Ehr (in *Tintin au Pays de l'Or Noir*) is cut from a word in the same slang meaning 'blabbermouth.'
5. The notion of interlanguage was coined to designate the language of learners who have not yet thoroughly grasped the language which they learn (L2) and still imprint this intermediary linguistic stage with elements and constructions pertaining to their original language (L1) (see Selinker 1992).

## Sources

*Lucky Luke*

*L'Amnésie des Dalton* (Morris, X. Fauche et J. Léturgie. Genève: Lucky Productions 1991).
*Chasse aux Fantômes* (Morris et L.H. van Banda. Genève: Lucky Productions 1992).
*Les Dalton à la Noce* (Morris, X. Fauche et J. Léturgie. Genève: Lucky Productions 1993).
*Le Pont sur le Mississippi* (Morris, X. Fauche et J. Léturgie. Genève: Lucky Productions 1994).
*Kid Lucky* (Morris, Pearce et J. Léturgie. Genève: Lucky Productions 1995).
*Les Cerveaux* (Morris. Genève: Lucky Productions 1996).
*O.K. Corral* (Morris, X. Fauche et E.Adam. Genève: Lucky Productions 1997).
*Oklahoma Jim* (Morris, Pearce et J. Léturgie. Genève: Lucky Productions 1997).
*Marcel Dalton* (Morris et E. de Groot. Genève: Lucky Productions 1998).
*Sarah Bernhardt* (Morris, X. Fauche et J. Léturgie. Rennes, France: Dargaud ed. 1999).
*La Belle et le Bête* (Morris, V. Leonardo et B. de Groot. Belgique: Lucky Comics 2000).
*Le Fil qui Chante* (Morris et E. Goscinny. Paris: Dargaud ed. 2000).
*Le Prophète* (Morris et P. Nordmann. Belgique: Lucky Comics 2000).
*L'Artiste Peintre* (Morris et B. de Groot. Belgique: Lucky Comics 2001).
*La Légende de l'Ouest* (Morris et P. Nordmann. Belgique: Lucky Comics 2002).

*Astérix*

*Astérix et Cléopatre* (A. Uderzo et R. Goscinny. St Maur, France: Dargaud ed. 1965).
*Astérix chez les Bretons* (A. Uderzo et R. Goscinny. St Maur, France: Dargaud ed. 1966).
*Astérix en Hispanie* (A.Uderzo et R. Goscinny. Neuilly/Seine, France: Dargaud ed. 1969).

*Astérix chez les Helvètes* (A. Uderzo et R. Goscinny. Neuilly/Seine, France: Dargaud ed. 1970).
*Astérix chez les Belges* (A.Uderzo et R.Goscinny. Neuilly/Seine, France: Dargaud ed. 1979).
*L'Odyssé d'Astérix* (A. Uderzo et R. Goscinny. Neuilly/Seine, France: Dargaud ed. 1981).
*Astérix chez Rahazade* (A. Uderzo et R. Goscinny. Paris: Albert René 1987).

Les Aventures de Tintin
*Les Aventures de Tintin au Pays des Soviets* (Hergé. Tournai: Casterman 1930/1981).
*Tintin en Amérique* (Hergé. Tournai: Casterman 1945).
*Le Lotus Bleu* (Hergé. Tournai: Casterman 1946).
*Tintin au Congo* (Hergé. Tournai: Casterman 1946).
*Tintin au Pays de l'Or Noir* (Hergé. Tournai: Casterman 1950).
*Les Bijoux de la Castafiore* (Hergé. Tournai: Casterman 1963).
*Tintin et les Picaros* (Hergé. Tournai: Casterman 1976).

Spirou
*Spirou le Journal d'un Ingénu* (E. Bravo. Belgique: Dupuis 2008a).
*Spirou 300e Album* (Belgique: Dupuis 2008b).
*Spirou Hebdomadaire* (Belgique: Dupuis 2005-2006).

Le Chat du Rabbin
*Le Chat du Rabbin: La Bar-Mitsva* (J. Sfar. Paris: Dargaud ed. 2002).
*Le Chat du Rabbin: Le Malka des Lions* (J. Sfar. Paris: Dargaud ed. 2002).
*Le Chat du Rabbin: L'Exode* (J. Sfar. Paris: Dargaud ed. 2003).
*Le Chat du Rabbin: Le Paradis Terrestre* (J. Sfar. Paris: Dargaud ed. 2005).
*Le Chat du Rabbin: Jerusalem d'Afrique* (J. Sfar. Paris: Dargaud ed. 2006).

Les Schtroumpfs
*L'Apprenti Schtroumpf: Trois Histoires de Schtroumpfs* (Peyo. Bruxelles: Dupuis 1977).
*Le Schtroumpfissme et Schtroumpfonie en Ut* (Peyo. Bruxelles: Dupuis 1978).

Others
*Le Dernier Round* (W. Vance et J. Van Hamme. Dargaud Benelux ed. 2007).
*Mickey* (weekly) 2005–2006.
*Picsou* (monthly) 2005–2006.
*Witch* (monthly) 2005–2006.
*Julie* (monthly) 2005–2006.

# References

Appadurai, A. (ed.) (2002) *Globalization*. Durham, NC: Duke University Press.
Ben-Rafael, E., Shoami, E., Amara, M., and Hecht, N. (2006) The symbolic construction of the public space: the case of Israel. *International Journal of Multilingualism* 3 (1): 7–28.
Ben-Rafael, M. (2007a) L'anglais dans la presse jeune francophone: danger ou ouverture. Paper delivered at the 5ième Réseau Français de Sociolinguistique international conference, Université de Picardie Jules Verne.

Ben-Rafael, M. (2007b) Linguistic purism versus language reality: the case of the French youth press. In Israel National Commission of UNESCO (ed.) *The Effect of Globalization on Center, Periphery and Multiculturalism*, pp. 30–39. Beit Berl: Beit Berl Academic College.

Ben-Rafael, M. (2008a) French: Tradition versus innovation as reflected in English borrowings. In J. Rosenhouse and R. Kowner (eds.) *Globally Speaking: Motives for Adopting English Vocabulary in Other Languages*, pp. 44–67. Clevedon: Multilingual Matters.

Ben-Rafael, M. (2008b) English in French comics. *World Englishes* 27 (3–4): 535–48.

Blanchet, P. (2007) La langue française, victime idéologique. *Lemensuel.net/la langue – française-victime.html*. Retrieved on 16 June 2008 from http://www.prefics.org/credilif/ travaux/IdeologieLingFr.pdf

Boly, J. (1979) *Chasse au Franglais-Petit Glossaire Franglais-Français*. Bruxelles: Louis Musin.

Bulot, T. (2007) Grammaire et parlers (de) jeunes – Quand la langue n'évolue plus... mais continue de changer. *Les Cahiers Pédagogiques*, no 453. Retrieved on 15 June 2008 from http://www.cahiers-pédagogiques.com/spip.php?article3076

Cholewka, N. (2000) U comme us et pratiques de la langue. Quelques aspects ici et maintenant. In B. Cerquiglini, J.-C. Corbeil, J.-M. Klinkenberg, and B. Peeters (eds.), *Le Français dans tous ses États*, pp. 305–23. Paris: Flammarion.

Crystal, D. (2003) *English as a Global Language*. Cambridge: Cambridge University Press.

Deniau, X. (1983) *La Francophonie*. Paris: Presses Universitaires de France, Collection 'Que sais-je?'

Doppagne, A. (1979) *Pour une Écologie de la Langue Française*. Bruxelles: Commission française de la culture de l'' aglomération de Bruxelles.

Etiemble, R. (1964/1991) *Parlez-vous Franglais?* Paris: Gallimard.

Forsdick, C. (2005) *The Francophone Bande Dessinée*. Amsterdam and New York: Rodopi.

Glick Schiller, N. (1999). Transnational nation-states and their citizens: the Asian experience. In P. Dicken, L. Kelley, K. Kong, H. Olds, and W. Yeung (eds.) *Glocalization and the Asia Pacific: Contested Territories*, pp. 143–55. London: Routledge.

Görlach, M. (2001) *A Dictionary of European Anglicisms: A Usage Dictionary of Anglicisms in Sixteen European Languages*. Oxford: Oxford University Press.

Görlach, M. (ed.) (2002) *English in Europe*. Oxford: Oxford University Press.

Hagège, C. (1987) *Le Français et les Siècles*. Paris: Odile Jacob.

Hagège, C. (1996) *Le Français, Histoire d'un Combat*. Boulogne: Editions Michel Hagège et La Cinquième Edition.

Höfler, M. (1982) *Dictionnaire des Anglicismes*. Paris: Larousse.

Humbley, J. (2002) French. In M. Görlach (ed.) *English in Europe*, pp. 108–27. Oxford: Oxford University Press.

Laroche-Claire, Y. (2004) *Evitez le Franglais, Parlez Français!* Paris: Albin Michel.

Le Cornec, J. (1981) *Quand le Français Perd son Latin: Nouvelle Défense et Illustration*. Paris: Les Belles Lettres.

Lederer, M. (1988) Les fausses traductions sources de contamination du français. In M. Pergnier (ed.) *Le Français en Contact avec l'Anglais*, pp. 119–29. Paris: Didier Erudition.

Lenoble-Pinson, M. (1991) *Anglicismes et Substituts Français*. Louvain-la-Neuve: Duculot.

Maurais, J. and Morris, M. A. (eds.) (2004) *Languages in a Globalising World*. Cambridge: Cambridge University Press.

McKinney, M. (ed.) (2008) *History and Politics in French-Language Comics and Graphic Novels*. Jackson: University Press of Mississippi.

Miller, A. (2007) *Reading Bande Dessinée: Critical Approaches to French-Language Comic Strip*. Chicago: Chicago University Press.

Pergnier, M. (1989) *Les Anglicismes*. Paris: Presses Universitaires de France.

Pieterse, J. N. (2000) Globalization as hybridization. In F.J. Lechner and J. Boli (eds.) *The Globalization Reader*, pp. 99–105. Oxford: Blackwell.

Pivot, B. (2004) *100 Mots à Sauver*. Paris: Albin Michel.

Selinker, L. (1992) *Rediscovering Interlanguage*. London and New York: Longman.

Rosenhouse, J. and Kowner, R. (eds) (2008) *Globally Speaking: Motives for Adopting English Vocabulary in Other Languages*. Clevedon: Multilingual Matters.

Truchot, C. (1990) *L'Anglais dans le Monde Contemporain*. Paris: Le Robert.

Vessels, J. E. (2010) *Drawing France: French Comics and the Republic*. Jackson: University Press of Mississippi.

Voirol, M. (1980) *Anglicismes et Anglomanie*. Paris: Victoires Editions, collection Métier Journaliste.

Walter, H. (1998) *Le Français dans tous les Sens*. Paris: Edition Robert Laffont.

Walter, H. (2001) *Honni Soit qui Mal y Pense. L'Incroyable Histoire d'Amour entre le Français et l'Anglais*. Paris: Edition Robert Laffont.

Yaguello, M. (2000) X comme XXL, la place des anglicismes. In B. Cerquiglini, J.-C. Corbeil, J.-M. Klinkenberg, and B. Peeters (eds.) *Le Français dans tous ses États*, pp. 353–62. Paris: Flammarion.

# 7
# To and Fro Dutch Dutch: Diachronic Language Variation in Flemish Comics

*Gert Meesters*

## 7.1 The language situation in Flanders: a brief introduction

In this chapter, I will discuss language evolution in two long-running Flemish comics: *Suske en Wiske* and *Jommeke*. I will show how comics can provide insight into the development of spoken language. My research is set in the context of the current interest in the recent evolution of Dutch in Flanders compared to the evolution of Dutch in the Netherlands.

Officially, Dutch is spoken by fifteen million people in the Netherlands and by six million people in Flanders, i.e. the northern part of Belgium. Linguists do not use the term Flemish as a name for the variant of Dutch spoken in Belgium, but tend to speak of Flemish Dutch or Belgian Dutch instead. The differences between Standard Dutch in the Netherlands and Standard Dutch in Belgium are rather limited. The clearest difference is the pronunciation. Speakers from one region may even find it difficult to understand speakers from another region. Lately, this has led to subtitling television fiction from Flanders on Dutch television and vice versa, which a lot of viewers think of as an exaggerated measure (see Vandekerckhove, De Houwer and Remael 2007). Furthermore, the national variants of Standard Dutch display some minor differences in vocabulary and very limited variation in syntax. In writing, Belgian and Dutch Standard Dutch look very much alike. Summing up, one could refer to the relationship between American and British English as a more or less comparable language situation.

Underneath the surface of this apparently minor variation in the standard varieties of Dutch, the language situation in the northern part of Belgium is far more complex. Since the sixteenth century, what is now known as Belgium has been politically separated from the Netherlands, with the exception of fifteen years in the early nineteenth century. In Belgium, French has mostly been used as the standard language for official occasions and education up to the twentieth century. Whereas in the Netherlands a standardized variant of Dutch has developed from the seventeenth century onwards, the need for a Standard Dutch in Belgium only became acute at the end of the nineteenth century, when language laws slowly began to assure the same rights for Dutch as for French. The legislation against the discrimination of Dutch in public life continued through the first half of the twentieth century.

Although Standard Dutch as it had developed in the Netherlands was officially adopted as the Belgian standard variey as well, most Belgian speakers of Dutch did not master this standard variant of the language. Until the second half of the twentieth century, many speakers used only a local dialect, which they had learned as their mother tongue. Especially in the 1960s and 1970s, Standard Dutch was promoted in the Flemish media, most notably by means of dedicated radio and television programs. The propaganda was often aimed against dialects and against the lexical influence from French upon Flemish Dutch.

Meanwhile, the need for a standardized language was also felt in everyday oral communication because of the rise in mobility. Dialects in Flanders differ greatly. There are three main groups, but even within those groups, the differences can be considerable. Therefore, Flemings who moved could not continue to use their dialect for local communication. For several reasons, they did not adopt Standard Dutch for conversations with their new neighbors, but a new colloquial variant developed, often dubbed *tussentaal*, 'in-between language,' because of the mixture of characteristics of Standard Dutch and of the dialects. Following up on other recent publications (e.g. Geeraerts 2003), I will use the name Colloquial Belgian Dutch (CBD) in this chapter. The dialect characteristics of CBD can be attributed mostly to the central Brabantian dialect group, which inspired some CBD morphosyntactic features, e.g. in the verbal system, the use of personal pronouns and a few phonological features, e.g. the deletion of /t/ at the end of small function words (see Taeldeman 2008).

A lot of recent research has been devoted to Colloquial Belgian Dutch. Topics include its status as a coherent language variant (Taeldeman 2008) or as a flexible set of more or less independent features (Plevoets 2008), its distribution over Flemish provinces (Taeldeman 2008) and the communicative situations in which this variant is used (Zenner, Geeraerts and Speelman 2009). In recent decades CBD has for instance become the mother tongue of a lot of young Belgian speakers of Dutch, slowly replacing the dialects that were previously used at home (De Caluwe 2009). The development of CBD has been described as a divergence from Dutch Standard Dutch, whereas Belgian Standard Dutch has mostly converged with Dutch Standard Dutch as a result of the zealous language-political activity in Belgium from the 1960s onwards in favor of Dutch Standard Dutch.

To conclude this introduction, the language situation in the Dutch-speaking part of Belgium can be summarized in a simplified way as follows: the importance of dialects is diminishing; the importance of CBD is rising. Belgian Standard Dutch is very close to Dutch Standard Dutch and is mostly used in writing or in very formal oral contexts.

## 7.2 Corpus, previous research, and hypotheses

One problem with diachronic research about Colloquial Belgian Dutch is the scarcity or laboriousness of older sources of spoken language. Radio or television recordings are often in Belgian Standard Dutch and if they are not, selecting and transcribing them is time-consuming. The *Corpus Gesproken Nederlands* (CGN) is a well-annotated corpus of spoken language with a workable quantity of CBD, but it is only synchronic material from around the year 2000.

In this chapter, I will use comics as an accessible source of colloquial Dutch. Most texts in comics are dialogues in text balloons. The dialogues are artificial in the sense that they are written to be read instead of spoken to be heard. One should therefore not equate text balloons with spoken language, but cartoonists who want to keep the language in their comics close to reality will have to use balloon texts that the reader readily accepts as language that could be spoken aloud. The captions can be equally colloquial at times, because some of the interventions of a narrator seem to belong to the register of a traditional oral storyteller. Other captions can belong to a different, more literary register.

In an earlier paper (Meesters 2000), I used the most popular Flemish comic in history, *Suske en Wiske* by Willy Vandersteen (see the example in Figure 7.1), to examine the evolution of Belgian Dutch in the second half of the twentieth century. *Suske en Wiske* was the comic that established the norm for Flemish comics just like *Superman* in the US or *Tintin* in the French-speaking part of Belgium. It is a family comic that has been published in the newspapers since its inception in 1945.

Panel 1: "Sidonie, give him the letter! Don't let me explode for a piece of paper!" Panel 2: "In heaven's name, don't be secretive, it's not the right moment!" – "But we don't have a letter, auntie! Lambik is lost!" Panel 3: "Fuse almost up – Balloon head leaving for eternity – Will cost lots – Price streetcar gone up."

*Figure 7.1*  A strip from the 1955 *Suske en Wiske* album *De dolle musketiers*, p. 27. Copyright: Standaard Uitgeverij, 2012.

It features a mixture of humor and adventure and is published at the rate of two strips a day. Every year, four or five comic books or albums are assembled from the newspaper strips. Vandersteen stopped making *Suske en Wiske* on a daily basis in 1970 and died in 1990, but the comic is still continued by a studio. It has never really become group work, though. Paul Geerts and later Marc Verhaegen took over both story and penciling. Lately, Peter Van Gucht writes the stories, while Luc Morjaeu draws them. Interestingly, all writers share the same regional background, which simplifies our research into language use in the series. So far, more than 250 stories of at least 46 pages have been published and more than 130 million books have been sold.

*Suske en Wiske*'s publishing history made it especially relevant (see De Ryck 1994). Immediately after its start, it became the most popular comic in Flanders. Soon the publisher started promoting *Suske en Wiske* in the Netherlands as well, although the real breakthrough there happened only in the 1970s. From 1953 onwards, *Suske en Wiske* was published in two different versions: one in the original Belgian Dutch language variant and a second one with adapted language for the Dutch market. The original version served as a starting point, but alterations in the Dutch version were rather frequent. Meesters (2000: 168–73) showed that up to 16 changes per 100 words were made. Half of those were replacements for content words (nouns, adjectives, verbs or expressions), while the other half consisted of changes in syntactic structures, interjections, morphology or function words (prepositions, conjunctions, etc.). Two parallel versions were published until 1964, when only one version remained: the version originally destined for the Dutch market. In twenty years, *Suske en Wiske* evolved from a comic with very local language (in early stories Vandersteen even used typical vocabulary from the dialect of Antwerp, his home town) to a comic with an internationally acceptable variant of Dutch. It is in itself telling that it was possible to promote an adapted version for the Netherlands to the only remaining version. In the years preceding the union, one could already see the numbers of changes from one version to another diminish radically (e.g. only six changes per 100 words in the album *Het Rijmende Paard* from 1963). The convergence of both versions could be completely attributed to the evolution of the original Belgian Dutch variant, which became more suitable for the readership in the Netherlands. The claimed convergence of Belgian Dutch with Dutch Dutch in the 1960s (see Geeraerts, Grondelaers and Speelman 1999) was thus mirrored in the *Suske en Wiske* books from that period.

In Meesters (2000), the development of the language use in *Suske en Wiske* beyond 1964 was also treated. At three moments in the comic's history, roughly 1955, 1975 and 1995, I checked the texts of three comics per moment in the best contemporary descriptive dictionary of the time, *Van Dale Groot Woordenboek van de Nederlandse taal* (Kruyskamp 1970), for regional labeling. Using *Van Dale* as a referee, I demonstrated that from 1955 to 1975, say the period of the most active media propaganda for Dutch Standard Dutch (see supra), the use of regional vocabulary diminished drastically, whereas its slight increase from 1975 to 1995 was too limited for statistically significant conclusions. These conclusions confirmed earlier research about the convergence of Dutch vocabulary in the 1960s and 1970s (Geeraerts, Grondelaers and Speelman 1999) and suggested a slight divergence from the 1970s onwards. Comics proved to be a reliable corpus for diachronic research on the evolution of language varieties.

In this chapter, I want to expand the research reported in Meesters (2000). Firstly, I aim to describe the further evolution of regional vocabulary in *Suske en Wiske* in the early twenty-first century, to see whether the slight divergence from 1975 to 1995 has continued afterwards. Secondly, the scope of the earlier research project expanded in two ways. In the last ten years, linguists have made an inventory of some grammatical features of CBD. Although some of these are still open to discussion, the list of features is clear enough to be able to involve them in my project. Also, I added another Flemish comic to the project: *De belevenissen van Jommeke* ('The adventures of Jommeke', by Jef Nys, created in 1955; see Figure 7.2).

Panel 1: "We're not drunk, professor, I wouldn't know what from?!" Panel 2: "We're not dreaming either, because I'm pinching my foreleg, I mean arm and I feel it, ouch!"

*Figure 7.2* Two panels from the 1960 *Jommeke* album *Purpere pillen*, p. 25. Copyright: Ballon Comics, 2012.

*Jommeke* is similar to *Suske en Wiske* in several ways. It is a mixture of adventure and humor, primarily meant for children. Since the switch from gags to long stories in 1959, it has become at least as popular in Flanders as *Suske en Wiske*. The author Jef Nys grew up in Antwerp, just like Willy Vandersteen. He died in 2009, but since 1972, the work on *Jommeke* had slowly been taken over by collaborators (no real team work, usually a single cartoonist or a team of a writer and an artist). These collaborators still ensure the steady release of new stories. Five or six books per year are published and as with *Suske en Wiske*, the total number of available books has exceeded 250.

There are some differences however, that make *Jommeke* an interesting complement to the research on *Suske en Wiske*. *Jommeke* was never popular outside of Flanders. It sold over 55 million copies in Flanders alone (population six million), but no serious attempts were made to reach the same popularity in the Netherlands or beyond linguistic borders. As a result, the authors of *Jommeke* never had to take Dutch readers into account, just Flemish ones. This difference in the audience may have influenced the language use in *Jommeke*. The hypothesis would be that *Jommeke* has preserved or cultivated more regional vocabulary and more features of CBD than *Suske en Wiske*. Besides the almost exclusive Flemish readership, there is another external indication of different language use in *Jommeke* than in *Suske en Wiske*. As late as 1984, *Jommeke* was severely criticized for its language in Flanders' biggest weekly *Humo* (Herten 1984). Part of the criticism dealt with spelling, but *Humo* mentioned the regional wording explicitly, implying that this language was a bad example for the kids who read the books. As a result of this kind of criticism, one important change in the texts of *Jommeke* was explicitly implemented: use of the Standard Dutch second person singular personal pronoun *jij* instead of *gij* (archaic in Standard Dutch, but normal in CBD) from 1989 onwards (see Bex 1995). The same magazine *Humo* admitted in a 1970 interview with author Willy Vandersteen that the language in *Suske en Wiske* had by then already clearly 'improved' (Guillaume 2005:78).

## 7.3 Methodological considerations

From Meesters (2000), I adopted some methodological principles. I focused on a few specific years as snapshots in time, to describe the evolution. Per chosen year, I selected three comics. I did not use them entirely as parts of my corpus, but I excerpted the first ten pages only. This approach makes sure that the corpus is not too dependent on the subject matter of individual stories. Of course, the theme may highly influence the vocabulary or the register of a particular story. It is therefore better to excerpt more stories than to take more pages of one story. The limit of ten pages is quite arbitrary, but our experience with comics corpora (see Meesters 2000, 2011) has shown that ten pages suffice as a sample of the complete comic for classic European albums of about 50 pages. After those ten pages, few new language elements are found. The main vocabulary related to the story has already been introduced. Compared to American or Asian comics, the number of words per page is relatively high. For both *Suske en Wiske* and *Jommeke*, ten pages means about 1,500 words, the extremes being 1,100 and 2,000. These exceptional, extreme results do not endanger comparisons, because each moment in time is represented by three stories, so the possibly disturbing effect of extremely wordy or extremely silent stories is flattened out. I have not chosen to count a fixed number of words per story. Some stories may be wordier than others, but it always takes about the same number of pages for a story (and the related vocabulary) to be introduced.

As for *Suske en Wiske*, the selection was simple: I used the same comics I used for Meesters (2000) for the years 1955, 1975, and 1995. I added 2010 as an additional sample period for comparison. With *Jommeke*,

I took three of the first books from 1959 and 1960, but for the other sample periods, I followed the *Suske en Wiske* timeline: 1975, 1995, and 2010. Obviously, a comparison between the two comics is only valuable when roughly the same sample periods are chosen. Because of the frequent reprints and the possibility of altered text in those, I used first editions or facsimile editions of those first editions only. The appendix shows the list of selected books of both series.

To find out how many lexical items in the comics can be considered regional, I reused the method I had tested in Meesters (2000): I checked the words for regional labeling in the biggest contemporary dictionary, *Van Dale Groot Woordenboek van de Nederlandse taal*. With *Suske en Wiske* in Meesters (2000), I had checked the latest edition (1992) that was available at the time of the publication of the comic that I was examining. The reasoning behind that approach was that regional labeling can change over time. However, this made it more difficult to use statistics such as Pearson's chi-square test, because our touchstone, the dictionary, was not completely stable. Moreover, the test samples showed that there were only very small individual differences in labeling in the different editions of the dictionary and these would not affect the proportional rise or fall of regional wording in the comics. Therefore, I used only the latest edition of *Van Dale Groot Woordenboek van de Nederlandse taal*, the fourteenth edition from 2005, to check the labeling of the words in all selected *Jommeke* books, as I did for the *Suske en Wiske* books from 2010.

Contrary to the approach in Meesters (2000), I kept the lexical regionalisms apart from the grammatical aspects. I took all grammatical aspects of CBD into account that were described in Taeldeman (2008). I counted the types per comic per year, but not the tokens. This means that I counted the use of the personal pronoun *gij* (type) only once, even if it occurred more than 50 times (tokens) in the corpus of the three *Suske en Wiske* books I selected for 1955. A token count would give too much weight to frequent words, whereas the presence of a less frequent regionalism is as telling for the regional character of the language use as an extremely frequent one. Moreover, words are often triggered by the theme of the story. Counting the tokens would therefore disturb the theme-neutral equilibrium we have been trying to establish.

## 7.4 Results: lexicon

Table 7.1 shows how regional the vocabulary of *Suske en Wiske* and *Jommeke* is according to *Van Dale Groot Woordenboek van de Nederlandse taal*.

At the starting point of the diachronic research (1955 for *Suske en Wiske* and 1959 for *Jommeke*), both series use a considerable amount of regionally labeled vocabulary (both more than 60 items in 30 pages (3 stories × 10 pages)). Later in their histories, both comics tend to use less regional vocabulary. The evolution is not entirely parallel, however. Both reach a stage where less than 10 items per 30 pages (3 × 10 pages) get a regional label in Van Dale, but *Suske en Wiske* reaches this stage in 1975, whereas *Jommeke* reaches it in 1995. Both comics have in common though that the number of regional items in the vocabulary seems to climb in recent years. *Jommeke* is more subject to this evolution than *Suske en Wiske*; *Jommeke* reached the level of more than 20 regional items (types) in 2010, whereas *Suske en Wiske* only slightly passed the 10 item mark.

While looking at these numbers, clear though they may seem, one should keep in mind that only the evolution from regionally marked to less regionally marked is statistically significant. For both comics, the evolution until 1975 is highly significant ($p<0.001$ with Pearson's chi-square test). The evolution of *Suske en Wiske* after 1975 seems non-existent. The value of p here seems to indicate that *Suske en Wiske* did not become more regionally marked. The later stages of the development in *Jommeke* are a bit different. Here the p-values (0.27 and 0.16 respectively for the transition 1975–1995 and 1995–2010) are far from the threshold for statistically significant differences ($p<0.05$), but p, the chance that there is no difference between the samples, is in both cases rather low: one out of four or less. This leads us to the hypothesis that larger samples would have given statistically significant results, for both intervals.

A larger sample can only be reached by adding other books to the last three snapshots in time of the *Jommeke* corpus. Taking more pages from the same books is no productive solution, since it would hardly increase the number of regionalisms (see supra). Selecting more stories would be possible, but might blur the precision of a sample in time.

*Table 7.1* Lexical regionalisms in *Suske en Wiske* and in *Jommeke*

| Year | Suske en Wiske | Jommeke |
|---|---|---|
| 1955/1959 | 62 | 68 |
| 1975 | 6 | 20 |
| 1995 | 12 | 8 |
| 2010 | 14 | 22 |

Depending on how many stories would be necessary to reach statistical significance, the moment in time (1975, 1995 or 2010) might be extended by a couple of years.

In order to test the hypothesis that the later stages in *Jommeke*'s development might display real changes, I added one more book to each moment in question (*De stenen aapjes* for 1975, *De mandoline van Caroline* for 1995 and *Schattenjagers in Bokrijk* for 2010). The amount of lexical regionalisms increased to a total number of 26, 11 and 28 respectively, with the p-value for the 1975–1995 interval diminishing to 0.19 and the p-value for 1995–2010 resulting in 0.12. Taking into account the rather equal distribution of the regionalisms over the four books that now constitute each sample in time and the diminishing p-values with the transition from three to four books, I was satisfied with the tendencies in the p-values and I have not added any more books to the *Jommeke* corpus.

## 7.5 Results: grammar

The preceding discussion illuminates the occurrence of lexical items across the chosen comics, and Table 7.2 shows the results for the grammatical features, which in fact reinforce the image Table 7.1 sketches out.

I did not perform statistical tests for table 7.2, because these are less useful with low numbers and empty cells. The low numbers are not caused by small samples, but are due to the limited number of possible grammatical regionalisms (types). Nevertheless, the evolution between the 1950s and the 1970s shows a clear drop in regionalisms. In *Suske en Wiske*, the situation remains stable on a very low level of regional marking. In *Jommeke*, the development from many regional characteristics to few or none takes another step in our timeline and seems to continue until 1995. Afterwards the regional features tend to reappear, although their presence is less striking than in 1959. In general, the similarities between Table 7.1 and Table 7.2 compensate for the lack of

*Table 7.2* Grammatical regionalisms in *Suske en Wiske* and in *Jommeke*

| Year | Suske en Wiske | Jommeke |
|---|---|---|
| 1955/1959 | 4 | 12 |
| 1975 | 1 | 7 |
| 1995 | 2 | 0 |
| 2010 | 1 | 3 |

statistical proof with table 7.2: when both tables show exactly the same evolution, the chances of coincidence diminish radically, although they may not disappear completely. When comparing both tables in more detail, some differences can be seen as well. First, *Suske en Wiske* used fewer grammatical features of Colloquial Belgian Dutch in the 1950s than *Jommeke*. With lexical regionalisms, both comics showed more or less the same image, but in 1955, *Suske en Wiske* used only one-third of the grammatical features that *Jommeke* used in 1959.

A second difference, one that should be considered with caution because of the small numbers, is the more hesitant reappearance of the grammatical features in *Jommeke* after 1995 compared to the resurgence of the lexical features. Part of this difference between lexical and grammatical features can be attributed to the slower evolution to a less regionally marked grammar. In 1975, *Jommeke* had still retained more than half of the grammatical regionalisms it contained in 1959. The number of grammatical features retained in 1975 is thus especially high.

## 7.6 Summarizing the lexical and grammatical features

To understand the differences between Table 7.2 and Table 7.1, we need to look at the distribution of the grammatical features in more detail (Table 7.3). The presence of a regional grammatical feature (type) is only marked with an x in the table. The number of tokens is usually rather restricted and the exact number is therefore considered irrelevant for this discussion. The low number of tokens entails that the absence of grammatical features where they are expected is always possible. Our corpus of comic books cannot be large enough to encounter these regionalisms or their Standard Dutch variant with certainty, since some of these grammatical features are rather rare in the language. The absence of a feature can therefore always be a coincidence and should not lead to deep analyses. This table does permit comparing the presence of specific grammatical features before and after the development towards a less regional language in the comics in question. As became clear from Tables 7.1 and 7.2, the language in *Suske en Wiske* became much less regional before 1975 whereas this change occurred in *Jommeke* between 1975 and 1995. The shaded cells mark the more regional moments based on Tables 7.1 and 7.2, i.e., 1955 for *Suske and Wiske* but 1959 and 1975 for *Jommeke*.

To clarify Table 7.3, I will briefly explain what the short descriptions of the regionalisms stand for.

- In Standard Dutch, the diminutive is formed by attaching *-je* or an allomorph to the stem of the word. The word *woord* ('word') thus becomes *woordje*. In CBD, diminutive formation with *-(e)ke* is preferred. This results in a form such as *woordeke*.
- Among the clearest markers of CBD are the second person personal pronoun *gij* (together with its unstressed variant *ge* and *u*, its form in cases other than the nominative) and the possessive pronoun *uw* in contexts that do not require politeness. These pronouns are rarely written, but are very frequent in everyday conversation. In the Netherlands, these forms are obsolete and are mostly used to refer to God. In Dutch and Belgian Standard Dutch, two different forms are used: *jij/je/jouw* for non-polite contexts and *u/uw* in polite contexts (the forms in *-w* being the possessive pronouns and *je* serving as unstressed variant of both personal and possessive pronoun).
- References to the future can be expressed in a number of ways in Dutch. The auxiliary *zullen* is often used in Standard Dutch, but speakers of Dutch mostly use the present tense. The use of the auxiliary *gaan* ('to go') with this future function is very limited in Standard Dutch, but generalized in CBD.
- 'Definite determiner before proper name' refers to the custom in CBD to use definite determiners before the proper name of people one knows (well). In CBD, it is normal to say *de Jommeke* ('the Jommeke').
- Articles, both definite and indefinite articles, are not inflected in Standard Dutch, but they are in CBD. I have also counted the deviant inflection of some other determiners, such as demonstrative pronouns, in this category.
- Dutch has very little case marking, but in some contexts, it is obligatory in Standard Dutch. In Table 7.3, one can see the inflection after prepositions of *voor*, *achter* ('front', 'back', e.g. *naar voren* 'to the front') and of numbers (e.g. *met tweeën* from *twee* 'two'). In CBD, this inflection tends to disappear.
- The function of the plural of the imperative has been taken over by the singular (the stem) in Standard Dutch, but the plural form in *-t* still exists in CBD.
- CBD also has slightly different rules for the inflection of adjectives. Neutral nouns preceded by a definite determiner normally entail a final *-e* /ə/ in the attributive adjective, but the *-e* remains absent in CBD in this context.
- Certain auxiliaries, such as *durven* ('to dare') or *beginnen* ('to begin') require a complementing infinitive to be preceded by *te* in Standard Dutch, but in CBD, they do not take this *te*.

Table 7.3 Distribution of grammatical regionalisms in *Suske en Wiske* and *Jommeke*, in order of decreasing type presence in subsections of the corpus

| Feature | Comic | 1955/59 | 1975 | 1995 | 2010 |
|---|---|---|---|---|---|
| Diminutive formation with -ke | Suske en Wiske | x | x | x | x |
|  | Jommeke | x | x |  | x |
| Personal pronoun *gij* | Suske en Wiske | x |  |  |  |
|  | Jommeke | x | x |  |  |
| Auxiliary *gaan* for future tense | Suske en Wiske |  |  |  |  |
|  | Jommeke | x | x |  | x |
| Definite article before proper name | Suske en Wiske | x |  |  |  |
|  | Jommeke | x |  |  |  |
| Deviant inflection of determiners | Suske en Wiske |  |  | x |  |
|  | Jommeke | x |  |  |  |
| No case marker with adverbs *voor, achter* or a number after preposition | Suske en Wiske |  |  |  |  |
|  | Jommeke | x | x |  |  |
| Plural imperative in -t | Suske en Wiske | x |  |  |  |
|  | Jommeke | x |  |  |  |
| Deviant inflection of adjectives | Suske en Wiske |  |  |  |  |
|  | Jommeke | x | x |  |  |
| Deviant verb construction after certain auxiliaries | Suske en Wiske |  |  |  |  |
|  | Jommeke | x |  |  | x |
| Conditional clause introduced by *moest* | Suske en Wiske |  |  |  |  |
|  | Jommeke | x | x |  |  |
| Deviant order in final verb group | Suske en Wiske |  |  |  |  |
|  | Jommeke | x |  |  |  |
| No partitive genitive after *iets* ('something') | Suske en Wiske |  |  |  |  |
|  | Jommeke | x |  |  |  |
| Adhortative construction *laat ons* ('let's') | Suske en Wiske |  |  |  |  |
|  | Jommeke |  |  | x |  |

- One of the possibilities to build a conditional clause in Dutch is by introducing it with the past verb form *mocht*. In CBD, *moest*, past form from another modal auxiliary, can be used instead.
- The next grammatical feature, the deviant word order in a final verb cluster, concerns the position of the past participle in such a cluster. In Dutch Standard Dutch, it can only be placed at the beginning or the end of a cluster. In Belgian Dutch, the past participle can be located between a finite verb or an infinitive and another infinitive, i.e. inside a verb cluster.
- The next to last feature, the absence of a partitive genitive after *iets* or *niets* is a feature of CBD that is quite logical given the history of the loss of morphological case marking in nouns and adjectives in Dutch, just like the *naar voren* case above. The obligatory -*s* that is added to the subsequent adjective (e.g. *iets moois*, from the adjective *mooi* 'beautiful') is a relic from the historically fuller case system.
- Finally, the equivalent of the adhortative construction *let us/let's* in English is different in Standard Dutch and in CBD. In Standard Dutch it is *laten we* ('let [simple present indicative first person] we'); in CBD the most common construction would be *laat ons* ('let [imperative] us').

## 7.7 Discussion

Let us take a closer look at the presence or absence of the features over time. Most features are of course found in the regional 'areas' of Table 7.3. But even there, the differences already hinted at in the discussion of Table 7.2 stand out. As for grammatical marking, the early *Jommeke* was clearly more regional than the *Suske en Wiske* albums that predate the *Jommeke* albums by five years. In the early *Jommeke* samples, almost every regional grammatical feature that was found in the entire corpus was present. Only one feature was not found in *Jommeke* in 1959. The regional aspect of the language of *Suske en Wiske* in the 1950s was mostly due to the regional vocabulary and less to grammar.

There are many possible explanations for this phenomenon. It may be due to the difference in experience between the two authors, because Vandersteen had been in the business ten years longer than Nys at this point; but it may also be a personal difference between the authors. The most viable hypothesis is to see this as evidence of Vandersteen's particular strategy with regional language. Some of his characters spoke a variant of Dutch close to Standard Dutch, especially characters with a lot of social prestige. More popular characters, such as the main characters of his comic, used more regional wordings in some situations than in others, showing that Vandersteen had the ability to vary the amount of regional elements. He often used regional wording as a funny element (De Ryck 1996: 28–29). This effect could be achieved because the regional words he used were often spoken, but rarely written. This gave them an awkward quality in the (written) texts of the comic. On the other hand, most grammatical features in the table rarely create or reinforce humor. They are usually much less obvious to the reader and therefore lack this inappropriate effect. One should keep in mind that the readership of these comics is mostly Flemish, all the authors are Flemish, and the Standard Dutch propaganda in Flanders was mostly aimed at the elimination of lexical regionalisms, which raised the awareness for lexical differences but not as much for grammatical variance. For *Jommeke*, Nys also adapted the language of the characters to the social and situational context, but the variation is less obvious and he used it less for humor. He seemed to be more concerned with natural language use as opposed to Standard Dutch, which he perceived as artificial (Bex 1995).

After the evolution towards less regional language use (i.e. the area without double borders in Table 7.3), most grammatical features of CBD disappear completely. The ones that do not go or stay away are the most interesting for our purposes. Our hypothesis would be that these remaining grammatical features can be very different, in the sense that speakers of CBD know that some of these features are clear markers of regional language, while with other features, they do not know this at all. This difference in awareness has multiple causes: salience, frequency, and education being the most important ones. Our hypothesis would be that the presence of clearly regional grammatical features indicates a will to cling to regional language, while the presence of less clear features operates on a more subconscious level. Assuming this hypothesis is correct, it is reasonable to expect all clear features to disappear in the post-regional area of the table. The less obvious features should show a much less clear and less radical development over time because these follow the evolution of language use in general.

The following grammatical features can be called clearly regional in the mind of most language users: diminutive formation with *-ke*, personal pronoun *gij*, deviant inflection of determiners, definite article before proper name. The other features are less clear, although there is some difference in this category as well: the conditional clause introduced by *moest* is more obvious as a regional element than the different verb construction after some auxiliaries.

Table 7.3 does not seem to confirm our hypothesis about the different behavior of clear and unclear features. Most 'unclear' features disappear completely in the post-regional area of the table. There are exceptions, like the future tense with *gaan* and the verb construction after some auxiliaries, which do reappear in 2010, but overall, these features are less present in the post-regional area than expected. The 'clear' ones display very dissimilar patterns. Two of them (*gij* and the definite article before a proper name) disappear completely in the post-regional area, while the deviant inflection of determiners is present in one sample. In contrast, the diminutive formation with *-ke* is present in all but one sample and is the most persistent grammatical feature of CBD in my corpus.

For the 'unclear' features, the limited presence in the post-regional area of the table can be explained by linking them to the evolution of regional language use in both comics in general, i.e. a strong evolution towards Dutch Standard Dutch at first and a very light tendency towards more regional language use in recent years. Because of the less salient character of these features, it does not seem illogical that they would follow the general evolution closely.

Even the major differences in the presence of the clearly regional grammatical features can be explained. Firstly, the personal pronoun *gij* is a litmus test of CBD. Using it implies a choice for regional language use; omitting it means an explicit move towards Standard Dutch. Therefore, once the decision has been made to replace *gij* by Standard Dutch *jij* and *u*, it is very difficult to return to the previous use of *gij*. The token value of *gij* can be seen in the evolution of both comics. When the language in *Suske en Wiske* began to conform more to Standard Dutch in the early 1960s, very few systematic regional aspects remained, but the use of *gij* continued until the unification of the Dutch and Belgian editions in 1964. In *Jommeke* also, the changing of *gij* to *jij* and *u* can be seen as the final step in an evolution: *Jommeke* only started using *jij* in 1989, 25 years later than *Suske en Wiske*. Apparently, it is all or nothing with this feature: either *gij* or *jij*. No comic from this corpus uses both. *Gij* and *u* can sometimes both be used, when *u* takes its Standard Dutch polite role in formal language.

The presence of regionally inflected determiners in *Suske en Wiske* in 1995 should not be exaggerated: it is only one form, a Flemish form of the indefinite article *een*: *ne*. It is a weird exceptional case, since *Suske en Wiske* did not even use these regionally inflected forms in the 1950s.

The ubiquitous category of diminutive formation in *-ke* is less different than it seems. In the regional area of the table, with double borders, these forms are used systematically with every noun. In the other area, the use of the diminutives in *-ke* in *Suske en Wiske* is mostly restricted to a few proper names (not counting the names of the main characters *Suske* and *Wiske*, originally diminutives as well), and a few isolated examples in *Jommeke*.

Although the distribution of these 'clear' grammatical features of CBD seemed chaotic at first, the evolution of their use proves to be very relevant. The difference between the presence of these features in the early areas (shown with double borders) and the later areas has gone from systematic to indicative. In the post-regional area of the table, the presence of the clear grammatical features is an identity marker: it is used consciously to lightly point to the Flemish origin of the comics, without 'overdoing' it and upsetting readers who are advocating the use of Standard Dutch. Generally speaking, from 1975 (*Suske en Wiske*) and 1995 (*Jommeke*) onwards, these clear markers make an appearance every now and then, but they are not used systematically anymore.

## 7.8 Conclusions

This chapter has shown that studying language use in comics can clarify the complicated history of Dutch in Flanders in recent decades. Belgian Standard Dutch and Dutch Standard Dutch have grown more alike in that period, as a result of the zealous language political activity in Belgium in favor of Dutch Standard Dutch from the 1960s onwards. Meanwhile Colloquial Belgian Dutch has developed as a more and more important variant for oral communication, instead of local dialects. This variant can be seen as a divergence from Dutch Standard Dutch, because CBD has taken over some of the functions of a standardized variant.

Our research on language use in two long-running Flemish comics, *Suske en Wiske* and *Jommeke*, allows for some interesting conclusions with respect to these evolutions. Firstly, comics can be a useful source of CBD. Of course, we have to take into account that the language in balloons is written and not spoken, but these balloons display a lot of the characteristics that have been attributed to this colloquial variant of Dutch in Flanders in the literature. The dialogues in the balloons also confirm the sociolinguistic variation demonstrated by Zenner, Geeraerts and Speelman (2009) with television programs, with characters changing their language use according to their interlocutor, although we did not go into that specifically because it would lead us too far along for the purpose of this chapter. The language material in comics has disadvantages as well. The number of authors is limited. Personal language use can therefore lead to hasty conclusions, as can be seen by comparing the very different statistics of both series. Moreover, comics corpora are labor-intensive because they require finding the (old) editions and excerpting from many books.

Secondly, the most interesting asset of our comics corpus is the possibility of doing diachronic research. The evolution in the language use in the comics, seen from the author's perspective, is both conscious and subconscious: sometimes an author makes a clear-cut decision about a change in the language, but most differences over time simply happen because of the evolution of language use in general. By comparing the language use in comics at regular intervals, one can see how – in the case at hand – comics witness a history of language use in the Dutch-speaking part of Belgium. They clearly illustrate the evolution from a colloquial language use with relatively many regional aspects to a less regionally influenced language use. Nowadays, the tendency seems to be to let more regional aspects appear, although one should not exaggerate this evolution.

We can ask ourselves if the language use in these two comics remains representative of colloquial language use in Flanders throughout their histories. One could think that the language use of these comics has gradually become more polished and thus less colloquial because of the sometimes harsh criticism of the language in comics. This hypothesis may be partly true for *Suske en Wiske*, since Vandersteen and his studio adapted the language use in order to conquer the Dutch market. In recent years, Flemings have sometimes deplored the 'artificial' language in this comic. But the hypothesis is certainly not valid for *Jommeke*, a comic that never had to take Dutch readers into account. Yet, Colloquial Belgian Dutch language use can contain more regional features than the language found in *Jommeke* in 2010. The regional elements in current day *Jommeke* (and *Suske en Wiske*) are token elements in the sense that only a few regional characteristics can be identified and their use is restricted: they hint at the regional background of the comic, serve to mark the regional identity, but the language has been polished in order to be written down and to conform to language political pressure. The diachronic evolution shows the general tendencies, but because of the polishing in the last forty years, the exact extent of the evolution in the general language use cannot be measured. Comics from the 1950s or the 1960s are a much more reliable source of colloquial language than the comics from the late twentieth or early twenty-first century. Of course, much depends on the comic itself. A comic series like *Urbanus* (by Willy Linthout and Urbanus, created in 1983), which has a famous television comedian in the title role, cultivates a very colloquial word use, much closer to dialect than

*Suske en Wiske* or *Jommeke* ever were. But this is an exception. In general, Flemish comics have evolved towards Standard Dutch.

The Dutch readership of *Suske en Wiske* can account for the small differences between *Suske en Wiske* and *Jommeke* since the 1970s. The fact that the readership of *Suske en Wiske* has been mainly Dutch since that decade prevents the comic from evolving like *Jommeke*. It keeps the dialogues closer to Standard Dutch. At several stages of our research, the publishing history of the comics proved to be of great value, because it explicitly demonstrates a number of findings about the history of Belgian Dutch in the literature. As explained in Meesters (2000), the evolution from two different editions for the Belgian and the Dutch markets to one single edition is in itself a very telling illustration of the overt convergence of Belgian Dutch towards Dutch Dutch in the twentieth century. Further, the 'correction' of the language use in earlier stories with important new editions (e.g. the replacement of *gij* by *jij* in the colored versions of early *Jommeke* books) confirms the evolutions that we have found by comparing the first editions.

## Appendix: Comics in the corpus

| Year | Suske en Wiske | Jommeke |
|---|---|---|
| 1955/1959 | De dolle musketiers | De jacht op een voetbal |
|  | De knokkersburcht | De koningin van Onderland |
|  | De speelgoedzaaier | Purpere pillen |
| 1975 | De nare varaan | De slaapkop |
|  | De bokkige bombardon | Choco ontvoerd |
|  | De raap van Rubens | De gekke wekker |
| 1995 | De vonkende vuurman | Het zevende zwaard |
|  | Het kostbare kader | De snoezige dino's |
|  | De razende race | Dinopolis |
| 2010 | De dartele draak | Krokodillentranen |
|  | De jokkende joker | Het oké-parfum |
|  | De rillende rots | Savooien op de Galapagos |

## References

Bex, J. (1995) Veertig jaar Jommeke: Jef Nys bekent – 'Ik kan geen strips lezen.' *Het Belang van Limburg*, 28 October 1995: 35.

De Caluwe, J. (2009) Tussentaal wordt omgangstaal in Vlaanderen. *Nederlandse Taalkunde* 14: 8–25.

De Ryck, R. (1994) *Willy Vandersteen. Bibliografie. Van Kitty Inno tot De Geuzen.* Antwerp: Standaard Uitgeverij.

De Ryck, R. (1996) *Van Glimlach tot Schaterlach. Humor bij Vandersteen*. Antwerp: Standaard Uitgeverij.

Geeraerts, D. (2003) Cultural models of linguistic standardization. In R. Dirven, R. Frank, and M. Pütz (eds.) *Cognitive Models in Language and Thought: Ideology, Metaphors and Meanings*, pp. 25–68. Berlin: Mouton de Gruyter.

Geeraerts, D., Grondelaers, S., and Speelman, D. (1999) *Convergentie en Divergentie in de Nederlandse Woordenschat. Een Onderzoek naar Kleding– en Voetbaltermen*. Amsterdam: Meertens Instituut.

Guillaume, L. (2005) *Willy Vandersteen. De Interviews, de Foto's*. Antwerp: Standaard Uitgeverij.

Herten, S. (1984) Een kleine bijdrage tot 25 jaar scheldproza: sfeervol bulshitten met Jef Nys. *Humo* 24 May 1984: 29–34.

Meesters, G. (2000) Convergentie en divergentie in de Nederlandse standaardtaal. Het stripverhaal *Suske en Wiske* als casus. *Nederlandse Taalkunde* 5: 164–76.

Meesters, G. (2011) La narration visuelle de l'Association. La différence que fait un Lapin. In ACME (ed.), *L'Association. Une utopie éditoriale et esthétique*. Brussels: Les impressions nouvelles.

Plevoets, K. (2008) *Tussen Spreek– en Standaardtaal: een Corpusgebaseerd Onderzoek naar de Situationele, Regionale en Sociale Verspreiding van Enkele Morfosyntactische Verschijnselen uit het Gesproken Belgisch-Nederlands*. Doctoral dissertation, K. U. Leuven.

Taeldeman, J. (2008) Zich stabiliserende kenmerken in Vlaamse tussentaal. *Taal en Tongval* 60: 26–50.

Vandekerckhove, R., De Houwer, A. and Remael, A. (2007) Intralinguale ondertiteling op de Vlaamse televisie: een spiegel voor de taalverhoudingen in Vlaanderen? In D. Sandra, R. Rymenans, P. Cuvelier and P. Van Petegem (eds.) *Tussen Taal, Spelling en Onderwijs: Essays bij het Emeritaat van Frans Daems*, pp. 71–83. Ghent: Academia Press.

Zenner, E., Geeraerts, D. and Speelman, D. (2009) Expeditie tussentaal: leeftijd, identiteit en context in 'Expeditie Robinson.' *Nederlandse Taalkunde* 14: 26–44.

<u>Van Dale dictionaries:</u>

Kruyskamp, C. and de Tollenaere, F. (1950) *Van Dale's Nieuw Groot Woordenboek van de Nederlandse taal*. 7th edition. The Hague: Martinus Nijhoff.

Kruyskamp, C. (1970) *Van Dale. Groot Woordenboek van de Nederlandse taal*. 9th edition. The Hague: Martinus Nijhoff.

Geerts, G. and Heestermans, H. (1992) *Van Dale. Groot Woordenboek van de Nederlandse taal*. 12th edition. Utrecht/Antwerp: Van Dale Lexicografie.

den Boon, T. and Geeraerts, D. (2005) *Van Dale. Groot Woordenboek van de Nederlandse taal*. 14th edition. Utrecht/Antwerp: Van Dale Lexicografie.

# 8
# Linguistic Codes and Character Identity in *Afro Samurai*

Frank Bramlett

## 8.1 Introduction

One of the guiding principles of linguistics is that people use different kinds of language in different social situations. This is true for every human being in the world who speaks, writes, or signs language. We speak one way at home and another way at worship services, one way at the night club and another in the classroom. Sometimes the differences between these styles are subtle, sometimes dramatic. A register is a variety of language that is associated with certain speech situations, whose parameters include participant identity, setting, and topic or purpose (Finegan 2004: 19). This chapter will examine how characters in both the anime and the manga versions of *Afro Samurai* use language and, under certain circumstances, shift registers.[1] In particular, the linguistic choices of the protagonist, Afro Samurai, and his sometime companion, Ninja Ninja, reveal a complex world in which a long list of discordant sociocultural and linguistic codes blend in the story of hero, demon, computer technology, hip-hop and R&B, and Japanese samurai warrior culture.

In explaining styles of drawing, Will Eisner (1985: 151) says that the cartoon image 'is the result of exaggeration and simplification.' This is doubtless true for animation, as well. In contrast to the cartoon, '[r]ealism is adherence to most of the detail. [...] Retention of detail begets believability because it is closest to what the reader actually sees. The cartoon is a form of impressionism' (Eisner 1985: 151). A linguistic investigation of language in comics, then, should consider the balance of 'realism' in the language the characters produce and the amount of linguistic 'exaggeration and simplification.' Since some shorthand visual representation is expected in most comics and animation, is it reasonable to expect some shorthand or skewed linguistic performance by the characters? If language is simplified and/or exaggerated, what kind of impact does this have on the amount of stereotype that readers find in the text? As Royal (2007:7) points out, comics artists 'may expose, either overtly or through tacit implication, certain recognized or even unconscious prejudices held by them and/or their readers. In comic art, there is always the all-too-real danger of negative stereotype and caricature, which strips others of any unique identity and dehumanizes by means of reductive iconography [...], features that have historically composed our visual discourse on the Other.' But what of linguistic production? How do comics artists navigate linguistic stereotype? What textual realities result from the limitations of language and shorthand drawn forms? Foster (2002: 168) interrogates images from underground comix and finds that the label of 'racist' or 'stereotype' is problematic. He argues for a contextualized study of images, not whether an image in isolation is or is not racist or a stereotype.

In *Afro Samurai*, the main character journeys through a vortex of sociocultural codes which is reflected in the admixture of linguistic codes that various characters produce. These codes include various styles or registers of English as well as very brief uses of Japanese. In the anime *Afro Samurai*, there is a great deal of written language in the form of calligraphy, found mostly on temple walls and temple doors. Likewise, in the manga *Afro Samurai*, almost all sound effects are rendered in both Japanese syllabary and English alphabet. My goal in this chapter is to trace the linguistic codes used in *Afro Samurai* and how those codes map onto particular characters, settings, or social actions. The analysis reveals that the two main characters regularly opt for a particular code and only in rare circumstances do they switch codes, whether that be a socially situated register within English or a bilingual code switch. Since code choice can tell us something about a character's identity, the final section of the chapter will explore the sociocultural implications of characters' code choice. Further, this chapter will question the role of linguistic stereotype and extend the discussion of identity formation from sociolinguistics into comics scholarship. In essence, the characters of Afro and Ninja Ninja are a kind of photographic negative of each other: they are very similar, but each is the reversal of the other, both in linguistic code and warrior code.

## 8.2  Samurais and ninjas, swords and justice

The setting of *Afro Samurai* resembles a wasteland, a dark and troubled place full of uncertainty and danger; this preponderance of the dismal is punctuated with brief moments of brightness, sunshine, and sometimes joy. It synthesizes the essence of seventeenth- and eighteenth-century samurai warrior culture and twenty-first century technology, featuring cell phones, elevators, RPG grenade launchers, cyborgs, and warrior-robots capable of laser-based warfare and flight, among other things.

Like the landscape, the story of Afro brims with sadness, loss, and revenge. As a young boy, Afro sees his father killed by a man named Justice, and he spends years training to become a samurai to avenge his father's death.[2] The situation is complicated by the presence of headbands – whoever wears the Number 1 headband is the most powerful warrior in the land, becoming something of a god. Whoever wears the Number 2 headband is, of course, the second most powerful warrior and has the opportunity to displace the wearer of the Number 1 headband and assume the title himself. The warrior with the Number 2 headband must fight any person who challenges him, always to the death. By definition, then, Number 2 is responsible for the deaths of tens or even hundreds of men just to keep his headband and challenge Number 1.[3]

As a young man, Afro has to make a decision that changes his life forever. He discovers that his own teacher wears the Number 2 headband; thus, he must fight and kill his teacher so that he can continue his journey of vengeance. Naturally, this wreaks havoc on his adopted family, those other students who live in the *dojo*. After he kills his teacher, and after he sees one of his best friends die, Afro experiences a kind of psychic rift and a new character appears, one who is visible and audible almost exclusively to Afro. This character, Ninja Ninja, is Afro's companion, advising him in matters ranging from battle to romance. Near the end of the anime, Ninja Ninja 'dies' in combat and Afro must continue alone in his quest. While there are several minor narrative differences between the anime (henceforth *AS–anime*) and the manga (henceforth *AS–manga*), the profoundness of loss and revenge remains constant in both.

## 8.3 The role of visual stereotype and images of blackness

In his introduction to Volume 2 of *Black Panther*, Priest (2001) writes of his hesitation in accepting the job of writing for a character who 'was, by any objective standards, dull. He had no powers. He had no witty speech pattern [...].' But more to the point, 'Panther was a black superhero,' and Priest was unsure how to 'do a book about a black king of a black nation who comes to a black neighborhood and not have it be a 'black' book.' Priest explains that 'Panther's ethnicity is certainly a component of the series, but it is not the central theme. We neither ignore it nor build our stories around it' (n.p.).

In discussing the historical development of artistic trends in comics, Harvey (1996:72) reviews the origin and maturation of Ebony White, one of Will Eisner's best-known characters from *The Spirit*. In June 1940, audiences were first introduced to Ebony White, who:

> appeared as a cab driver. A young black man, his eyes roll in that stereotypical expression of wide-eyed fright as he drives by Wildwood Cemetery. Everything about Ebony is a stereotype of his race: his large eyeballs, pink big-lipped 'mushmouf,' his linguistic mutilation of pronunciation and diction, his comic costume [...], his low-comedy behavior. Years later, Eisner would feel no little embarrassment at having perpetuated such a racial caricature. [...] Eisner's latter-day discomfort with the character arose from his realization that he had employed a racial stereotype, often in stereotypical fashion. But this portrayal was [...] a consequence of an unwitting insensitivity to the feelings of other races rather than a desire to persecute.

As Harvey suggests in his history, a great deal of change in comics occurred over the course of the twentieth century, and scholars continue to debate how images of blackness function.

Some scholars argue that certain comics maintain the stereotypes found so commonly in the early twentieth century, even though these stereotypes may come in different packages. Scott (2006: 310) argues passionately that superhero comics are always already invested in white supremacy: 'If blackness is a conspiracy [against whiteness], could one who wears it realistically assume access to it in order to render it art? Is perhaps the conflation of the visuality of black bodies with the experience of white supremacist views of reality that which characterizes a "black experience"?' She goes on to say that the culture of superhero comics is so biased against blackness that it is difficult to escape 'this conspiracy we call blackness, this crisis of representation that so marginalizes people who wear that skin that their own bodies elide narration altogether' (p. 312). If the hegemony of whiteness inhabits the black experience so completely, then what hope do we have of articulating a different vision, a comics without the conspiracy? Scott's question about black and white may be present in *Afro Samurai*, but it is even further complicated by the role of the Japanese influence. If Afro journeys through a vortex of sociocultural codes, then the question is whether these codes sit together comfortably like oil and water or they blend together into a new compound, a liquid of unique characteristics. In other words, if the codes blend together so well that they appear seamless, does this obviate the potential for racist caricature?

## 8.4 The verbal meets the visual: linguistic codes and characterization

Images of blackness have been explored extensively, in comics scholarship as in other fields. To a lesser degree, so has the language of blackness. Scholars of nineteenth-century literature have mapped out the role of African American English in antebellum and postbellum texts. Jones (1999: 184) explores dialect literature (e.g., Mark Twain, George Washington Cable, Paul Dunbar): 'written dialects are, at best, gestures toward a spoken reality that the reader can bring to the text from his or her recollection of heard experience.'

For late twentieth-century media, scholars who investigate stereotype have analyzed movies like *Shaft*, *Jungle Fever*, and *Bamboozled*, among many others. Most often, these studies have addressed black stereotypes vis-à-vis white/Anglo ideologies and sociocultural and political constructs. For *Afro Samurai*, the question is complicated by the presence of Japanese culture, language, and ideologies. Relations between African Americans and Asian Americans, especially through hip-hop, show an uncertainty about the state of affairs. The outlook regarding stereotypes in comics ranges from outright reproduction of white supremacist/racist codes (Scott 2006) to a cautious optimism about the future (Wang 2006; Whaley 2006).

Several kinds of English are in evidence throughout both anime and manga *Afro Samurai*. The audience encounters General American English, African American English, and some Japanese. Davenport (1997:25) surveys superhero comics featuring black characters and finds that 'most non-superhero Blacks within the comics sampled used Black English, while the superheroes themselves spoke standard English (Black Goliath, Brotherman, Captain Marvel, Icon, Night Thrasher, Nightwatch, and Storm).' Overall, this is true for both the anime and manga *AS*.

In her book-length introduction, Green (2002) explores the relationship between language used in the black community, what she and many linguists term 'African American English' or AAE, and English as it is commonly used across the United States, what she terms 'General American English' or GAE. Because AAE is a type of English, it shares an extraordinarily high number of linguistic features with other kinds of English. There are, of course, identifiable differences, and these differences are well-documented. However, AAE is much more than a linguistic system. Like all language varieties, understanding AAE means looking at the world through the lens of experiences shared in the communities that use it.

Morgan (2002: 65) defines African American English as 'the language, discourse and interactional styles and usage of those socialized in the [Black] speech community. It functions within the political context as both a stigmatized sign and an authenticating sign.' It is this stigmatized and authenticating sign that so captures the imagination of scholars writing about representations of race in literature, media, and comics. Green (2002: 214) discusses the role that language plays in creating images of blackness in media: 'Linguistic features from the area of speech events as well as from other parts of the grammar contribute to creating images of blackness, [and] images of socioeconomic, and social and ethnic class. [Furthermore, much] can be learned from inaccurate uses of linguistic features.' However, Green cautions that the 'question with regard to language in minstrel performance is not about accuracy of representation because the focus was not on depicting authentic language or other features, but instead it was on exaggerating stereotypes and creating grotesque figures. The minstrel representations perpetuated the stereotypes on which they were based, and these stereotypes, some reflected through language, were carried over into early films' (p. 202). Gavin Jones (1999: 184) agrees: 'Where dialect literature is concerned, however, one point becomes apparent immediately: ideas of authenticity have definite limits as criteria for aesthetic evaluation.'

The blend of African American culture and Asian American culture has received a growing amount of scholarly attention. Wang (2006) explores the role of hip-hop and the complications that arise when Asian Americans perform this musical genre. Likewise, Whaley (2006) interrogates 'Black American and Asian American cultural, political, and social crossroads' (p. 191) and finds that these 'forms of cultural production – that is, visual and hip-hop culture – can exist as a formidable site of transformation within the realm of representation and in social relations' (p. 191). Wang seems less optimistic, since he concludes that 'shared cultural habits [like hip-hop in Black American and Asian American communities] do not erase historical enmities, and they cannot resolve larger structural inequities' (p. 160). However, he does argue that these shared cultural habits create 'the opening of communication, without which the possibility for change and transformation cannot exist' (p. 160).

I do not wish to speculate on the presence/absence of minstrelsy in *Afro Samurai*, that is, whether Afro may play the stereotype of the 'buck' and whether Ninja Ninja may play the stereotype of the 'coon' (Green 2002: 201; Strömberg 2003). In fact, it may be a moot point; traces of minstrelsy from the nineteenth-century Jim Crow performances in New York City and elsewhere (Lhamon 1998: 207) have found their modern reflex in hip-hop performances by artists like M.C. Hammer (p. 218). Instead, my purpose is to discuss the linguistic codes, how characters use those codes, and their possible signification for the development of character identity in *Afro Samurai*, taking into consideration the fact that Japan is the physical location and cultural backdrop yet English is the primary linguistic code.

Russell's take is especially germane here because his discussion locates forms and functions of race in Japan: 'Japanese literary and visual representations of blacks rely heavily on imaginary Western conventions. Such representations function to familiarize Japanese with the black Other, to preserve its alienness by ascribing to it certain standardized traits which mark it as Other but which also serve the reflexive function of allowing Japanese to meditate on their racial and cultural identity in the face of challenges by Western modernity, cultural authority and power' (1991: 4). Even though Green (2002) and Jones (1999) articulate the limitations of 'authenticity' in media representations of blackness through AAE, audiences may still respond to those representations through their understanding of linguistic authenticity, their expectations – stereotyped or not – of linguistic performance. In any case, the question of linguistic authenticity plays a central role in twenty-first century media. However audiences may perceive the authenticity of characters, the very notion of identity lies at the heart of the matter. Bucholtz (2003: 408) makes the case for thinking of identity not as a stable glomeration but as an ever-emergent, ever-renewed formation; authenticity should be understood from a social standpoint as a process of authentication, that 'identity formation is closely tailored to its context.' Later in this chapter, the question of linguistic production and its relationship to identity will be explored for Ninja Ninja and Afro Samurai.

## 8.5 Analysis: Afro Samurai's linguistic production

An examination of all of Afro's contributions is beyond the scope of this chapter, so this analysis focuses exclusively on his use of language in the 'present time' of the story, excluding the scenes containing his contributions as a child and as a teenager. Strictly speaking, this time period refers to Afro's utterances produced after Ninja Ninja enters the narrative. Afro remains silent through most of the five anime episodes and the two manga volumes. Appendix 8.1 summarizes Afro's contributions in the anime, as does Appendix 8.2 for the manga.

In the anime, the majority of Afro's utterances are directed at Ninja Ninja, though he also directs several utterances to Okiku/Otsuru and to Jinno. This proportion might suggest that Afro and Ninja spend quality time together, sharing ideas and building their 'friendship.' The content, brevity, and infrequency of these utterances, though, reveal a different picture. Of the eight utterances directed at Ninja, half of them consist of the directive 'Shut up.' Afro plainly demonstrates his unwillingness to engage in conversational interaction with Ninja while simultaneously indicating that Ninja should stop trying to keep the conversation going, seemingly about anything at all. Three of the four remaining utterances are less rude, hinting that while Afro does not wish to engage Ninja in the present, he might be willing to engage him (or at least listen to him) later on: 'Suit yourself' and 'Do as you please' have enough ambiguity built into them that Ninja interprets them in positive, neutral, or negative ways.

Regarding linguistic features at or below the sentence level, Afro's language in both the anime and the manga cleaves rigorously to what Green (2002) calls General American English. In the anime, Afro uses standard contractions: *it is* becomes *it's* and *I will* becomes *I'll*. On the other hand, at times Afro chooses not to contract but to use full separate forms; e.g., *you will* does not contract to *you'll*. In the manga, Afro's speech shows more features of casual register: he is slightly more likely to contract, e.g., *It's*, *Where's*, and *won't*. He also uses casual linguistic features to express his opinion or emotion about a situation, e.g., *Oh man*. One distinct difference is that Afro uses the word *Shit!* in the manga form but does not use any swear words in the anime.

One last point to make about Afro's language usage is that in the anime, he uses three bilingual code switches. At the tea house, he uses the Japanese word *oishii*, which means something like 'delicious,' to express his satisfaction with the lemonade he has just finished. Later on, he switches to the Japanese word *katajikenai* in a conversation with Otsuru/Okiku. This term indicates two concepts at once. First, it means 'thank you' in the sense that he expresses his gratitude that Okiku/Otsuru fed him, protected him, and healed his wounds after he was hit by a poisoned arrow in battle. But uttering *katajikenai* means that Afro humbles himself before Okiku, expressing gratitude and humility at the same time. The spirit of this utterance seems diametrically opposed to the indifference and open hostility Afro channels toward Ninja Ninja.

The final Japanese expression that Afro utters in *AS–anime* is *namu*. This word signals a prayer to the Abida Buddha, romanized to something like *Namu Amida Butsu*. Uttering this word helps the supplicant to enter into paradise. Since Afro produces this word after he takes the Number 1 headband from his enemy, Justice, the audience understands that the battle is over and Afro has successfully avenged his father's death.

In *AS–manga*, adult Afro's speech patterns very closely resemble those in *AS–anime*. He speaks rarely, and when he does, the turns/utterances are short. Several of the utterances are nonlinguistic in that they consist of laughter or non-words (these have been excluded from the analysis). One significant difference between the anime and the manga, however, is that Afro speaks to a wider variety of characters in the manga. The impact of this difference is that he directs proportionately fewer utterances to any individual character, Ninja Ninja and O-sachi included. As in the anime, Afro's speech in the manga largely conforms to the GAE register.

## 8.6 Analysis: Ninja Ninja's linguistic production

The rarity of Afro's utterances lends itself to charting the time, place, and interlocutor for each one. In stark contrast, Ninja Ninja is by far the most talkative of all the characters in *AS–anime*, and this is largely true for *AS–manga* as well. Ninja's only conversation partner is Afro, which is understandable since Ninja exists only in Afro's mind. The single exception occurs in the manga, when Ninja realizes that one other character, one of Afro's enemies, can see and hear him. All of Ninja's conversations are long (measured in time but not necessarily in number of turns), and most of his turns per conversation are long (measured in time and in number of words per turn). Appendix 8.3 contains three representative samples of Ninja's discourse in *AS–anime*, as does Appendix 8.4 for *AS–manga*.

Ninja's use of register shows more variety than Afro's use. In the anime, Ninja uses pronunciation style frequently associated with AAE. He often uses a monophthongal [ɐ] or [a] rather than the diphthongal [aj] in words like the personal pronoun *I* and the word *find*. His speech shows a high degree of r-lessness, e.g., pronouncing the phrase *for sure* without postvocalic [r]. Another very common feature in both the anime and the manga is Ninja's use of the alveolar nasal [n] rather than the velar nasal [ŋ] in words like *somethin'* for *something*. In the manga, spelling conventions indicate Afro's pronunciation to a certain degree. The pronoun *you* is sometimes spelled *ya*. The phrase *all right* is spelled *aight*. The words *about* and *appreciation* are shortened to *'bout* and *'ppreciation*, respectively.

Contractions are frequent: e.g., *you'll, it's, might've,* and *y'know*, among others. In the rare circumstances when Afro contracts, he uses enclitics only, but when Ninja contracts, he also uses proclitics, which are very rare in both *AS–anime* and *AS–manga*. Further, at the word level, Ninja contracts more frequently than Afro, e.g., *I am going to* is produced as *I'm gon*. In the anime, Ninja's speech shows occasional copula deletion: 'You one lucky dog, Afro!' In both the anime and the manga, he shows variability in the *be* conjugation: 'You're no normal man, is ya?'

At the discourse level in *AS–anime*, Ninja reads dialect: 'Excuse me for probing but what you just did back there. Was it absolutely necessary?' Morgan (2002: 265) explains that reading dialect 'occurs when members of the African-American community contrast or otherwise highlight obvious features of AAE and [GAE] in an unsubtle and unambiguous manner to make a point [which] may or may not be a negative one.' Ninja pushes Afro to reflect on how many people he has killed in his quest.

In both the anime and the manga, Ninja employs rhetorical rhyming. Smitherman (1977: 3) demonstrates rhetorical rhyming with a quote from Jesse Jackson: 'Africa would if Africa could. America could if America would. But Africa cain't and America ain't.' Note that for some speakers of AAE and some speakers of English in the American South, the word *can't* rhymes with the word *paint*, hence the spelling change of *can't* to *cain't* on analogy with *ain't*. When Ninja tries to warn Afro of impending danger in the anime, he uses rhyme for discourse effect: 'He don't want your autograph more like your epitaph.' In the manga, he observes that Afro is 'maxin' and relaxin'' as he is recovering from his battle wounds under O-sachi's care.

Like Afro, Ninja employs bilingual code switches, one each in the anime and the manga. In the manga, Ninja refers to Afro's *katana*, a type of sword used by samurais. In the anime, Ninja uses the word *mamasan* to convince Afro should be interested romantically (or at least sexually) in Okiku/Otsuru. However, unlike his use of the word *katana*, Ninja's use of this term does not meet a dictionary definition. The *Oxford English Dictionary* describes *mamasan* as a combination of the Japanese word *mama* and the honorific suffix *-san*, referring to 'a matron in a position of authority, specifically one in charge of a geisha house; the mistress of a bar.' The *New Partridge Dictionary of Slang and Unconventional English* offers a similar gloss: 'a woman whose age demands respect, especially a brothel madam.' Both sources stipulate the advent of the word as post-1945.

Ninja uses *mamasan* to refer to Okiku/Otsuru, but as far as the narrative goes, she does not fit the dictionary definitions. She is a young woman (at least a couple of years younger than Afro), and there is no indication that she is a brothel madam or a geisha house manager. In fact, Ninja seems to use the term in a positive sense: Okiku is 'practically marriage material.' On the other hand, traditional gender role ideologies are strongly reinforced here. Okiku is marriage material seemingly because she serves Afro: she gives him medical attention and cooks for him. There is no unambiguous positive reference to her intelligence, though, so the 'compliment' that Ninja seems to be paying may be a double-edged sword. Ironically, although he seems to know all about the dangers that Afro faces, Ninja never warns Afro that Okiku/Otsuru is a ninja spy who works for Afro's enemies.

Ninja uses linguistic innovation in the anime. For instance, he uses the syntactic frame *to get one's X on* to comment on Afro's convalescence: 'I see somebody's getting their little resort on!' Further, Ninja blends the word *homeboy* with the word *android* to yield the innovation (morphological blend) *homedroid*. (Arguably, Ninja's use of *mamasan* also fits the category of lexical innovation.) Lastly, Ninja uses *ass* words ('lanky ass'; 'stank ass'), and he explains to Afro where they are when he says 'This is it, Muthafucka, Bhava-Agra, the city of gods. No. 1 resides within' (Volume 2, page 104, panel 1). Spears (1999) explains the grammar of *ass* words (pp. 234–38) and challenges the assumption of 'so-called obscenity' in some expressions in AAE, preferring the term 'uncensored speech' to describe the register (p. 248).

## 8.7 Discussion

In general, both Afro and Ninja maintain similar discourse tendencies across *AS–anime* and *AS–manga*. That is to say, Ninja is highly talkative in those scenes where he appears, and he employs a register in which AAE figures prominently. Afro is highly taciturn in those few scenes where he speaks, and he employs a register in which GAE figures prominently. Afro speaks slightly more in the manga than in the anime; he speaks to a larger number of characters in the manga and his turns are slightly longer (though the amount of data is small, so the differences probably are not statistically significant). Afro's use of Japanese in the manga is restricted to proper names of people and places; he does not use bilingual code switches. He uses a noticeably small number of contractions in the manga: e.g., 'I'll keep fighting' (Volume 2, page 114, panel 1). Only once does he substitute an alveolar nasal [n] for a velar nasal [ŋ]: 'Thanks for puttin' me up' (Volume 2, page 28, panel 3). Afro uses GAE for the majority of his utterances, but he also uses a more formal register that makes him sound as much a bureaucrat as a friendly conversationalist.

Likewise, Ninja's speech in the manga differs slightly from the anime. He speaks less in the manga, though not markedly so. In the anime, Ninja speaks only to Afro; however, in the manga, he speaks to Afro and to one other character (Takimoto Kougansai). He uses some formal features of AAE in the manga, but not at the same frequency as in the anime. These features include pronunciation, word forms, and sentence structure, but they also include discourse features. Notable are his use of *ass* words ('lanky ass' in the anime and 'big ass' in the manga) and rhetorical rhyming ('autograph/epitaph' in the anime and 'maxin' and relaxin'' in the manga [Volume 2, page 14, panel 2]). One last difference of import is that while Ninja uses Japanese in an innovative way in the anime (*mamasans*) he uses standard Japanese once in the manga (*katana*, a type of sword used by samurais [Volume 2, page 97, panel 4]).

Ninja performs at least two kinds of functions in the narrative, and his linguistic production provides strong evidence of this. Ninja's primary function is that of *advisor/counselor*; he always has advice for Afro, usually relating to battle strategy, but he also counsels him regarding friendship and romance. Ninja's advice is almost always about Afro's safety, that it is best to retreat or otherwise avoid the fight. Ninja also tries to help Afro retain some semblance of humanity. He concedes that Afro's road to vengeance means not having any emotional availability, but Ninja knows that Afro's connection to other humans is very fragile, so Ninja convinces him that a sexual connection with Okiku/Otsuru is better than no connection at all.

The second major role that Ninja plays in *AS–anime* is acting as a *source of humor*, in particular by offering redundant/unnecessary narration, clowning, and a distortion of priorities. Ninja narrates for Afro and the viewer the possibilities that await Afro as he journeys toward his goal. He also injects levity into an otherwise terrifying possibility, that Afro goes to his almost certain death. Ninja pays attention to a somewhat trivial detail (Afro's hair style, his 'fro) in a humorous way rather than what really matters (Afro could die if the mountain exploded).

When Ninja cautions Afro about the two swords in this passage, it is redundant and/or unnecessary. Afro clearly sees that the swords are drawn, as does the viewer:

> Ninja Ninja: Aaww. Watch it! He bout to bust a two sword move on you. Uh oh. Watch your left. No no no no no. Watch your right. (1:12:18 – 1:12:33)

This narration seems redundant because the action on screen is unambiguously equivalent to Ninja's description. Further, Ninja's confusing admonition about the direction of attack serves no purpose: it does not solve a problem; instead, it states the obvious and perhaps interferes with Afro's concentration in a very dangerous sword fight. One phonetic indicator that this is meant to be funny is Ninja's articulation of the [w] in *sword*.

In contrast to Ninja, Afro functions in the narrative as a character who is emotionally cold, almost inhuman. His last utterance to Ninja in *AS–anime* demonstrates this. In this scene, Ninja and Afro are having their last interaction of the story. As is his wont, Ninja advises Afro about the danger of his opponent, who in this case carries two swords:

Ninja Ninja: [laughs] Like I told you. Might not make it this time baby. Didn't I say to be careful? [some omitted] So go on take care of business. Whack that punk. Even if he is your old pal. You of all people should understand that. Don't forget your goal. Your opponent's dual sworded. You got to make twice as many moves in order to win.
Afro: The force of the blow is twice as powerful for the double-handed grip.
Ninja Ninja: [laughs] Okay if you say so. You just don't listen to people do you. Alright dog. You're on your own from here. I guess this is goodbye. (1:34:25–1:35:48)

The coldness, the inhumanity that Afro wields contrasts dramatically with Ninja's heartfelt speech. As Ninja is dying, Afro chooses not to comfort him or say goodbye to him but to disagree with him about a principle of sword fighting: whether two swords or one sword is better. The length of this utterance surpasses all of his other turns, and it evinces his 'heightened' willingness to engage Ninja in talk. Simultaneously, it solidifies Afro's cold, distant attitude toward Ninja. Other utterances indicate Afro's principles and approaches to life and the way of the samurai warrior. After killing one of his adversaries, Afro says simply, 'My aim is only to move forward.'

## 8.8 Social roles, language codes, and identity

In their research on language and identity, Bucholtz and Hall (2005) articulate several principles by which speakers performatively produce their identity within interpersonal interaction; they summarize their program in a single sentence: '*Identity is the social positioning of the self and other*' (p. 586, italics in original). While all the principles are relevant for this discussion, one stands out in particular. Bucholtz and Hall explain that in the process of interactively constructing our identity, we use particular linguistic forms that invoke particular (ranges of) meanings: this is the characteristic of indexicality. One aspect of indexicality in *Afro Samurai* is 'the use of linguistic structures and systems that are ideologically associated with specific personas and groups' (p. 594). For Afro and Ninja, the use of different registers (GAE for Afro and AAE for Ninja) means that they are attempting to index different identities. While GAE and AAE are types of English, they are different enough in the minds of native English speakers that people who speak one code are socially and culturally different from people who speak the other one.

The indexing of identity in *AS–anime* becomes crystal clear after Afro drinks his lemonade in the tea house; he uses a Japanese word *oishii* 'delicious' and then switches to an English word *refreshing*. In that one conversational turn, Afro signals membership in the English speaking community and in the Japanese speaking community. Likewise, when Ninja reads dialect when commenting on Afro's 'killing spree' early in the anime, he signals membership in the African American English community and also signals knowledge of the General American English community. This pragmatic usage of dialect indicates that Ninja can speak GAE when he wants to or needs to.

In both the manga and the anime, Afro's linguistic performances contrast in a dramatic fashion with Ninja's. Afro communicates in minimal ways linguistically, many times solely with facial expression, hand gestures, and silence. He most often uses General American English with a rare code switch to Japanese. In contrast, Ninja is talkative; we might say that from Afro's standpoint, Ninja is garrulous to the point of annoyance. In *AS–anime*, Ninja frequently uses AAE with occasional use of GAE, primarily for pragmatic impact. In *AS–manga*, there are fewer observable AAE features in Ninja's speech, but their presence harmonizes with the code distribution found in *AS–anime*.

Revenge is a constant, guiding trope by which Afro measures his success, his progress, his path. Afro's behaviors and world view reflect a single-minded journey on a path of vengeance; even when he has the opportunity to turn away from that path, he kills his own teacher in memory of his father who died wearing the Number 1 headband. Afro's conversational interaction can be described as minimal at best and, when directed at Ninja, hostile. Afro's linguistic production thus indexes his world view, reflexively producing and reproducing an approach to life that depends on attitudes of coldness, social distance, a lack of compassion, and a willingness to kill.

Ninja Ninja's verbal performance, like his physical characteristics, is similar to a photographic negative of Afro's verbal performance. His linguistic contributions are fluent and generous; the content of his utterances demonstrates that he has Afro's best interests at heart. In fact, he repeatedly attempts to persuade Afro that the path of vengeance is not the right one at all. Ninja's linguistic production, then, indexes his world view, reflexively producing and reproducing an approach to life that indicates warmth, connection, sexuality, and an unwillingness to fight. Granted, as Afro gets closer to his goal, Ninja relents and encourages him to win his battles rather than avoiding them.

## 8.9 Conclusions

Characters who are authentic use authentic language. This means that the character will produce speech that conforms to a range of linguistic features deemed acceptable by an audience capable of judging; a character's linguistic production will index a (reasonably?) authentic identity. The problem here is the question of audience: who knows whether a black character is using authentic language? (See Beers Fägersten, this volume.) Who knows whether a character who speaks Irish English is using authentic language? (See Walshe, this volume.) If a character does not use authentic language, then the believability of that character fails, unless of course the audience has an expectation of linguistic performance that belies realistic use. This is the role of ideology and stereotype. It seems reasonable to claim that white audiences with expectations of stereotyped speech in the mouths of black characters find those characters who produce stereotyped speech to be believable. Green (2002) makes a similar point: 'the linguistic description [of AAE] can be used to determine the extent to which the representation of black speech […] is authentic, but a rating of "authentic" does not necessarily mean that the character will be perceived as being positive or negative' (p. 200).

Russell (1991: 21) concludes his exploration of the black Other in Japan by focusing on how a Japanese audience responds to images of blackness:

> Japanese representation of blacks tends to be condescending and to debase, dehumanize, exoticize, and peripheralize the black Other, who at once serves as a symbolic counterpoint to modernity,

rationalism, and civility—and as an uncomfortable reminder of the insecurities and ambiguity of Japanese racial and cultural identity vis-à-vis an idealized West.

Doubtless, though, it is more complicated than that. As Singer (2002) argues about superhero comics, this genre can 'perpetuate stereotypes, either through token characters who exist purely to signify racial clichés or through a far more subtle system of absence and erasure that serves to obscure minority groups even as the writers pay lip service to diversity. However, [...] these comics also demonstrate that the concepts of double-consciousness and divided identity remain artistically viable techniques for representing race, as valid in the popular culture of today as they were in Du Bois's study nearly one hundred years ago' (p. 118).

The Afro/Ninja split of personality, as well as the samurai/ninja split of warrior codes, could very well symbolize a sort of dichotomy between nostalgic reflex (samurai culture) and a lingering stereotype of the minstrel. The blend of history and contemporary could also represent an advance in understanding, appreciation, and even celebration of the Other writ large as instantiated in the black Other in Japan. This situation is further complicated by the audience: Japanese audiences and US audiences bring varied sociocultural codes and cultural models, and their reading of the story and the characters and the distribution of linguistic codes will certainly differ to some degree.

Kiuchi (2009) argues for the cultural bridge, the nexus of US mainstream, African American, and Japanese cultures embodied by the enka singer Jero, born Jerome White, raised in Pennsylvania and now living and working in Japan. Admitting that 'his Japanese blood may have made it easier for many Japanese to accept [Jero] more readily,' Kiuchi explains that 'he is popular not because of his blackness [but] because he has been able to successfully execute a convergence of his African American and Japanese heritages and a convergence between people of various generations together' (p. 527). This is likewise true for other situations when more than two identities are in play. Chun (2001: 61) finds that some Korean American speakers of English 'negotiate their identities and jointly (re)construct the links between language and social categories' when they use General American English and AAE. In other words, using these linguistic strategies 'allows [a speaker] to project a uniquely Korean American male identity in the context of complex historical, cultural, and political relationships that Korean

American men have with both African Americans and European Americans' (Chun 2001: 53).

As Wilson explains in his (2002) translation of Musashi's *The Book of Five Rings*, 'the Zen Buddhist insistence on absolute personal experience and transcendence of the interfering self' is a foundational principle of the samurai warrior's code of behavior (p. 27). The entire scope of *Afro Samurai* constructs a narrative that embodies the struggle of the warrior who must avenge his father's death (personal experience) while simultaneously paying a high price for this goal (loss of friends, loss of love, extreme asceticism: transcending the interfering self). Likewise, the social codes and the linguistic codes involved in *Afro Samurai* function together to help construct the viewer's sense of the warrior. The tensions between the two types of codes provide and reflect motivations for the narrative and for character choice. In the anime version, the ostensible death of Ninja Ninja, along with Afro's ultimate victory against Justice, leaves the viewer aware of an array of possibilities. One suggests a resolution of conflicts, both external and internal. It suggests an integration of sorts of social codes and linguistic codes into one character, yielding a 'whole' Afro Samurai. Another suggests that the warrior has succeeded in transcending the interfering self (Ninja Ninja, the moral compass, the link to humanity, has died). In *AS–manga*, Ninja Ninja exists as a counterweight to Afro's tendencies toward revenge, even after Afro takes the Number 1 headband and assumes his place as a god in the story.

The blending of social codes and linguistic codes creates a space where stereotypes are employed in complex ways. These stereotypes, these expectations, often result from ideologies, social and linguistic alike. As Chun (2001: 62–3) argues, '[it] is through the mechanisms of everyday talk that the dominance of whiteness is both maintained and resisted. And it is through the racialized imaginings of language enacted in such talk that whiteness is constructed and distinct from – and yet related to – other social identities.' Brown (2006) investigates the attitudes and ideologies of a young black female in an ethnographic study, and he argues that 'the use of African American English [is] an unambiguous act of ethnic identity' (p. 597). Further, Brown paraphrases Coupland and Jaworski (2004: 34) to suggest that his research informant's linguistic production matches a multiplication of linguistic repertoires, which 'might be seen as increasingly accessible to many speakers in the post-modern, post-industrial world' (p. 607). Brown goes on to explain that speakers' awareness of their own linguistic choices suggests that they can 'both reinvigorate and contest dominant [language] ideologies' (p. 607). The supply of linguistic codes wielded by Afro and Ninja Ninja together both maintain and overturn the viewer's sense of language use, manipulating the stereotype for a variety of purposes. As far as comics goes, most representations are 'shorthand' and substitute for something 'real.' It is normal for physical representations to be skewed, so it should not be surprising for linguistic representations to be skewed to some extent as well.

## 8.10 Future directions

The discourse contributions shared between Afro Samurai and Ninja Ninja suggest further research questions. For instance, we need more research on the globalization of comics: how comics artists from different parts of the world are taking what they see from global culture (especially movies, television, and internet media) and rendering a mélange that can speak universal truths. From a linguistic standpoint, we should study the range and depth of African American English in print and web forms, especially as it is used by characters like Rachel Rage, who appears in comic books published by Olde Towne Comix. Further, what is the impact of African American English on readers/viewers of new black comics? As new generations of comics readers grow up, how does their social reality mesh with the stereotypes? In what ways will AAE have a positive impact on readers on the global stage and the production of blackness? (See Beers Fägersten, this volume.)

*Afro Samurai* reveals a rich, complex social and cultural artifact that problematizes what we often consider to be the basic scenario in the US: that white majority culture engages in hegemonic practices in ways that diminish and undermine African American culture and language practices. While this system of oppression continues to be the case in many ways, what we also see is that the social, cultural, artistic, and linguistic contributions made by African Americans to US culture are being celebrated and extended into East Asian cultures, in this case, Japan. While the question may still be about black and white, it now also includes the global. The impact that African American culture and African American English have on the world stage is greater now than ever before.

## Notes

1. An early version of this chapter concentrating solely on the anime version of *Afro Samurai* was presented at the *2008 International Comic Arts Forum*

    in Chicago. I wish to thank two anonymous reviewers at Palgrave for
    encouraging me to extend the analysis and argument to the manga.
 2. Afro seems to live his life according to samurai principles, especially those
    articulated by Musashi in 'The Way of Walking Alone.' In the anime *Afro
    Samurai*, various characters comment on sword fighting principles resonant
    of, especially, 'The Fire Chapter' and 'The Emptiness Chapter' in Musashi's
    *The Book of Five Rings*.
 3. I use the masculine pronouns here purposefully. In *Afro Samurai*, there are no
    female samurais and it is not clear that women can even train for that status.
    In the anime, one female character (Okiku/Otsuru) does become a ninja, and
    although she demonstrates a measure of combat skill, her primary role is espionage. One (unexpected?) outcome of her assignment is that her childhood
    friendship with Afro is rekindled and becomes an adult, romantic love affair,
    consummated sexually just before Okiku dies in battle. While the question of
    racial stereotype in *Afro Samurai* is under discussion here, the stereotyped role
    of gender is, in my opinion, much clearer.

Appendix 8.1 Adult Afro's utterances in AS-*anime*

| (Hr:)Min:Sec* | Utterance | Interlocutor | (Hr:)Min:Sec | Utterance | Interlocutor |
|---|---|---|---|---|---|
| 8:59 | Nothing personal. It's just revenge. | dead warrior(s) | 1:00:13 | Do as you please. | Ninja Ninja |
| 13:00 | Lemonade. Ice cold. | bartender | | | |
| 14:16–14:32 | Oishii.** Refreshing. //*** Thanks. | bartender | 1:06:34 | Shut up. | Ninja Ninja |
| 15:13 | Shut up. | Ninja Ninja | 1:11:51 | Suit yourself. | Ninja Ninja |
| 16:48 | My aim is only to move forward. | dead warrior | 1:30:31 | Jinno? | Jinno |
| 34:31–34:48 | Thanks for everything. // Tomorrow. // Katajikenai. | Okiku/Otsuru | 1:35:16 | The force of the blow is twice as powerful for the double-handed grip. | Ninja Ninja |
| 39:17 | Sorry. | Okiku/Otsuru | 1:43:15 | Hiro. Master. Otsuru. Otsuru. | internal monologue |

| | | | | | |
|---|---|---|---|---|---|
| 40:31–41:11 | I do. // East. | Okiku/Otsuru | 1:49:23 | Finished. | Justice/himself (?) |
| 42:41–43:03 | Get out of my way. Otsuru. // Otsuru! | Okiku/Otsuru | 1:51:48 | Namu. | Prayer to Amida Buddha |
| 48:22 | Shut up. | Ninja Ninja | 1:53:14–1:53:42 | Jinno. // | Jinno |
| 53:02 | I'll see you then. | Ninja Ninja | | You will only die again my friend. // | |
| 54:10 | Shut up. | Ninja Ninja | | Alright. Let's do this. | |

\* All times are taken from a Sony DVD display in hour, minutes, and seconds; e.g. 66 minutes 34 seconds is represented as 1:06:34. The number in the left-hand column indicates the onset of Afro's turn. A time range is given when Afro makes at least two linguistic contributions to a conversation; the second number in the range indicates the end of his final turn, ignoring turns from other speakers which follow in those conversations.
\*\* My thanks to Ed McNamara and Reiko Take Lakouta for giving me insight into the uses of Japanese language in *Afro Samurai*. Any errors in interpretation are my own.
\*\*\* The place holder // indicates that other speakers' turns are omitted and suggests that a (variable) amount of time passes between Afro's turns. In any given conversation with Ninja Ninja, Afro takes one and only one linguistic turn; thus, only multiturn conversations with other characters use the place holder symbol, specifically, the bartender, Otsuru/Okiku, and Jinno.

Appendix 8.2  Adult Afro's utterances* in *AS–manga*, Volumes 1 and 2

| Vol.Pg | [Panel] | Utterance | Interlocutor | Vol.Pg | [Panel] | Utterance | Interlocutor |
|---|---|---|---|---|---|---|---|
| 1.30 | {1} | It's just revenge. | dying warrior | 2.28 | {3} | 'Thanks for puttin' me up. | O-sachi |
| 1.33 | {4} | Which way? | Ninja Ninja | 2.28 | {7} | No. | O-sachi |
| 1.34 | {1} | Where's No. 1? | Ninja Ninja | 2.29 | {3} | Stand down. | O-sachi |
| 1.50 | {2} | No. | Shichigoro | 2.29 | {5} | Stop! | O-sachi |
| 1.53 | {4} | You're awfully knowledgeable. How do you know all this? | Oi-chan | 2.85 | {2} | Brother Jin. | Jinno |
| 1.97 | {6} | Mt. Sumeru… | Ninja Ninja | 2.95 | {7} | Goodbye, Brother Jin… | Jinno |
| 1.102 | {5} | Mt. Sumeru… | Ninja Ninja | 2.105 | {6} | No. 1.… | Ninja/himself |
| 1.130 | {2} | Shit! | Himself? | 2.113 | {6} | I won't do either. | Takimoto Kougansai |
| 2.9 | {6} | Oh man… | O-sachi/himself | 2.114 | {1} | I'll keep fighting, until I find him. | Takimoto Kougansai |
| 2.12 | {4} | Oh man. | Doctor/himself | 2.115 | {5} | Father? | Afrodroid |
| 2.25 | {2} | Yeah. | O-sachi | 2.154 | {6} | Do as you like. | warrior |
| 2.25 | {8} | Yeah. | O-sachi | 2.155 | {1} | I … if you want to kill me, kill me. | warrior |
| 2.28 | {2} | A long time ago this man sliced my father's head off right in front of me. | O-sachi | 2.162 | {1} | I'll just keep moving forward. | Ninja Ninja and Takimoto Kougansai |

* Several of Afro's utterances were nonlinguistic productions written in speech balloons, like '!!' or 'Grr!' Included in this table are only those utterances that unequivocally count as words.

*Appendix 8.3* Excerpts from Ninja Ninja's utterances in AS-*enime*

| (Hr:)Min:Sec* | Utterance | Interlocutor |
|---|---|---|
| 14:36–15:20<br><br>Afro and Ninja Ninja are walking on the Low Down East Path | Hey Afro. Yo. Pro Boy. You sure this the right way to go? These cliffs don't look so safe man. And one other thing. Uh don't look now but I think you got a stalker. A crazy fan boy from the tea house. And uh he don't want your autograph more like your epitaph. Now I suggest you keep stepping because this one is a stone killer. For real. Yo. Didn't you hear me? Get going. I'm trying to pull your coat. Oh. All you gonna do is ignore me? Alright uh alright. You want to blow it for me and you huh?<br>//**<br>Well fine. Go on and get your lanky ass killed then. But don't say I ain't warned you. Later! | Afro |
| 39:37–40:18<br><br>Afro and Ninja Ninja sit on the porch discussing Otsuru. | Y yo yo yo yo Afro. Can I ask you something? Now I know you cast away your feelings and all but does that mean casting away your manhood too? Oh mamasans don't get much better than old girl. She's sweet. Pretty. Knows her first aid. Can broil a mean fish dinner. I call that practically marriage material. Come on now. Confess. You want to hit that booty. I just want to see my boy happy you know? Seriously man. Now don't you think there's something vaguely familiar about that woman? Look. If you ain't gonna knock them boots I will. | Afro |
| 47:42–48:22<br><br>Afro and Ninja Ninja approach Mt. Shumi in a canoe for the final battles | I do not like the looks of that thing. See that smoke oozing out? It ain't natural. Well what I'm saying is this bitch just one thunder from (her) and she can blow. That's the end of your fro bro. BAM. Ha ha. Hee hee hee. By the way. I- I'm trying to imagine you without a fro. Think you look cute with little braids. [laughs] Hey hey hey. | Afro |

* All times are taken from a Sony DVD display in hour, minutes, and seconds; e.g. 66 minutes 34 seconds is represented as 1:06:34. The first number indicates the onset of Ninja Ninja's first turn in the conversation and the second number indicates the end of his final turn in each conversation.
** The place holder // indicates a (variable) amount of time passes, with Afro remaining silent or producing a short utterance.

*Appendix 8.4* Excerpts from Ninja Ninja's utterances in *AS-manga*, Volumes 1 and 2

| Vol.Pg | {panel} | Utterance | Interlocutor |
|---|---|---|---|
| 1.33 | {2} | Heyyyyy! | Afro |
|  |  | >* |  |
| Ninja advises Afro after a battle. | {3} | Yo! |  |
|  |  | > |  |
|  | {5} | Damn, man, you never change, do ya? You mean you can't even give me a simple friendly greeting after all this time? |  |
|  |  | > |  |
|  | {6} | Y'know you need me if you ever gon' make it to where No. 1's hiding on Mt. Sumeru. How 'bout showing a little 'ppreciation? |  |
| 1.34 | {2} | Damn, man. Fine, then… | Afro |
|  |  | > |  |
|  | {3} | Uh… that way… maybe? |  |
|  |  | > |  |
|  | {4} | Well how 'bout that? Not a single word of thanks, as usual. |  |
|  |  | > |  |
|  | {5} | I don't know why I even bother. |  |
|  |  | > |  |
|  | {6} | Damn… |  |
| 1.97 | {1} | 'Eyyy. Yo. No. 2! Over here. | Afro |
|  |  | > |  |
| Ninja advises Afro after a battle. | {3} | Yo! It's been ages! |  |
|  |  | > |  |
|  | {4} | Yo yo! I was worried you might've been caught in that big-ass explosion. Damn! Didn't want to be waitin' around here for nothing. Heh heh heh. |  |
|  |  | > |  |

|  |  |  |  |
|---|---|---|---|
|  | [5] | Yeah, man. Aight, look, we gotta go by boat from here. If you just go up this river for a few days, you'll get to a ravine. It's right by Mt. Sumeru, where No. 1 is. Finally comin' to the end of your journey, m'man. > |  |
|  | [7] | Yeah. But ... |  |
| 1.98 | [1] | To get to Mt. Sumeru, first you gotta pass through Tecchisen, the hideout for those damn old motherfuckers. And there's no doubt they've set up somethin' nasty for you to fall into. > | Afro |
|  | [2] | No. 2 you know you've only survived this long on luck. But that ain't gonna do you much good from here on out. You sure you wanna keep going? > |  |
|  | [5] | Heh heh heh ... So that's how it's gonna be, huh? Well, don't say I din't warn ya. |  |
| 2.135 | [6] | That's it man. Keep it goin'. > | Afro |
| Ninja observes Afro in battle and talks to another character for the first time. | [7] | Huh?! Who ... You mean me?! > | Takimoto Kougansai |
| 2.136 | [1] | Ohoho! Man, you've made a guy happy! So you can see me! Heh heh heh. > | Takimoto Kougansai |
|  | [2] | But I don't suppose who I am is all that important. What's more important right now is that you can see me. > |  |
|  | [3] | You're no normal man, is ya? Heheheh. > |  |
|  | [6] | O ho ho ... Nice, man! I think we might dig each other. |  |

\* The angle bracket > indicates that a shift of at least one panel has occurred and that a (variable) amount of time passes between Ninja's utterances.

## References

*Afro Samurai*. (2006) Created by Takashi Okazaki. Directed by Fuminori Kizaki. Starring Samuel L. Jackson, Kelly Hu, and Ron Perlman. Studio Gonzo, Bushwazee Films, Samurai Project, Spike TV, and Funimation.
Brown, D. W. (2006) Girls and guys, ghetto and bougie: metapragmatics, ideology and the management of social identities. *Journal of Sociolinguistics* 10 (5): 596–610.
Bucholtz, M. (2003) Sociolinguistic nostalgia and the authentication of identity. *Journal of Sociolinguistics* 7 (3): 398–416.
Bucholtz, M. and Hall, K. (2005) Identity and interaction: a sociocultural linguistic approach. *Discourse & Society* 7 (4–5): 585–614.
Chaney, M. A. (2004) Coloring whiteness and blackvoice minstrelsy: representations of race and place in *Static Shock*, *King of the Hill*, and *South Park*. *Journal of Popular Film and Television* 31 (4): 167–75.
Chun, E. (2001) The construction of White, Black, and Korean American identities through African American Vernacular English. *Journal of Linguistic Anthropology* 11 (1): 52–64.
Coupland, N. and Jaworski, A. (2004) Sociolinguistic perspectives on metalanguage: reflexivity, evaluation and ideology. In A. Jaworski, N. Coupland, and D. Galasinski (eds.) *Metalanguage: Social and Ideological Perspectives*, pp. 15–51. New York: Mouton de Gruyter.
Davenport, C. (1997) Black is the color of my comic book character: an examination of ethnic stereotypes. *Inks: Cartoon and Comic Art Studies* 4 (1): 20–28.
Eisner, W. (1985) *Comics & Sequential Art*. Expanded edition: print and computer. Tamarac, FL: Poorhouse Press.
Finegan, E. (2004) American English and its distinctiveness. In E. Finegan and J. Rickford (eds.) *Language in the USA: Themes for the Twenty-first Century*, pp. 18–38. Cambridge: Cambridge University Press.
Foster, W. H., III. (2002) The image of Blacks (African Americans) in underground comix: new liberal agenda or same racist stereotypes? *International Journal of Comic Art* 4 (2): 168–85.
Green, L. (2002) *African American English: A Linguistic Introduction*. Cambridge: Cambridge University Press.
Harvey, R. C. (1996) *The Art of the Comic Book: An Aesthetic History*. Jackson: University Press of Mississippi.
Jones, G. (1999) *Strange Talk: The Politics of Dialect Literature in the Gilded Age*. Berkeley and Los Angeles: University of California Press.
Kiuchi, Y. (2009) An alternative African American image in Japan: Jero as the cross-generational bridge between Japan and the United States. *The Journal of Popular Culture* 42 (3): 515–29.
Lhamon, W. T., Jr. (1998) *Raising Cain: Blackface Performance from Jim Crow to Hip Hop*. Cambridge, MA: Harvard University Press.
Morgan, M. (1999) More than a mood or an attitude: discourse and verbal genres in African American culture. In S. Mufwene, J. R. Rickford, G. Bailey, and J. Baugh (eds.) *African-American English: Structure, History and Use* 251–281. London and New York: Routledge.
Morgan, M. (2002) *Language, Discourse and Power in African American Culture*. Cambridge: Cambridge University Press.

Musashi, M. (2002) *The Book of Five Rings*. [Trans. William Scott Wilson.] Tokyo: Kodansha International Ltd.
Okazaki, T. (2008) *Afro Samurai*. Vols. 1 and 2. [Trans. Greg Moore.] New York: Tor/Seven Seas.
Peterson, J. (2004) Linguistic identity and community in American literature. In E. Finegan and J. Rickford (eds.) *Language in the USA: Themes for the Twenty-first Century*, pp. 430–44. Cambridge: Cambridge University Press.
Priest, C. (2001) Introduction. *Black Panther: The Client*, Vol. 2. New York: Marvel Comics (not paginated).
Royal, D. P. (2007) Introduction: coloring America: multiethnic engagements with graphic narrative. *MELUS* 32 (3): 7–22.
Russell, J. (1991) Race and reflexivity: the black other in contemporary Japanese mass culture. *Cultural Anthropology* 6 (1): 3–25.
Scott, A. B. (2006) Superpower vs supernatural: black superheroes and the quest for a mutant reality. *Journal of Visual Culture* 5 (3): 295–314.
Singer, M. (2002) Black skins and white masks: comic books and the secret of race. *African American Review* 36 (1): 107–19.
Smitherman, G. (1977) *Talkin and Testifyin: The Language of Black America*. Detroit: Wayne State University Press.
Spears, A. K. (1999) African American language use: ideology and so-called obscenity. In S. Mufwene, J. R. Rickford, G. Bailey, and J. Baugh (eds.) *African-American English: Structure, History and Use*, pp. 226–50. London and New York: Routledge.
Strömberg, F. (2003) *Black Images in the Comics: A Visual History*. Seattle: Fantagraphics Books.
Wang, O. (2006) These are the breaks: hip-hop and AfroAsian cultural (dis)connections. In H. Raphael-Hernandez and S. Steen (eds.) *AfroAsian Encounters*, pp. 146–64. New York: New York University Press.
Whaley, D. (2006) Black bodies/yellow masks: the orientalist aesthetic in hip-hop and Black visual culture. In H. Raphael-Hernandez and S. Steen (eds.) *AfroAsian Encounters*, pp.188–202. New York: New York University Press.
Wilson, W.S. (2002) Introduction. In M. Musashi, *The Book of Five Rings*, pp. 14–31. Tokyo: Kodansha.
Wood, J. (2006) The yellow Negro. In H. M. McFerson (ed.) *Blacks and Asians: Crossings, Conflict and Commonality*, pp. 463–83. Durham, NC: Carolina Academic Press.

# 9
# Pocho Politics: Language, Identity, and Discourse in Lalo Alcaraz's *La Cucaracha*

Carla Breidenbach

## 9.1 Introduction

In *Your Brain on Latino Comics*, Aldama (2009: 105–06) states: 'In reading the signposts laid out by Latino comic book and comic strip author-artists, we are cued to reexperience or reconstruct our core selves in complex and specifically directed ways: ways that direct us to realize a fuller experience of U.S. ethnicity – specifically, Latino and Latina identity.' For Gee (1999), Discourses (with a capital 'D') are more than just language; they are social practices influenced by cultural and social models surrounding us. He argues that Discourses are 'identity kits' incorporating 'specific devices (i.e., ways with words, deeds, thoughts, values, actions, interactions, objects, tools, and technologies) in terms of which you can enact a specific identity and engage in specific activities associated with that identity' (2001:719–20). Comic strip and political cartoon artist Lalo Alcaraz, a self-proclaimed 'pocho' Spanglish-speaking Chicano, uses his comic strip *La Cucaracha* to enact his Chicano identity and political ideologies from within a post-Chicano (Wegner 2007) Discourse, the combination I name 'pocho politics.'

An analysis of 163 cartoon strips collected between September 2007 and May 2010 shows language choice and language ideologies as two features intertwined throughout seven major themes (Latino holidays and culture, immigration, language issues, political pundits, presidential politics, social issues, and self-hate mail), all of which are the target of Alcaraz's political commentary and social criticism. Close linguistic analysis reveals four important facets identified in the comic strip *La Cucaracha*: Mock Spanish (Hill 1998, 2008), fake Spanish accents, code switching, and language ideology. The comic strip is written mostly in English, peppered with some Spanish words. Not simply used to reach a wider audience, English is also the dominant language for many Chicanos (Fought 2003). Spanish indicates a linguistic landscape, situating the reader in a Chicano barrio in East Los Angeles. Code switching also appears. This mixture of English and Spanish has often been referred to by the derogatory terms 'pocho' or 'Spanglish' (Fought 2003; Lipski 2008). In Alcaraz's view, however, 'pocho' and 'Spanglish' are terms to be celebrated and reclaimed, reminders of a heritage of which Chicanos should be proud, not ashamed.

Alcaraz uses political discourse, narrative and rhetorical techniques, and the linguistic codes of Spanish and English to make his mark as a Chicano comic strip artist. This chapter examines how language choice (Spanish and English) and language ideologies are used in *La Cucaracha* to enable him to enact his specific identity as a socio-political Chicano, his own special brand of 'pocho politics,' situated from within a post-Chicano Discourse.

## 9.2 Constituting identity in *La Cucaracha*

Alcaraz uses text and image interaction as well as only image in his comic strips and single panel cartoons. In his discussion of Alcaraz's comic strip from 2002 to 2003, Wegner (2007) excludes analysis of image alone: because 'of the verbal humor involved, the pictures serve to identify speakers and time comedic revelation' (Harvey 2002: 80, quoted in Wegner 2007: 232). Since some of Alcaraz's most politically important recent cartoons are solely image, some of the comic strips in this chapter are analyzed as a 'mode of visual art' (Wegner 2007: 232). In the present analysis of *La Cucaracha*, 33 of the 163 cartoons (20%) were primarily images lacking words or simple captions. A typical comic strip is usually three to four frames in black and white during the week and, on Sundays, they stand alone to 'crank up his attacks' (Wegner 2007: 238). Since Wegner's analysis, stand-alone editorial cartoons like 'LAPD' (Alcaraz 2007), 'Viva Obama' (Alcaraz 2008), 'Little Judge Lopez' (Alcaraz 2009), and 'Stop Arizona SB 1070' (Alcaraz 2010) have garnered national attention and play an important role in this analysis.

The cartoon can constitute a 'graphic jibe,' and cartoonists are artists who use their ability to draw 'as a means of making statements, usually of a somewhat derisory nature, about the absurdities and incongruities (real or imagined) of human behavior' (Geipel 1972:20). In *La Cucaracha*, Alcaraz uses humor as suggested by Rybacki and Rybacki (1991) 'to expose social problems, to confront societal taboos, and to vent frustration' (Shultz and Germeroth 1998: 230). As Boskin (1979) (paraphrased by Shultz and Germeroth 1998: 230) contends, humor 'simultaneously is a form of social control that allows society to regress and a form of cultural release that allows a society to aggress.' More specifically, according to Marín-Arrese (2008: 9), political humor takes aim at social groups, institutions, and other power holders, often to undermine the 'legitimacy' of those institutions. Indeed, the stance that Alcaraz takes is one of an attacker, and his comic strips and cartoons are, as Geipel suggests for cartoons in general, 'an aggressive medium, an offensive weapon whose effect can be devastating' (Geipel 1972: 21). As Marín-Arrese states: 'The aim of a political cartoon is to launch an attack against a specific target by profiling specific actions and/or characteristics of the targets and depicting them as politically incompetent and/or morally wrong' (2008: 9). Not only is this especially relevant in Chicano satire (Wegner 2007), this also places Alcaraz within the long-standing tradition of Mexican political satirists like late nineteenth-century political cartoonist/satirist José Guadalupe Posada, who paid great attention to social problems, and more recently, Abel Quezada, who before his death in 1991 was a leading Mexican political cartoonist (Alba 1967). In the US, Alcaraz's work holds similarities to Garry Trudeau, creator of *Doonesbury*. In fact, Alcaraz has jokingly called his strip 'Doonesbarrio' on occasion.[1] Like Trudeau, Alcaraz 'draws truth' (Newton 2007: 81), that is, he draws what he knows; he speaks from his own experience as a Chicano situated in a post-Chicano Discourse.

## 9.3 A post-Chicano Discourse

Gee (1999, 2001) distinguishes between Discourse (with a big 'D'), which involves more than just language, and discourse (little 'd') which pertains to just language in use (i.e., conversations, strings of speech). Gee (2001) writes that '[a] Discourse integrates ways of talking, listening, writing, reading, acting, interacting, believing, valuing, and feeling (and using various objects, symbols, images, tools, and technologies) in the service of enacting meaningful socially situated identities and activities' (p. 719). Social and cultural experiences make up identities and determine how individuals act in society. These cultural models 'inform the social practices in which people in a Discourse engage' (Gee 2001: 720). Due to different experiences, people may or may not relate to another's Discourse. In order to be able to relate to something, one must find it relevant. Some scholars refer to this phenomenon through relevance theory based on the work of Sperber and Wilson (Yus Ramos 1998, Forceville 2005), while other scholars prefer the term 'multiliteracies' (El Refaie 2009).

In *La Cucaracha*, the ingredients for the cultural models constituting Alcaraz's pocho politics are found in what Wegner (2007) describes as 'the post-Chicano.' Wegner situates Alcaraz in what he calls the post-Chicano period which follows the civil rights and Chicano movements:

> I use the term *post*-Chicano to describe the political, social, and cultural experience of the Mexican American community since the end of the civil rights and Chicano movements. This aesthetic has largely been developed by artists and critical thinkers, like Lalo Alcaraz, who were born after 1960; those who came to maturity in the age of Reaganomics, or later, and experienced the change from urban industrialization to deindustrialization, from essential notions of Chicano-ness to meta-narratives of Chicano-ness, without any nostalgic allegiance to the past but with an understanding of the harsh postmodern realities of the present. (Wegner 2007: 231)

Alcaraz has also situated himself within the post-Chicano period by saying, 'I don't claim I'm doing the classic Chicano art, but I'm doing it in the spirit of the Chicano art of the 60's and 70's' (Basu 2002). Born in San Diego, California, in 1964 to Mexican immigrant parents, Alcaraz (Eduardo López) was aware of and active in establishing his identity as a Chicano from an early age by working with Chicano artists who showed him 'how to combine politics and art' (Basu 2002). As a child of Mexican immigrant parents, he experienced the feeling of not being 'Mexican' enough or not being 'American' enough (Wegner 2007: 233), and he also experienced language troubles. Through his work, Alcaraz has continued to establish his identity as a figure from the Chicano community with which to be reckoned, early on co-founding the satirical magazine *POCHO* and touring with the comedy troupe Chicano Secret Service, and most recently as a guest speaker/artist, and host of the radio show *Pocho Hour of Power*.[2] These are cultural models that inform Alcaraz's post-Chicano Discourse.

## 9.4 Lalo Alcaraz as a Latino comic strip author-artist

Aldama (2009) describes Latino visual and written works, comic strips included, 'as part of a Latin American (and Third world) tradition of unmaking of dominant systems of representation' (2009: 12). The choice to 'represent and reframe all aspects of our everyday world' is a common thread in the works of Latino comic book and comic strip author-artists (2009: 17). According to Aldama, 'In choosing to reframe real and fictional worlds, they also ask the reader-viewer to look upon this newly reframed object from a new angle of vision; in so doing, they amplify the reader-viewer's everyday cognitive and emotive activities' (2009: 83). This is an explicit process of guidance and education for the reader, one which leaves virtually 'no guesswork' (Aldama 2009: 97) as to what the author's vision is. This is not to say that all readers will pick up on the messages or interactions between text and image, nor will a work appeal to all readers (Aldama 2009: 103). Nevertheless, Aldama points out that though authors might guide them in a particular way, it is the readers who have the last say.

In contrast to other Latino comic book and strip author-artists, like Gus Arriola, creator of *Gordo*, and Hector Cantú and Carlos Castellanos, creators of *Baldo*, Alcaraz revels in taking on topics of political controversy. Aldama (2009: 25) includes Alcaraz as an author-artist who 'chooses social satire as his preferred form for conveying values' and whose topics often include issues such as bilingualism and taking 'shots' at English-only proponents. At the same time Alcaraz 'complicates the representational map' by poking fun at Latinos, thereby drawing criticism from some members of the Latino community (p. 26).

## 9.5 Chicano

Helpful to truly understanding the Discourse of *La Cucaracha* is familiarity with the term *Chicano* and what it means to be *Chicano* in the US. *Chicano* is viewed differently according to people's ideologies shaped by their social and cultural models. Thus, *Chicano* has many definitions and appeared widely in the media during the 1960s and 1970s. While the origin and meaning of *Chicano* is 'elusive' (Stephenson 1969), the simplest definition is a person of Mexican American heritage born in the United States (Fought 2003). Fought (2003: 38–40) has identified several specific and intertwined conflicts and oppositions within the US Mexican American community regarding the term *Chicano*. First, it signifies a division between people born into the community and immigrants. In Fought's interviews, one informant mentioned that he approved of code switching because it distinguished Chicanos from Mexicans born in Mexico (2003: 40). Second, Chicano identity may signify a degree of assimilation whereby conventional success equals the adoption of habits of Anglo culture, dress, speech, etc., which often conflicts with ethnic pride and stokes fears of selling out. Believing Chicano to be equivalent to assimilation, another of Fought's informants preferred the term 'Mexicana' (2003: 41). *Chicano* is also used to distinguish between gang members and non-gang members. Fought speculates that 'part of the resentment of immigrants toward [US-born] Chicanos may be due to the fact that most gang members in the area are [US-born], so that a negative stereotype of native born Latinos is perpetuated among the immigrants' (2003: 41). *Chicano* has also been linked to images of competition for job resources and members of a depressed underclass and desperate immigrants (Vigil 1938, cited by Fought 2003: 41–42).

In Mexico, *Chicano* is usually perceived to be a derogatory term (Lipski 2008). In the United States, for many, usually those with conservative ideologies (including many Mexican Americans, such as Eddie's father in *La Cucaracha*) and those whose beliefs about Chicanos are influenced by negative and stereotypical portrayals in the media, *Chicano* is a word loaded with connotations of lower-class, lazy, trouble-making, Spanglish-speaking, *pocho* (caught between two cultures, the American and the Mexican, but belonging to neither or a traitor to one), *pachuco* (a Chicano zoot-suiter from the 1940s), *cholo* (gang-bangers). Mexican Americans who reject this term do so to distance themselves from what they believe to be a poor image. For others, like Alcaraz, who tend to be more liberal in their political views and/or participated in or have knowledge of the Chicano movement, *Chicano* usually carries with it a connotation of political awareness, referring to the ideologies and fights for civil rights of the Chicano movement in the 1960s and 1970s.

## 9.6 Language ideologies

Ideologies, based on cultural and social models, are an important part of Discourse. For Hill (2008: 34), an ideology 'suggests a way of thinking or a perspective saturated with political and economic interest.' In Gal and Irvine's (2000) discussion of language ideologies, ideological aspects are 'the ideas with which participants and observers frame their understanding of linguistic varieties and map those understandings onto people, events, and activities that are significant to them.' They call these 'conceptual schemes ideologies' because 'they are suffused with the political and moral issues pervading the particular sociolinguistic field and are subject to the interests of their bearers' social position' (Gal and Irvine 2000: 35). Hill (2008: 33) believes that 'linguistic ideologies shape and constrain discourse (talk and text), and thus shape and constrain the reproduction of other kinds of ideologies, such as ideologies of gender, race, and class.' According to Hill, 'Linguistic ideologies persist not only because they have a certain internal coherence, and because they resonate with other cultural ideas, but because they support and reassert the interests of many (but not all) of those who share them' (2008: 34).

Ideologies are often not available to people's awareness and reflection. People rarely question the ideologies with which they have been raised and which have been instilled in them by various state institutions, especially when it comes to language. Bourdieu (1977) attributes this to people being in a state of *doxa*, of unquestioning acceptance of the social order as an order of nature. What makes Alcaraz's work so powerful is that he does question the *doxa*, the status quo, and infuses the comic strip with his own ideology of pocho politics. He lampoons the hypocrisies and corruptions of not only the Anglo society but also the Latino and, more specifically, the Chicano (Mexican American) communities. This is what makes people uncomfortable. According to Shultz and Germeroth (1998), it has been demonstrated that people from the United States are 'notoriously uncomfortable with argument and divisiveness.' They feel as if they are being attacked. Readers even question whether his cartoons should be placed with the cartoons, as their sharp political edge would indicate that they are editorial in nature.[3] At times, newspapers (*Los Angeles Times*, *Houston Chronicle*) have pulled *La Cucaracha* from circulation due to reader complaints.

## 9.7 Language choice and identity

As Aldama (2009: 48) states in his discussion of Latino comic book author-artists and cultural diversity, 'Diversity is not just skin deep; it is also how one self-identifies and how one uses language to announce group affiliation.' In *La Cucaracha*, language is a huge part of Alcaraz's Chicano identity. This section identifies the theoretical linguistic concepts which are the base of the language issues analyzed in *La Cucaracha*: Chicano English, pocho, Spanglish, code switching, and Mock Spanish.

### 9.7.1 Chicano English

Though Alcaraz's strips are written mostly in English, the primary language of communication between the characters, it is not clear whether the English dialect being spoken is Chicano English or another variety of English. Fought (2003: 1) defines Chicano English as 'a nonstandard variety of English, influenced by contact with Spanish, and spoken as a native dialect by both and bilingual and monolingual speakers.' Chicano English, Fought reminds us, is a 'cultural marker, a reminder of linguistic history, and a fertile field for the study of language contact phenomena and linguistic identity issues' (2003: 2). A continuum exists among speakers, ranging from monolingual in Chicano English to being bidialectal in Standard and Chicano English to completely bilingual in English and Spanish (Fought 2003: 18). Chicano English may be marked by certain syntactic (negative concord, habitual 'be,' nonstandard verb forms), semantic (*like, all, barely*), and phonetic features ([ɪ]/[i], intonation) (Fought 2003). Chicano English speakers may exhibit features of one without the other (e.g., phonetic features but no syntactic features). Its users may or may not use or know Spanish. In her field notes, Fought makes the following observation about Los Angeles:

> Spanish language and Latino culture permeate the environment. It is impossible to live and work there without at least occasionally hearing Spanish spoken. Many of my speakers are bilingual, speaking Spanish in varying degrees and fluent English. For most of them, the dialect of English they speak is Chicano English. (2003: 18)

Because the comic strip takes place in East Los Angeles, a predominately Chicano community, and all of the main characters are Chicano, it is not unreasonable to guess that the English being spoken may be a variety of Chicano English, especially by Cuco since he so strongly identifies as a Chicano activist.[4] It is also possible that they are speaking a more mainstream, broader California English dialect.

### 9.7.2 Pocho

Field (1994) formally calls the mixed Spanish and English language 'Pocho,' and provides a very detailed analysis and portrait of speakers whose varying degrees of linguistic abilities inform their use of Spanish or English. What results is the creation of a new language variety and identity. Instead of looking at *pocho* as a positive term like Field and Alcaraz, there are many negative images about *pocho* as a term to describe both the speaker and his language. Under these circumstances, pocho is someone who has rejected or forgotten their heritage, including their language, resulting in the 'jumble' of Spanish and/or English, proficient at neither, barely comprehensible in both. Often, if one language becomes dominant, it is English. However, this view does not recognize that, for Chicanos who have lived for generations in the United States, as is the case in much of the Southwest, speaking mainly English is part of a natural process of language acquisition and shift among generations. The further a speaker is from belonging to the first generation, the more English is used (Lipski 2008). However, as English replaces Spanish in many arenas like school, work, and friendship, Spanish can play a role in the home language. Thus, even if Chicanos know very little Spanish, they will recognize familiar household terms, terms of endearment, family relations, names of people and places (Torres 2007), or 'the linguistic landscape' (Ben-Rafael 2008), and probably some expletives. In that respect, *La Cucaracha* reflects the Chicano norm as these are arenas and contexts where Spanish appears in the strip. The term *pocho* is used by Alcaraz with pride to celebrate the Chicano culture and language which has surged and flourished due to the mix of both cultures.

### 9.7.3 Spanglish

*Spanglish* is another term commonly attributed to the mix of English and Spanish, though, while *pocho* can refer to language or identity, *Spanglish* is used solely to discuss the language hybrid which can take the form of code switching, word blends, and creation of new words. It is spoken mostly by Latinos born or living in the United States. Critics of Spanglish (e.g. Octavio Paz) see it as a destruction of the purity of the Spanish language and a sign that speakers of Spanglish may not be competent in either language. Lipski (2008) writes that Spanglish is usually not associated with whiter backgrounds like the Southern Cone countries (Chile or Argentina) or Spain, 'thus suggesting an element of racism coupled with xenophobia that deplores any sort of linguistic and cultural hybridity' (p. 39). However, 'some U.S. Latino political and social activists have even adopted Spanglish as a positive affirmation of ethnolinguistic identity' (pp. 38–39). Ilan Stavans, professor of Latin American and Latino culture who has collaborated with Alcaraz says: 'Spanglish is proof that Latinos have a culture that is made up of two parts. It's not that you are Latino or American. You live on the hyphen, in between. That's what Spanglish is all about, a middle ground.'[5] Alcaraz is a self-identified Spanglish speaker: 'We don't live neatly in two worlds. I teach my kids Spanish, yet my wife and I speak English to each other. Spanglish is its own unique point of view. It's more of an empowering thing to us, to say we have a legitimate culture.' In Alcaraz's comic strip, Spanglish appears mostly in the form of code switching.

### 9.7.4 Code switching

Code switching is the use of more than one language in one speech situation. Rather than being a random mix of two languages, in this case English and Spanish, code switching is rule-governed and has very specific uses. It is either marked (situation/arena is not expected for a certain language choice) or unmarked (expected, predictable for the speakers) (Myers-Scotton 1988). Code switching can be thought of as a result of stable bilingualism, where there exists a main and embedded language (Myers-Scotton 1988) or the creation of 'an entirely new, mixed language' of which there is a continuum (Field 1994: 87). There are linguistic (morphological, syntactic) constraints on possible, acceptable code switching (Timm 1975; Smith 2007). Code switching is used to express identity, solidarity, and emotion (Romero 1988). It can also show cultural aspects (Torres 2007), when there is simply not the word or right phrasing for the concept, called emblematic switches or gaps (Jonsson 2010: 1304). In her interviews with California Mexican Americans, Fought's informants viewed code switching positively because it serves as an identity marker for Chicanos, not Mexican immigrants (2003: 209). She found code switching occurred fairly rarely when the interviews were conducted in English (usually only emblematic switches), but occurred more frequently when the interviews were conducted in Spanish. Fought mentions the possibility that English may be the higher prestige variety as well as an expression of assimilation into mainstream Anglo society. For many speakers in Fought's study, the importance of Spanish was symbolic of pride in their heritage. To provide a contrast between Anglo and Latino (Chicano) culture and to highlight ethnic identity, Alcaraz uses emblematic code switching from English to Spanish.

### 9.7.5 Mock Spanish

*Mock Spanish* functions as a highly stylized and specific type of pseudo-Spanish, and according to Hill (1998, 2008) the term refers to a form of subtle racism in American English, used by Anglos to degrade Latinos. Hill attributes this to an air of cultural conformity, where white people are not comfortable when public figures 'exhibit styles and expressions that are distant from White norms' (2008: 23). In Mock Spanish, the imitation of certain authentic Spanish features is used in controlled ways by the speaker to 'represent' the whole Spanish language through mimicry and parody. In doing so, they indirectly index racist stereotypes of Latinos (e.g., lazy, corrupt, always at a fiesta, hyper-sexual, dumb) (Hill 1998, 2008). Mock Spanish is found in a very informal register and is generally used for the particular purpose of joking in informal social settings or even very public ones. Hill uses the example of 'El Rushbo,' the nickname Rush Limbaugh, right-wing talk show host, has given to himself in order to present an image of a likeable jokester. Hill

(2008: 147) writes that 'Mock Spanish associates the Spanish language irrevocably with the non-serious, the casual, the laid-back, the humorous, the vulgar,' 'joking' and 'insult,' not 'gravitas or sophistication.' It also relegates native Spanish fluency to the realm of the 'un-American' where 'to pronounce Spanish place names or the names of public figures' risks being 'stuffy, effete, p.c., even ridiculous' (2008: 148).

When speakers use Mock Spanish, Hill (2008) calls this 'linguistic appropriation,' meaning that 'speakers of the target language (the group doing the borrowing) adopt resources from the donor language, and then try to deny these to members of the donor language community.' Speakers of Mock Spanish impart their own (primarily negative) meanings to the 'borrowed' Spanish words for their own benefits (laughter, advertisement, propaganda), which often excludes or makes fun of the Spanish-speaking population. Hill (1998: 62) identifies four major strategies for the 'incorporation' of Spanish-language materials into English. These strategies represent a pragmatic continuum which ranges from 'merely jocular' at one end and 'obscene insult' at the other:

(1) *semantic derogation*, the borrowing of neutral or positive Spanish words that function in Mock Spanish in a jocular and/or pejorative sense, e.g., *Adios cucaracha!*, an advertisement for a bug extermination company on a bus bench in an upscale neighborhood where 'adios' is a kiss-off instead of a friendly 'goodbye' (elaborated in Hill 2008: 135);
(2) *euphemism*, e.g., *Casa de Pee Pee* for restroom;
(3) *affixation*, e.g., consisting of *el* + word + *-o*, like *el cheapo*; and
(4) *hyperanglicization*, using a broad American English accent to pronounce Spanish words.

A fifth overt strategy, not considered Mock Spanish by Hill, but in line with Chun's (2004) discussion of Mock Asian, is the exaggerated and contrived imitation of a Spanish speaker's accent (e.g., 'Somtheeng deeferent').

Not all scholars agree that 'mocking' can go only in one direction, that is, be used only by Anglos to exert their power over minority groups (Zentella 2003; Chun 2004; Breidenbach 2007; Barrett 2006). Multiple strategies, such as linguistic reappropriation (re-direction of negative language use), can be used to 'resist hegemonic and racist notions of language' (Zentella 2003: 55). Mock Spanish functions as a form of resistance and sometimes self-deprecation by using stereotypes of Latinos, the stereotypical accent and the affixation of *el* + word + *-o* (e.g., *El Jerko, Drinko de Cinco*).

Alcaraz's strategy of resistance includes using stereotypes typically considered part of Mock Spanish and inverting them to shed an ironic light on them. He often does this when his strip is dedicated to answering 'self-hate mail' from Latino readers. His 'self-hate mailers' complain of his use of stereotypes, which they believe to be demeaning to the Mexican American community, while often spouting typical stereotypical nonsense or circular logic themselves. Most Mock Spanish images represented by the media and society are based on stereotypes. Hall (1997) makes an excellent point when he says that stereotyping fixes the meanings that we give to groups. We are bombarded with a limited range of characteristics that are supposed to inform us, on the basis of racial categories, about what people can do, how they act, how they dress, what they eat, etc. In short, a stereotype 'represents' a culture. Contesting a stereotype means increasing the diversity of images in the media.

Unfortunately, the images of Latinos presented by the media are not diverse and stem from a long history of turbulence between Chicanos and mainstream Anglo society since at least the nineteenth century (De Leon 1983). These media images of Latinos often give a perception of a threat to stability and unity of the country due to the 'roles' which Latinos fill in the media. What is presented by the media are the sensational, attention-grabbing headlines of gang-banging, illiteracy, illegal immigration, and drug-smuggling (Fought 2003). Is it not, after all, much more interesting to watch the seedy side of life, instead of hearing the 'boring' success stories, or stories that simply have nothing to do with the 'idea' of Latinos created by society? Latinos are lawyers, bankers, teachers, laborers, students, athletes, doctors, just as other Americans. As Stuart Hall says, we must unhinge and dislodge everyday assumptions and tear apart the images presented to us. This is what Alcaraz does and what Wegner (2007) refers to when he says Alcaraz 'normalizes' Chicanos.

## 9.8 The data

To provide a background for language choice and language ideologies in *La Cucaracha*, I collected and analyzed a total of 163 comic strips and editorial cartoons of *La Cucaracha* from September 2007 to May 2010, divided into seven main themes (see Table 9.1). From the total number of comic strips and editorial cartoons, nine were then closely analyzed for language choice and ideologies.

Table 9.1  Seven main themes in *La Cucaracha* 2007–10

| Theme | Number of cartoons | % |
| --- | --- | --- |
| Latino holidays and culture | 44 | 27 |
| Immigration | 25 | 15 |
| Social issues | 24 | 15 |
| Presidential politics | 23 | 14 |
| Political pundits | 20 | 12 |
| Language issues | 17 | 10 |
| Self-hate mail | 10 | 6 |

### 9.8.1  Analysis of the data: themes and techniques used in *La Cucaracha*

Table 9.1 shows the number of cartoons/comic strips in the seven main themes: *immigration* (25); *language issues* (17) (English only, Anglicization of names, Spanish heritage language learners); *political pundits* (20) (Rush Limbaugh, Lou Dobbs); *social issues* (24) (education, jobs/economy, guns, the 2010 Census, health care); *presidential politics* (23) (Barack Obama, George W. Bush, Sarah Palin, Sonia Sotomayor); *Latino holidays and culture* (44) (cultural icons, family issues, Latino Heritage month, Day of the Dead, Cinco de Mayo); and *self-hate mail* (10).

The number of cartoons/comic strips within each category reflects current events salient during a given time period. For example, *immigration* maintained a steady presence, but saw a surge in 2009 and 2010 when in 2009 questions about 'who is an immigrant in the United States of all immigrants?' and the meaning of 'illegality' circulated in the news. In 2010, the Arizona law SB 1070 inspired several cartoons. In 2009, the 'Anglicization' of names resulted in seven cartoons dedicated to the topic.

*Latino holidays and culture* figured prominently, especially in 2008 and 2009 with strips dedicated to Cinco de Mayo, Día de los Muertos (Day of the Dead), and Latino Heritage month. Strips dedicated to politics were most numerous during the presidential election of 2008, with caricatures of George W. Bush and Barack Obama's Democratic opposition candidate, Hillary Clinton. The cartoon 'Viva Obama' with Barack Obama shown in Emiliano Zapata's revolutionary garb, also featured in 2008. In 2009, six strips were dedicated to Congressional hearings for the appointment of Supreme Court nominee Sonia Sotomayor. Social issues addressed in 2009 included the recession, unemployment, Obama's health care plan, and the erasure of César Chavez from history textbooks in Texas schools. In 2010, social issues continued with education, the right to bear arms in public, and the 2010 Census.

## 9.9 Language choice as identity and language ideology in *La Cucaracha*

This section explains how two specific elements of Alcaraz's Discourse, language choice and language ideologies, are used in *La Cucaracha* to enact his identity as a sociopolitical Chicano. One of the seven categories identified was *language issues*, which included strips specifically dedicated to English-only, Anglicization of names, and heritage learners of Spanish. However, 39 of the strips across theme categories contained some aspect of code switching (including Spanglish), Mock Spanish, and language ideologies. Specific examples from each of these aspects will be discussed below.

### 9.9.1 Language choice: code switching

In contrast to other Latino comics' characters (The Whip, El Dorado), Alcaraz's characters do not speak heavily accented Spanglish. *La Cucaracha* is written predominately in English but Alcaraz switches from English to Spanish for two main reasons: to highlight ethnic identity and to indicate the linguistic landscape of a Chicano barrio. Most of the code switches in Alcaraz's strip which occur intrasententially are emblematic, used to highlight ethnic identity, as in (1) below, or are tag phrases such as interjections (2, 3, 4). Intersentential code switching also appears (4 and 5). In example (1), Alcaraz uses emblematic code switching when Eddie's conservative Republican dad, Ernesto, asks for a *chancla* (flip-flop) to spank Eddie with instead of the traditional Anglo belt, paddle, or wooden spoon when he finds out Eddie is working for the Census. At emotional moments, especially ones associated with anger or surprise, Alcaraz mixes Spanish and English interjections, like when characters say 'Que what?!' or 'holy frijoles!' when confronted with some unbelievable news (2 and 3). Example (4) contains intersentential switches from Spanish to English. Ernesto typically uses Spanish when he becomes very emotional about a political topic. The uses of code switching are consistent with the requirement that no grammatical rule in either language be violated, particularly after the point of the switch. For example, in (2) the presence of fulcrum words allows switches; e.g., conjunctions and complementizers like Spanish *que* and English *that* (Lipski 2008: 55). In each case, the uses are also consistent with permitted switches as described by Timm (1975).

(1) Hi, pop. I'm working for the **2010 Census. Maria!** Fetch my shotgun! But I'm your **son. Maria!** Fetch my **chancla!** (March 19, 2010)

(2) Cuco! There's a guy at the door. He says he's the new manager of our comic strip and he's here to 'revamp operations.' Que what?! (November 10, 2009)

(3) **Holy frijoles!** This neighborhood sure is getting gentrified ... (October 5, 2009)

(4) **¿Qué? Estúpidos federales! ¡Obama es un idiota!** Which language is primarily spoken in your household? None of your business. (March 23, 2010)

Example (5) and Figure 9.1 demonstrate intrasentential emblematic code switching from English to Spanish. When Alcaraz's nemesis rightwing anti-immigrant conservative news host Lou Dobbs left CNN amid controversy, he became the topic of 'Bug Blog' by Cuco Rocha.[6]

*Figure 9.1* Bug Blog – Lou Dobbs Quits CNN. All images by Lalo Alcaraz. Used by permission.

Figure 9.1 and example (5) are from the first in a series of five strips (November 30, 2009 to December 4, 2009). Cuco writes:

(5) Today's topic to **taco** about: **Lou Dobbs Quits CNN (11/30/ 2009)** Lou Dobbs left CNN because the news network wouldn't let him magically pull fake anti-immigrant facts out of his **sombrero**.

He continues 'As a public service, the Bug Blog debuts: JOBBS FOR DOBBS.' This is because 'No one wants Lou to join the ranks of the unemployed in front of Home Depot.' In the last panel, Dobbs is shown with other day laborers looking for work in front of the Home Depot, wearing a hat commonly worn by day laborers and an anti-Mexicans T-shirt. Behind him, the day laborers look on, one in shock and the other arms crossed and perhaps ready to beat Dobbs up. The Spanish word *taco* is used in place of *talk*. A *taco* is usually used to refer to the Mexican food made from a tortilla containing meat, cheese, and beans, but here, Alcaraz uses an ethnic word to highlight that Cuco is a Chicano blog writer. In the following sentence, the Spanish *sombrero* is used in place of the English *hat*. This is significant on a number of levels. On his program, Dobbs' rhetoric frequently indicated his dislike of undocumented Latino immigrants, specifically Mexicans, and the (lack of) US immigration policy. *Sombrero* is used in the English idiomatic expression 'pull something out of a hat' when one wishes to indicate that something materialized from nothing, like a magician seemingly pulls a rabbit out of his hat where there was nothing before. The substitution of *sombrero* for *hat* indicates Chicano ethnic identity and, in a twist of irony, is from the language belonging to immigrant group most disliked by Dobbs.

Spanish lexical items also serve as linguistic signposts indicating to the reader that the characters live in a Spanish barrio where Spanglish is spoken. Table 9.2 shows a summary of the place names which appeared in the comic strips analyzed.

Table 9.2 Spanish used for place names

| | |
|---|---|
| Pancho's Retirement Villa | Donas (donut shop) |
| Barriobucks Café | Barrio Bugle |
| Peludo Arms (apartment complex) | Chuy's diner |

These place names have English grammatical structure such as English possessive formation and word order for nouns and adjectives while also containing Spanish lexical elements. The English possessive morpheme 's stands in place of Spanish possessive formation with the preposition *de* ('of'). For example, 'Chuy's Diner' would be 'El restaurante de Chuy.' English word order (adjective + noun) also takes precedence over the Spanish word order (noun + adjective) such as in 'Peludo [furry/fuzzy] Arms.' These examples might be considered Spanglish and Mock Spanish by others (Hill 1998, 2008). Alcaraz, who identifies as a Spanglish-speaking Chicano, probably uses these as a Spanglish technique. Spanish also appears in the names given to the main characters: Pepe, Chepe, Cuco Rocha, Eddie (Anglo shorthand for Eduardo) Lopez, Vero Varela, María (Eddie's mother), and Ernesto (Eddie's father).

### 9.9.2 Re-appropriation of Mock Spanish words and stereotypes

In *La Cucaracha*, elements such as in Table 9.3 can be considered Mock Spanish (Hill 1998, 2008) because they contain the affixation technique of *el* + word + *-o* and/or invoke a (seemingly) negative stereotype. However, Alcaraz often re-appropriates or re-directs this stereotype to lampoon the majority Anglo culture or the myths which exist about Chicanos. Sometimes, what is Mock Spanish for Hill appears to be Spanglish for Alcaraz, though Hill stresses that Mock Spanish is different from Spanglish. In addition, a Mock Spanish accent for Hill is a broad American English accent, not a more overtly racist parodic imitation. However, in her discussion of Mock Asian, Chun (2004) categorizes intentional mockery as part of a mock language. Chun (2004) also recognizes 'legitimate mockery' where speakers can re-appropriate, or use the Mock language themselves because of their insider-group status. The Mock Spanish in Alcaraz's work can be analyzed as a re-appropriation of Mock Spanish: it is more acceptable for him to use it since he as an insider to the cultural group, Chicano/Latino.

Figure 9.2 exemplifies the Mock Spanish accent and linguistic stereotypes of Mexicans in the US. It also pays homage to the first Mexican American comic strip Gordo, created by Gus Arriola, who died in early 2008.[7] Arriola's cartoon portrayed a humble Mexican bean farmer turned tour guide, first appearing with a heavy Mexican Spanish accented English, but later, due to criticism from Mexican American readers who disliked the stereotypical speech and persona being portrayed, lost his accent and some weight and became socially aware (Aldama 2009: 72–73). Arriola's strip was not political; he wished to reach as wide an audience as possible, and especially reach those who knew very little about Mexican culture. Arriola considered Gordo to be his alter ego, as Alcaraz considers Cuco his. Here Alcaraz puts Eddie in the shoes (or literally, at the desk) of Arriola at his job at the *Barrio Bugle*. He simultaneously casts Eddie in the position of Peter Parker (whose alter ego is Spider-Man) working for publisher J. Jonah Jameson at the *Daily Bugle*.

*Table 9.3* Mock Spanish words or phrases in *La Cucaracha*

El Rushbo
El Taco Cart Guy
Drinko de Cinco
Cinco de Marcho
It's Finger-Lickin' bueno [KFC slogan]
Dear El Jerko

*Figure 9.2* Lopez! We need a new comic strip! All images by Lalo Alcaraz. Used by permission.

'**Lopez!** We need a new comic strip!' shouts his boss. Dressed in a shirt and tie reminiscent of J. Jonah Jameson, scowling and shaking the newspaper in his hand, he appears in front of Eddie's desk announcing 'I hate all the comic strips we run! Why can't they be more like "**GORDO**"?!' Eddie morphs into a Gus Arriola/Gordo *campesino* type while pondering this, his thoughts portrayed in a thought bubble with a Mock Spanish exaggerated accent 'Meester jefe ... He-ees muy loco.' In this panel, his boss is now a Mexican American sporting a sombrero, a furry oversized mustache and *guayabera* (style of shirt). The boss points to Eddie and finishes, just like J. Jonah Jameson would to Peter Parker, with 'And get a real tie!!' The most obvious marker of the stereotypical accent is the elongated [i] in 'meester' for 'mister' and 'ees' for 'is.' This phonetic element has been found in Chicanos' speech (Mendoza-Denton 2008; Fought 2003), but not in this exaggerated Speedy Gonzalez-like portrayal. Alcaraz's use of this stereotypical language reminds us how Mexicans (Mexican immigrants *and* Chicanos) were/are viewed as far as speech is concerned: that they use uneducated, poor speech stemming from difficulties with English pronunciation, or speech with a heavy Mexican accent.

Figure 9.3 functions as a platform for several themes (immigration, commercialization of Latino holidays, and Latino's buying power) and techniques (news from a media source and Mock Spanish) in *La Cucaracha*.

In the final panel as the car tootles out of sight, the news reporter announces the name of the new commerce-oriented holiday: '**Cinco de Marcho.**' Discussed is the commercialization of Latino holidays, specifically 'el Día de los Reyes', January 6, the day many Latinos open gifts, instead of the Anglo custom of December 25. Latinos have an increasingly large buying power, leaving retailers wanting to take advantage of this group. Alcaraz incorporates two Mock Spanish stereotypes: undocumented immigrants and alcohol-laden fiestas. The word play

*Figure 9.3* Cinco de Marcho. All images by Lalo Alcaraz. Used by permission.

with 'Cinco de Marcho' is both an allusion to the Cinco de Mayo holiday celebrated by mainstream US society and beer companies as well as an allusion to the polar-opposite sentiment, the desire of that same faction of people in the US mainstream society to kick all undocumented immigrants from Mexico out of the country.

### 9.9.3 Self-hate mail

A favorite topic/theme of Alcaraz's is general reader hate mail and especially 'self-hate mail from Latinos.' Alcaraz pokes fun at Latino readers who disagree with his politics and ideology, declaring them as 'self-haters' or those who hate their 'Chicanoness,' while also poking fun at himself, as the long-suffering cartoonist who must deal with these inane objections. While in 2009, most of the 'hate mail' received decried Alcaraz's support for and sympathetic portrayal of undocumented citizens, in 2010, the 'self-hate mail' highlights Alcaraz's portrayal of Latinos in his strip. In several cartoons titled 'La Cucaracha answers reader mail from LATINOS, aka 'SELF-HATE MAIL,'' Mexican Americans who disagree with Alcaraz's politics and use of stereotypes write in. Alcaraz characterizes the discourse of these 'self-haters' with Mock Spanish, circular logic, hasty generalizations, self-contradiction, and irony, and he plays with stereotypes to represent the absurdity of their complaints. The term 'self-hate' implies that these Mexican Americans are not proud of their heritage the way Alcaraz portrays it, but prefer a 'whitewashed' or assimilated version. Figure 9.4 simultaneously shows Alcaraz's self-identification as a Chicano and criticizes ignorance.

Reader 'Sharon Sanchez,' with her Anglo first name and Spanish surname, believing she has the upper hand but unfamiliar with the term *Chicano*, self-validates her hasty and unfounded conclusion that the creator of the strip is not Latino because she believes that Latinos are portrayed negatively in the strip. In the last panel, a triumphant

*Pocho Identity in* La Cucaracha 229

*Figure 9.4* Self-hate mail. All images by Lalo Alcaraz. Used by permission.

Sharon proclaims, 'I knew he was a Norwegian from Chicago!' The humorous definition comes from the faulty logic of the reader's misanalysis of *Chicano* as a derivational blend from *Chica(go)* + *No(rwegian)*. She accuses Cuco of painting a negative picture of Hispanics because he is not one himself. In his response, Cuco agrees, but on the grounds that he does not identify with Hispanic as a generic term, preferring to self-identify as Chicano. Seemingly assimilated 'Sharon,' on the other hand, is so far out of touch with her own roots that she cannot even recognize the term or history behind *Chicano*. Through the 'self-hate' mail, Alcaraz exposes fallacies of stereotypes held by non-Latinos and Latinos alike while making his critics' faulty logic look foolish.

### 9.9.4 Leftist Chicano ideology

Arising from the big 'D' Discourse of a politically conscious Chicano, Alcaraz's strips reflect his views on immigration and presidential politics. He often uses narrative devices such as billboards, radio broadcasts emitting from a tiny car, and television newscasts which present seemingly neutral, or 'Fair and Balanced' news in the style of television stations like FOX News. In actuality, these 'neutral devices' provide Alcaraz a platform for ironic commentary sympathetic to the left and the Chicano community.

### 9.9.5 Presidential elections

In Figure 9.5, the 'neutral' news source is anything but; we see Alcaraz's opinion of the candidates and their politicking. As Cuco and Eddie begin their journey, the radio reports: 'All three major U.S. presidential candidates hope to capture the Latino vote this fall/apparently by scooping up large numbers of Latinos caught by the new U.S./Mexico border fence which all three voted for.' The humor lies in the error in reasoning and

*Figure 9.5* The Latino Vote. All images by Lalo Alcaraz. Used by permission.

*Figure 9.6* *Viva Obama*: Barack Obama as Emiliano Zapata. All images by Lalo Alcaraz. Used by permission.

self-contradiction of the candidates running in the 2008 Presidential election. Though later Alcaraz celebrates Obama's victory, here he has no problem recognizing the fact that all three candidates were at one point in favor of a fence being erected along the entire border of the United States and Mexico as a measure for keeping the undocumented out. As the campaign progressed, the candidates tried hard to get the Latino vote, each one appealing to the Latino community in a variety of ways (promises of reform, support for amnesty), none of which have materialized as of this writing. Alcaraz, as a politically aware Chicano, however, did not forget their earlier decisions to exclude that very same community.

'Viva Obama' (2008) may be the most famous (or famously talked about) Alcaraz cartoon to date, a 'grassroots runaway hit,' especially for many young Latinos (interview with Lalo Alcaraz in *Aztlán* Fall 2009). The winner of the 2008 US Presidential election, Barack Obama is portrayed as Mexican revolutionist Emiliano Zapata. This stand-alone editorial cartoon depends on direct indexical emblems: e.g., the sombrero, the rifle and ammunition, the military horseman's *charro* suit, all associated with Zapata.

The phrase 'Viva Obama' (Long Live Obama) parallels a Mexican *corrido* (folk song) ¡Viva el pueblo proletario! (Long live the proletariat / the people! – Jaeck 2001: 46). The *corrido* extols Zapata as the hero of land reform and of the proletariat. Jaeck writes that 'Corridos were an exemplary form of communication, their words and lyrics flying on the wings of the human spirit from one end of the nation to the other, seemingly faster than a telegraph transmission' and that 'a good revolutionary song speaks of love – the soul of the people' (p. 41). *Viva Obama* is Alcaraz's song to the people of the United States to celebrate that he is the people's candidate who will bring revolutionary change to the country.

### 9.9.6 Immigration

In Figure 9.7, the recurring comical figure of 'Miles Standish,' a symbol for the 'generic Anglo pilgrims,' Alcaraz's examples of the original 'illegal aliens' to enter the United States, stands on a podium in a Native American village, in the midst of reading a proclamation, when he is interrupted by a protesting Cuco, who is decked out in Native American dress. Cuco questions angrily: 'Why won't you show us your birth certificate?!' Cuco is accompanied by Native Americans holding protest signs reading 'You are not a real American' and 'Show us your papers.' These slogans allude to protest signs used by groups of Americans protesting 'illegal' immigration and by conservative factions such as the Minute Men (self-appointed border patrollers). The irony is that their ancestors were once 'illegal,' a point which seems forgotten in the mass media. Gee (1999) has written that part of being 'a real something' is the ability to act that part one hundred percent, which, as this cartoon suggests, anyone other than Native Americans cannot do. Alcaraz claims that they are the original 'real Americans' and the rest of us are immigrants. In this case, Alcaraz uses his alter ego, rabble-rousing Cuco, to resist anti-Chicano rhetoric and to relay his opinion of the hypocrisy of such statements.

232  *Carla Breidenbach*

*Figure 9.7* An early request for a birth certificate. All images by Lalo Alcaraz. Used by permission.

## 9.10 Language ideologies

### 9.10.1 English only

Figure 9.8 belongs to a series of seven comic strips (9–19 November 2009) that mirror a story in the news during July 2009 when Larry Whitten, a 63-year-old manager from Texas, was brought in to help a struggling hotel in Taos, New Mexico. In giving his new rules, he forbade his employees to speak Spanish in his presence and he also 'ordered some to Anglicize their names.' He said that changing the pronunciation of the employee's first name (formerly pronounced with a Spanish accent) was not racist: 'I'm not doing it for any other reason than for the satisfaction of my guests, because people calling from all over America don't know the Spanish accents or the Spanish culture or Spanish anything.'[8] Alcaraz does not share this sentiment and lampoons the incident. Over at the boys' apartment, Peludo Arms, Eddie and Cuco discover that the comic strip has hired a new manager to 'revamp operations,' says Eddie. 'QUE WHAT?!' hollers Cuco. A middle-aged manager type holding a sign which reads 'Lou Dobbsville' (an allusion to anti-immigrant Lou Dobbs) enters and says 'Howdy, I'm here to save this shabby comic strip. Where can I hang this here new title?'

In Figure 9.8, the caption in the first panel reads 'New manager Mel Gregvin is shaking up things at the comic strip "La Cucaracha."' Underneath the caption we see Mel pointing to Eddie and Cuco and informing them that 'Readers are turned off by your Hi-Spanic names,

Pocho Identity in La Cucaracha   233

*Figure 9.8*   Anglicizing Spanish-language names. All images by Lalo Alcaraz. Used by permission.

so I'm renamin' the lot of ya...' In the next panels, the captions read: '**Bruiser** the fry cook' is now '**Rachel Ray**,' a cultural reference to one of the most popular and commercialized Anglo Food Network Stars. Mel Gregvin embodies an ignorant (racist?) ideology of the English only policy to an incredible degree. The insistence to Anglicize or not use native pronunciation is part and parcel of erasing someone's identity and forcing them to assume a new 'American' (United States) identity under the guise of helpfulness. It also indicates a discomfort and indignation that many Anglos feel when encountering a 'foreign' language in America. It is 'un-American' to speak in other languages than English, and especially un-American to speak with an accent native to that country. Accents are not good for business. However, Mel Gregvin's accent contains features stereotypically associated with an 'ignorant country (hick) accent' such as the phonetic features [n] for [ŋ] in 'renamin' and semantic phrases 'the lot of ya' and 'yup.' Not only is this 'country' accent used to poke fun at the manager, but it also shows that to discriminate against a 'foreign' accent is silly since *everyone* has an accent. However, this linguistic phobia continues to live on in the United States.

### 9.10.2   Spanish for heritage speakers

In Figure 9.9, Alcaraz parodies the cultural list phenomenon 'Stuff White People Like' and highlights an issue quite prevalent in the Chicano community concerning education and bilingualism.[9] In Figure 9.9, Vero, Eddie's girlfriend, is a Chicana who for any number of possible reasons, ended up with a low grade in high school Spanish. As she strolls down the street to the Instituto de Español Barcelona, she happily thinks 'I'm paying, so I'll be sure to get a good grade!' The punchline in the last panel invites the reader to walk in Vero's shoes to sympathize with her about the experience. Many Latinos, whether Spanish is their first or

*Figure 9.9* Stuff Latinos Like #2. All images by Lalo Alcaraz. Used by permission.

second language, tend not do well in their high school Spanish classes for a variety of reasons. A Chicano may be second, third, or fourth generation (or later), and may be acquainted with Spanish to varying degrees. Normally, by the third and fourth generations, much of the language production has been lost, even though they may hear it from grandparents or other recently arrived family members. What Spanish they do know is often a nonstandard dialect of Mexican Spanish or Chicano Spanish, not the standard taught in schools. Thus, when Chicanos arrive in their Spanish classes they are confronted with two major obstacles. First, many Chicanos believe since they already know Spanish they will get an easy 'A,' but in reality, the Standard Spanish they are learning is far from what they have heard previously at home (Fought 2003: 156–57, Valdés and Geoffrion-Vinci 1998). Second, many teachers who speak and teach the Standard often hold Chicano Spanish or other nonstandard dialects in low regard (Peñalosa 1972; Valdés and Geoffrion-Vinci 1998: 473), and have negative opinions of Chicanos, which in turn demoralizes Chicano students (Peñalosa 1972). Sometimes, teachers see these students fooling around and not doing their homework and not doing well on tests (Matute-Bianchi 1986). This may be partly because, as Matute-Bianchi (1986) and Malagón (2010) suggest, Chicano students are uninterested and unmotivated because they do not want to assimilate themselves with other school factions, preferring to go against the status quo. It may also be that if the teacher is a nonnative speaker of Spanish, Chicano students may doubt the teacher's expertise and make this clear by not performing well in class (Merino, Trueba and Samaniego 1993). However, it is often the case that Chicano students are simply much more unfamiliar with the language than is generally believed (Valdés and Geoffrion-Vinci 1998; Fought 2003).

## 9.11 Conclusion

This chapter has examined how language choice (Spanish and English) and language ideologies are used in Alcaraz's *La Cucaracha* to enable him to enact his identity as a sociopolitical Chicano. As a political cartoonist, Alcaraz chooses to represent and take to heart the concerns and issues facing the Chicano community in the United States. He presents a multiplicity of voices and pointedly breaks down negative stereotypes of the Chicano community. In doing so, he incorporates Discourses which enable him to enact his Chicano identity and political ideologies from within a post-Chicano framework, infusing his strip with his special 'pocho politics.' The analysis of the comic strips has identified language choice and language ideologies as two techniques used by Alcaraz to enact his Chicano identity. The dominant language of the strip is English, perhaps a parallel to the mainstream society in which Chicanos live. The use of Spanish proper nouns and place names, on the other hand, situates the characters in a Chicano barrio. The appearance of Spanish code switching and Spanglish within the text has the special purpose of demonstrating the mix of two cultures, the Anglo and the Chicano. The use of Mock Spanish words and stereotypes mark a transformation of the negative images of Latinos into positive ones. In this way, Alcaraz reclaims what has been 'stolen.' The themes of the cartoons (e.g., English only, presidential politics, and immigration) provide a window into his political ideologies, sympathetic to the left and Chicanos. Always mindful of his audience, Alcaraz employs political discourse, narrative and rhetorical techniques, and varied linguistic choices to make his special mark as a Chicano comic strip artist by incorporating topics and discourses which are not only relevant to the Chicano community, but also provide the general readership with insights into perspective about Chicanos and their role in US society.

## Notes

1. http://www.allbusiness.com/services/business-services-miscellaneous-business/4694566-1.html.
2. Alcaraz hosts a radio talk show Friday afternoons from 4 to 5pm called 'Pocho hour of power' http://www.pocho.com/lalobio/lalobio2009.php
3. Jennings, A. (2010) Comic is not a good fit. Retrieved on 2 May 2010 from http://www.ivpressonline.com/articles/2010/04/15/voice_of_the_people/voice01-04-15-10.txt
4. No explicit syntactic features of Chicano English such as double negatives are present in the characters' speech. However, this does not exclude the possibility that if we were to hear the characters speak, we may hear certain

phonetic features identified with Chicano English. Alcaraz has an editor who checks the comic strip for content as well as grammatical errors. It may also be Alcaraz who chooses to use a Standard English variety in his strips.
5. Kong, D. 'Spanglish' creeps into the mainstream Retrieved on 6 July 2010 fromhttp://www.ampersandcom.com/GeorgeLeposky/SpanglishCreepsintoth eMainstream.htm
6. Altman, A. and Fletcher, D. (2009) Departing CNN Anchor Lou Dobbs. Retrieved on 22 April 2010 from http://www.time.com/time/printout/ 0,8816,1938382,0.html
7. Buchanan, W. (2008) http://articles.sfgate.com/2008-02-03/bay-area/ 20872771_1_mexican-culture-strip-chronicle
8. Dabovich, M. (2009) Hotel owner tells Hispanic workers to change names. Retrieved on 11 November 2009 from http://www.myfoxdc.com/dpp/news/ dpg_Hotel_Owner_Tells _Hispanic_Workers_to_Change_Names
9. Lander, C. (2010) Stuff white people like. Retrieved on 27April 2010 from http://stuffwhitepeoplelike.com

## References

Alba, V. (1967) The Mexican revolution and the cartoon. *Comparative Studies in Society and History* 9 (2): 121–36.
Aldama, F. (2009) *Your Brain on Latino Comics*. Austin: University of Texas Press.
Barrett, R. (2006) Language ideology and racial inequality: competing functions of Spanish in an Anglo-owned Mexican restaurant. *Language in Society* 35 (2): 163–204.
Basu, S. (2002) Lalo Alcaraz interview. Retrieved on 7 July 2010 from http://www.willamette. edu/~sbasu/polixxx/laloalcaraz/Interview.htm
Ben-Rafael, M. (2008) English in French comics. *World Englishes* 27 (3–4): 535–48.
Boskin, J. (1979). *Humor and Social Change in Twentieth-Century America*. Boston: Trustees of the Public Library of the City of Boston.
Bourdieu, P. (1977) *Outline of a Theory of Practice*. Cambridge: Cambridge University Press.
Breidenbach, C. (2007) *Deconstructing Mock Spanish: A Multidisciplinary Analysis of Mock Spanish as Racism, Humor, and Insult*. Doctoral dissertation, University of South Carolina.
Chun, E. (2004) Ideologies of legitimate mockery: Margaret Cho's revoicings of mock Asian. *Pragmatics* 14 (2–3): 263–89.
De León, A. (1983) *They Called Them Greasers. Anglo Attitudes Toward Mexicans In Texas, 1821–1900*. Austin: University of Texas Press.
El Refaie, E. (2009) Multiliteracies: how readers interpret political cartoons. *Visual Communication* 8 (2): 181–205.
Field, F. (1994) Caught in the middle: the case of Pocho and the mixed language continuum. *General Linguistics* 34 (2): 85–105.
Forceville, C. (2005) Addressing an audience: time, place, and genre in Peter Van Straaten's calendar cartoons. *Humor: International Journal of Humor Research* 18 (3): 247–78.
Fought, C. (2003) *Chicano English in Context*. Basingstoke and New York: Palgrave Macmillan.

Gal, S. and Irvine, J. (2000) Language ideology and linguistic differentiation. In P. Kroskrity (ed.) *Regimes of Language: Ideologies, Polities, and Identities*, pp. 35–84. Santa Fe: School of American Research Press.

Gee, J. (1999) *An Introduction to Discourse Analysis: Theory and Method*. London and New York: Routledge.

Gee, J. (2001) Reading as situated language: a sociocognitive perspective. *Journal of Adolescent & Adult Literacy* 44 (8): 714–25.

Geipel, J. (1972) *The Cartoon: A Short History of Graphic Comedy and Satire*. London: David & Charles.

Hall, S. (1997) *Representation. Cultural Representations and Signifying Practices*. London: Sage Publications Ltd.

Harvey, R.C. (2002). A couple chidings and some New Lang Syne: The year in review. *Comics Journal*. 250: 75–82. (There is no issue number, posted online in Feb. 2003.)

Hill, J. (1998) Language, race, and white public space. *American Anthropologist* 100 (3): 680–89.

Hill, J. (2008) *The Everyday Language of White Racism*. Chichester: Wiley-Blackwell.

Jaeck, L. (2001) Viva México/Viva la Revolución. One hundred years of popular/ protest songs: the heartbeat of a collective identity. *Ciencia Ergo Sum* 8 (1): 41–49.

Jonsson, C. (2010) Functions of code-switching in bilingual theater: an analysis of three Chicano plays. *Journal of Pragmatics* 42 (5): 1296–1310.

Lipski, J. (2008) *Varieties of Spanish in the United States*. Washington, DC: Georgetown University Press.

Malagón, M. (2010) Trenches under the pipeline: the educational trajectories of Chicano male continuation high school students. *UC Los Angeles: UCLA Center for the Study of Women*. Retrieved on 22 July 2010 from http://www.escholarship.org/uc/item/73q6f2fx.

Marín-Arrese, J. I. (2008) Cognition and culture in political cartoons. *Intercultural Pragmatics* 5 (1): 1–18.

Matute-Bianchi, M. (1986) Ethnic identities and patterns of school success and failure among Mexican-Descent and Japanese-American Students in a California high school: An ethnographic analysis. *American Journal of Education* 95 (1): 233–55.

Mendoza-Denton, N. (2008) *Homegirls: Language and Cultural Practice among Latina Youth Gangs*. Oxford: Blackwell.

Merino, B., Trueba, H., and Samaniego, F. (1993) *Language and Culture in Learning: Teaching Spanish to Native Speakers of Spanish*. London: The Falmer Press.

Myers-Scotton, C. (1988) Code-switching as indexical of social negotiations. In M. Heller (ed.) *Code-switching*, pp. 151–86. Berlin: Mouton de Gruyter.

Newton, J. (2007) Trudeau draws truth. *Critical Studies in Media Communication* 24 (1): 81–85.

Peñalosa, F. (1972) Chicano multilingualism and multiglossia. *Aztlán: A Journal of Chicano Studies* 3 (2): 215–22.

Romero, M. (1988) Chicano discourse about language use. *Language Problems and Language Planning* 12 (2): 110–29.

Rybacki, K. and Rybacki, D. (1991) *Communication Criticism: Approaches and Genres*. Belmont, CA: Wadsworth Publishing Company.

Shultz, K. and Germeroth, D. (1998) Should we laugh or should we cry? John Callahan's humor as a tool to change societal attitudes towards disability. *The Howard Journal of Communications* 9 (3): 229–44.

Smith, D. (2007) Spanish/English bilingual children in the southeastern USA: convergence and codeswitching. *Bilingual Review* 28 (2): 99–108.

Stephenson, E. (1969) Chicano: origin and meaning. *American Speech* 44 (3): 225–30.

Timm, L. A. (1975) Spanish-English code switching: el porqué y how-not-to. *Romance Philology* 28 (4): 473–82.

Torres, L. (2007) In the contact zone: code-switching strategies by Latino/a writers. *MELUS* 32 (1): 75–96.

Valdés, G. and Geoffrion-Vinci, M. (1998) Chicano Spanish: the problem of the 'underdeveloped' code in bilingual repertoires. *The Modern Language Journal* 82 (4): 473–501.

Vigil, J.D. (1938) *Barrio Gangs: Street Life and Identity in Southern California*. Reprinted 1988, Mexican American monographs, no. 12. Austin: University of Texas Press.

Wegner, K. (2007) La Cucaracha 'normalizes' Chicanos. *International Journal of Comic Art* 9 (1): 231–61.

Yus Ramos, F. (1998) Relevance theory and media discourse: a verbal-visual model of communication. *Poetics* 25 (5): 293–309.

Zentella, A. C. (2003) 'José can you see': Latino responses to racist discourse. In D. Sommer (ed.) *Bilingual Aesthetics*, pp. 51–66. Durham, NC: Duke University Press.

# 10
# The Use of English in the Swedish-Language Comic Strip *Rocky*

Kristy Beers Fägersten

## 10.1 Introduction

The Swedish comic strip *Rocky* first appeared in 1998 in the national edition of *Metro*, a free newspaper distributed at points of public transportation, targeting morning commuters. Since this humble debut, *Rocky* has so far been published in a total of 17 collected volumes, been adapted for the stage, made into a documentary series of short films, and inspired a franchise of products such as t-shirts, skateboards, and calendars. *Rocky* is currently featured as the headlining comic strip of Sweden's largest national newspaper, *Dagens Nyheter*, and has found distribution in Norway, Finland, Denmark, France, Italy, Slovenia and the United States. The strip's writer, Martin Kellerman (born in 1973), has been awarded a number of prizes for *Rocky*, including the Urhunden award in 2000 from Seriefrämjandet (an association for the promotion and support of comic strips and artists) for his first published volume, *Rocky I*; the Bern's Award in 2009 by Swedish PEN (an international association for Poets, Essayists and Novelists) for his contemporary depiction of Stockholm; and, most recently, the prestigious Bellman Award of 2010, an honorary distinction including a stipend awarded by the City of Stockholm.

*Rocky* has thus developed into what its publisher, Kartago, likens to an empire, attributing it the status of an institution in Swedish culture. The motivations for the awards and accolades bestowed on Kellerman and his comic strip seem to confirm this assessment: *Rocky* is praised by critics for its accurate and humorous portrayal of the day-to-day life of a young, single man in modern-day Stockholm. Indeed, the comic chronicles the daily life of Kellerman's alter-ego, Rocky, a 30-something single man, albeit in the gestalt of a dog, living and working as a comic strip writer in

Stockholm. *Rocky* is thus largely autobiographical, representing a public diary of Kellerman's own life and, notably, an expository documentation of the lives of his friends, also represented by animal figures. Kellerman has even admitted in a 2005 interview to being a 'parasite' with regard to his friends, relying on them for inspiration: 'They generate a lot of the material. [...] My friends are just such perfect cartoon characters. A lot of times they say things and all I have to do is write it down. Their personalities match and complement each other so well, it's impossible not to write it down' (MacDonald 2005). Kellerman's practice of documenting the social interaction of his network of friends in *Rocky* helps explain its profile as a dialogue-driven comic strip. In another 2005 interview, Kellerman aligned himself with other dialogue-heavy comic strip artists, citing inspiration from Peter Bagge, Joe Matt, and Robert Crumb (Kinn 2005), whose *Fritz the Cat* also featured anthropomorphized animals as regular strip characters. In yet another interview from 2005, Kellerman explained the conversation-driven aspect of the comic strip as a deliberate choice, claiming he feels that he's 'cheating' if he makes a strip 'with only a few words in it' (Spurgeon 2005).

It is precisely the verbosity of the *Rocky* cast of characters that makes the strip an obvious target for linguistic analysis. But perhaps most interesting to an international audience is the fact that this Swedish-language comic strip is characterized by the interlocutors' frequent use of English. To some extent, then, *Rocky* can be considered accessible to even non-Swedish speaking scholars of linguistics and comics.

In this chapter, I analyze the use of English in *Rocky* in terms of a distinct linguistic code with discursive and humorous functions. I approach the use of English (1) as a manifestation of the assimilation of English in Sweden, (2) as an indicator of in-group identity and (3) as a source of humor in the Swedish comic strip medium.

Since Swedish is the dominant language of *Rocky*, switches to English are indeed discursively significant, and not only reflect the in-group linguistic norms shared by the *Rocky* characters – and, presumably, by their real-life counterparts – but can also reflect or even introduce a similar linguistic behavior among the wider Swedish reading public. I argue that the use of English among the *Rocky* characters reflects a stylistic choice, symbolizing assimilation, cultural alignment, and in-group membership, or aspirations thereto. The establishment of a native/nonnative English speaker opposition as well as the appropriation of a hip-hop vernacular are recurring features in *Rocky*. The occurrence of code switching as the deliberate use of nonstandard English and the reciting of song lyrics serve as salient hallmarks of in-group/out-group opposition. Furthermore, the use of English itself is responsible for the opposition, and code switches are exploited as sources of humor because they capitalize on linguistic incongruities in group member identities.

In an effort to reveal why English is actually available as a valid code choice both in Swedish society broadly and within the speech community represented in *Rocky*, I review the status and role of English in Sweden. In section 10.3, I prepare for the linguistic analysis of English usage in Swedish-language *Rocky* by outlining theoretical tenets of code switching and crossing. In section 10.4, I present linguistic analyses of selected *Rocky* comic strips, focusing on examples of Swedish, nonnative English, and hip-hop English. In this section, I also discuss code switches to English as indicators of an in-group/out-group opposition and realizations of humor. In section 10.5, I summarize my findings considering the influence of *Rocky* on modern Swedish and the role of English in Sweden.

## 10.2  English in Sweden

Ever since the end of World War II, Sweden, like the rest of the world, has been exposed to a steadily increasing influence from the English-speaking world, an influence that takes many forms and uses many different channels and which has undoubtedly left its mark on the Swedish language (Ljung 1986:25). The Scandinavian and Northern European countries such as The Netherlands and Luxembourg are often identified as exemplary nations in terms of their successful assimilation of English (Haugen 1987; Labrie and Quell 1997; Phillipson 1992). Scandinavians in particular have been recognized as having a high level of English proficiency (Ferguson 1994), and it has been suggested that in the Scandinavian countries, English has attained the status of a second language 'rather than a foreign language, as the number of domains where English is becoming indispensible in Scandinavia is increasing constantly' (Phillipson 1992: 25).

In Sweden, English is used not only as a lingua franca in international contexts, but also intranationally, as Swedes can be observed incorporating English words and phrases in their Swedish communication with each other (Sharp 2001, 2007). This practice of code switching, attended to in more detail in Sections 10.3 and 10.4, reflects the powerful influence English has historically exerted on the Swedish language and, by extension, on Swedish culture.

These sociolinguistic trends continue to hold true even 25 years later, for a variety of reasons. First, exposure to the English-speaking world and its subsequent influence continues to increase. Taking broadcast television as an example, only two public television channels were available in Sweden until the 1980s (and no English-language channels, though the Swedish-language channels did occasionally broadcast English-language films) when the cable network was expanded. Now, almost three decades later, exposure to the English-speaking world via television and film is commonplace. The subsequent influence on Swedish culture is palpable, as the *Rocky* strips featured in this paper will illustrate.

Second, the 'forms' and 'channels' of this influence are in no way less numerous today: quite the contrary, with the expansion of cable networks, the advent of the internet, and further developments in information technology. Passive exposure to English increases, as do the possibilities to actively seek out English-language press, programming, or other varieties of input.

Finally, the 'mark' left on the Swedish language (and on the Swedish speech community) may be even more obvious now than 25 years ago, as the evolution of Swedish increasingly figures as a subject of popular interest: in the past ten years, a number of television series (*Värsta språket*; *I love språk*), radio programs (*Språket*) and magazines (*Språktidningen*) dedicated to discussions about Swedish have been enthusiastically met by wide audiences.

The use of English in Sweden is a matter of fact, but not an uncontroversial one. On a global scale, English is often accused of being a 'language killer' (Graddol 1996; Josephson 2004). Its use in non-English speaking countries tends to eclipse the status of other languages, and minority languages or national languages of smaller countries are particularly vulnerable to this fate. In terms of population, Sweden, at just over nine million people, ranks among the smaller countries of Europe; the nation's tradition of deliberate incorporation of English especially in the domains of education, trade, and business (Berg, Hult, and King 2001) reflects an awareness of this status and a conscious effort to contribute to its competitive edge internationally (Gunnarsson 2004; Haugen 1987; Hollqvist 1984; Truchot 1997). It has been claimed that, in the future, English may serve as the only language to be used in high-status domains in Sweden, and might even be adopted as the official language of the Swedish government (Hyltenstam 1999).

While the traditional strategy of assimilating English has not been without economic and social advantages for Sweden and its citizens, it has also caused concern over the fate of Swedish (Holm 2006; Josephson 2004; Teleman 1992; Westman 1996). It is a valid concern, too: the more contact Swedes have with English, the easier it is for them to transition to this language, and to an ever-increasing degree. Research reveals, in fact, that the use of English in Scandinavia is not limited to the elite or within high-status domains, but rather has come to characterize social interaction in low-status domains as well (Sharp 2001, 2007). For example, Hult's research on the use of English in southern Sweden shows that it is 'in the process of being appropriated and integrated with daily interaction in public and interpersonal domains [and] can be appropriated for use together with Swedish for expressive purposes' (2003: 60). Preisler (1999) observed a similar development in Denmark, leading him to conclude that the use of English in both high- and low-status domains exerts different influences on the native linguistic system. Specifically, Preisler uses the expressions 'English from above' and 'English from below.' English from above is provided by 'the hegemonic culture for the purposes of international communication.' This can be exemplified by Sweden's policy of including English in the school curriculum beginning at the elementary levels, and promoting the use of English in the domains of trade and industry. The native linguistic system is minimally affected as influence is often limited to the use of loanwords. On the other hand, the influence of English from below, that is, 'the informal – active or passive – use of English' is attributable to 'the desire to symbolize subcultural identity or affiliation, and peer group solidarity' (1999: 241, 246). The native linguistic system is more vulnerable to influences of English from below, as evidenced by lengthier code switches in low-status domains (see, for example Sharp 2001, 2007).

In Sweden, the influence of English from below can be attributed to the daily and prominent exposure to English in Sweden via popular culture media such as television, film, radio, internet, video games, printed press, and music. It is important to note that, like many Scandinavian and Northern European countries, Sweden's imported television programs and films are not dubbed. Swedes are thus regularly exposed to original-language programming, of which the majority is imported English-language films and television series. Furthermore, popular music broadcasting in Sweden features predominantly English-language songs, and native music productions are often recorded in English. Swedish websites frequently contain English texts or translations, or are entirely in English. The widespread exposure to English both from above and from below thus serves to secure it as a valid code for communication in Sweden, resulting in an increased use of English such that, in ever-increasing domains, it approaches that of Swedish. In general, it can be said that Swedes have each language at their disposal as a communicative tool, confirming previous assessments of the status of English as a second language in Sweden (Ferguson 1994; Josephson 2004; Phillipson 1992). For this reason, it is increasingly common for the languages to co-exist in one and the same communicative context, logically resulting in code switching.

## 10.3 English in Sweden: code switching and crossing

Having established the prominence of English in Sweden in Section 10.2, I now briefly present some basic tenets of crossing and code switching. While there are many terms and overlapping concepts in the bilingualism and language contact literature, code switching can for simplicity's sake be considered the umbrella term for the phenomenon of the simultaneous use of two languages in one conversational exchange. Code switching can be further distinguished as intrasentential or intersentential, depending on where the switch occurs in terms of clausal orientation. In terms of predictability, code switching predominantly occurs among bilingual speakers sharing the same or overlapping linguistic repertoires (Grosjean 1982; Li Wei 2005). The shared linguistic background of interlocutors automatically establishes them as members of an in-group, which code switching serves to confirm. Thus code switching frequently functions as a communicative strategy for achieving social goals, including to signal interpersonal relationships (Blom and Gumperz 1972), to redefine social roles (Myers-Scotton 1988), or to manage social relations (Auer 1988).

Similar to code switching, crossing (or 'code crossing', Rampton 1995) is also a socially motivated phenomenon, occurring among speakers with access to two or more linguistic systems. The critical difference between code switching and crossing lies in the status of the speaker as a legitimate member of the speech community associated with each language. Crossing 'is concerned with switching into languages that are not generally thought to belong to you. This kind of switching, in which there is a distinct sense of movement across social or ethnic boundaries, raises issues of social legitimacy that participants need to negotiate' (Rampton 1995: 280).

The prominent role of English in Sweden and its consequent spread from high-status to low-status domains has helped encourage an ideological shift in the view of English as a foreign language to English a second language. This shift is significant, as it reflects both the ever-increasing use of English across domains and the progression beyond the approach to English as merely a source for lexical borrowing to English as a valid, viable code for communication. In the context of analyzing the use of English in Sweden, the distinction between foreign language and second language is potentially significant: on the one hand, proposing the status of English as a *second language* in Sweden enables a perspective of Sweden as a Swedish-English bilingual speech community. This, in turn, sets the stage for an application of a code switching framework for analyzing the simultaneous use of English and Swedish, as code switching commonly characterizes bilingual speech communities (Auer 1988; Blom and Gumperz 1972; Myers-Scotton 1988; Li Wei 2005).

On the other hand, maintaining that English is a *foreign language* in Sweden invites an analysis of the use of English in Sweden as crossing. The status of English as a foreign language in no way precludes bilingual abilities among Swedes; on the contrary, like code switching, crossing assumes a bilingual linguistic repertoire. Crossing is thus a type of code switching, but one that entails a lack of rights to, belonging to, or ownership of a particular language. While it is not productive to categorize English conclusively as a second or foreign language in Sweden, it is useful to bear in mind the distinction as well as that between code switching and crossing, so as to better understand and interpret examples of English usage and the subsequent social implications.

## 10.4 English in *Rocky*

In this section, I consider examples of the use of English in *Rocky*, analyzing code switching and crossing as indicators of in-group/out-group membership with humorous overtones. The examples represent Swedish language usage and two categories of English usage: both nonnative English and hip-hop English.

### 10.4.1 Swedish, nonnative English

In the early publication years of *Rocky*, that is, 1998–2000, there are very few examples of the use of English. Perhaps this is not so surprising as any new publication of this sort initially seeks to find a readership and thus may be keen not to alienate a potential audience by such linguistic means. There is evidence, however, that Kellerman ignored the risk of alienating his audience. While being featured in Metro, *Rocky* was repeatedly bounced from various other local newspapers due to reader complaints of impropriety. For this reason, the conspicuous absence of English may instead point to Kellerman's own developing assimilation of English and subsequent penchant for code switching. Example 1 illustrates the oblique use of English from one of the earliest strips of 1998.

In Example 1, Rocky is calling an American, English-speaking acquaintance in New York, to inquire about the possibility of accommodation while visiting. The text of the strip is in Swedish, but in the first panel, there is an indication that the telephone conversation actually takes place in English, with Rocky speaking what is labeled in the strip as 'school English,' presumably referring to a rudimentary variety. The complete translation is provided below; in all examples, the English translations of Swedish are mine; code switches to English appear **in bold**; standardizations of Swedish-spelled English appear in [square brackets]; extralinguistic notes appear in (parentheses); cultural explanations appear in italics between /*slanted brackets*/.

Example 1

(1) Rocky: Hello? Marcus? This is Rocky! From Sweden! Hiiii! How is it going?* (*school English)
(2) Rocky: Yeah, it's been two years now. But you know I only get in touch when I need something. Ha ha ha! No, seriously...
(3) Rocky: I'll be in New York soon, and I need a place to stay... Really?! That's great!
(4) Rocky (to Tiger): What a guy! We've met one time, and he's letting me stay with him for free as long as I want!
(5) Tiger (thinking): Good! Then maybe I won't have to sell one of your kidneys to pay my phone bill...

The strip in Example 1 leaves the actual form of English used by Rocky to the reader's imagination. Nevertheless, the explicit mention of Rocky's use of English establishes it as an accessible code for him, validating and, in turn, confirmed by subsequent uses of English.

The use of translated English in *Rocky* is limited to the storyline of Rocky's New York *séjour*, initiated in the strip in Figure 10.1 and

Figure 10.1  Rocky calls New York. All figures copyright Kellerman/Kartago, 2008. Used by permission.

documented in the first months of publication, in 1998. In all subsequent strips, code switches to English remain untranslated. Much like the reference to 'school English,' however, the quality of English used by Rocky or other characters is usually conveyed somehow, for example, by creative phonetic representations of Swedish accent, or by more subtle grammatical, semantic or pragmatic deviations from idiomatic English. The English conversation in Example 2 illustrates such deviations, thereby establishing a contrast in quality between native and nonnative English.

Example 2

(1) Duck: How nice to run into you on a Monday! This is a friend from the USA.
(2) Rocky: **How do you do?**
(3) Rocky: **My name is Rocky, what is your name? How do you like Sweden, do you love it?**
(4) Rooster: I dunno yet. The people are kind of strange and uptight.
(5) Rocky: **Yes, we are wery (sic) strange and uptight. This is because of the German monk Martin Luther King who told us to work hard and feel bad about it.**
(6) Rooster: But it's a nice country, you have free health care. That's so weird!
(7) Rocky: **It's not free, we pay for it with tax! And we think it is funny and very sexy to pay the tax!**

In example 2, Rocky is introduced to an American friend (a rooster) of a mutual friend (a duck). Rocky then initiates a conversation, beginning with the formulaic greeting, 'How do you do?' which comes across as a rather formal expression for this social context. Without waiting for

*Figure 10.2* Rocky meets an American rooster. All figures copyright Kellerman/Kartago, 2008. Used by permission.

a response to his greeting, Rocky continues his turn in the second panel, stating his own name, asking the rooster's name, and then posing additional, back-to-back questions about how the rooster is experiencing Sweden. There is an awkwardness to this sequence of utterances, deftly conveyed by Rocky's unsophisticated verbosity in the form of persistent questioning and both lexical and structural repetition. In the third panel, Rocky's strategy of repetition persists with his appropriation of the rooster's phrase 'strange and uptight.' Furthermore, there is an indication of a mispronunciation typical of Swedish speakers of English, namely the use of the approximant /w/ instead of the voiced fricative /v/ represented by the spelling of 'very' as 'wery.' Finally, the fourth panel contains a variety of features typical of nonnative speech. In addition to further examples of lexical repetition in 'free,' 'pay,' and 'tax,' there is hyperarticulation in 'it is' (in contrast to the previous use of 'it's'), an overuse of the definite article in 'the tax,' and, finally, semantic incongruity in the utterance 'We think it is funny and very sexy to pay the tax.' This description of paying taxes as 'funny' and 'very sexy' can be understood as an infelicitous evaluation, as taxation usually conjures up negative associations. Rocky's utterance thus confirms his nonnative status both in form and function: the odd lexical combination in fact reflects a decidedly nonnative (i.e., non-American) attitude towards taxation.

The overall awkwardness of Rocky's use of English is even more palpable in juxtaposition with the rooster's, whose own use of English is more colloquial and idiomatic. Although the rooster says very little in comparison to Rocky, native fluency is suggested by the use of an combined form ('dunno') and contractions ('it's' and 'that's'), as well as the hedged ('I dunno' and 'kind of') expression of a negative evaluation ('strange and uptight'), countered by the pragmatically felicitous use of 'but' followed by a positive evaluation ('nice country' and 'free healthcare'), and concluded with the slang expression, 'That's so weird!'

The concise representation of two very different varieties of English is evidence of Kellerman's own linguistic aptitude, awareness, and proficiency. Clearly he has the grammatical and communicative competence to produce correct and pragmatically appropriate English. His deliberate decision to voice Rocky with a Swedish, nonnative variety of English reflects a move to align Rocky – and thus himself – with a Swedish, and, significantly, non-American, identity. This particular strip thus serves as an acknowledgment of, and suggests a self-consciousness towards, the Swedish nonnative speaker variety of English. Rocky's awkward English certainly makes a mockery of Swedish English in general, but it is delivered in a self-deprecating manner, as Rocky himself is the perpetrator.

As an example of in-group mockery, the portrayal is accepted as a pragmatic move to assert national, in-group identity.

In this section's final example, the theme of mocking the Swedish variety of English is revisited. In Example 3, Rocky is conversing with his friend, a rat, and the rat's girlfriend, a cat. This strip is part of an ongoing storyline, where Rocky and the rat are living temporarily in Berlin, Germany. The cat is visiting, and has recently complained that Rocky and the rat only talk nonsense and trivialities. The strip begins with Rocky addressing this accusation.

Example 3

(1) Rocky: We can talk seriously, but at four in the morning after four bottles of wine, we're not exactly Adaktusson in *Are You Smarter than a Fifth-Grader*! /*Aduktusson is a Swedish television journalist*/

(2) Cat: But you guys babble all day long, too! You're so afraid of things turning serious that it's pathetic!

(3) Rocky: Jeez, there were 30,000 serious assholes talking seriously in Copenhagen for three weeks, what'd we get out of that? Nothing! Reinfeldt raised his hand and was all, '[**Uhh, we in Sweden think it's very cold and wet all the time. We plan to set up a goal for 2012 to order the poor countries to let us come there in the wintertime!**]

(4) Rat: They're all, 'No, we're talking about lowering the temperature on Earth and that…'

(5) Rocky: Ookay, right. Then it's nothing … Is anyone headed north?

(6) Rat: Share a jumbo-jet? I've got room for 748 if anyone's going my way?

In this strip, it is Sweden's Prime Minister Fredrik Reinfeldt who is targeted by Rocky as a speaker of Swedish, nonnative English. The aspect

*Figure 10.3* Rocky speaks of serious topics. All figures copyright Kellerman/Kartago, 2008. Used by permission.

of nonnativeness is represented primarily by the use of Swedish spelling to approximate a nonnative accent; I have standardized Kellerman's English spelling in my transcript. Similarly to the strip in Figure 10.2, this strip capitalizes on in-group membership. Here, however, in-group membership can be understood to exist on two levels. First, there is an in-group consisting of Rocky and the rat; second, there is an in-group consisting of the Swedish population, all of whom are potential speakers of the Swedish, nonnative variety of English. On the one hand, Rocky's use of the personal pronouns 'we' ('vi') to refer to himself and the rat and 'they' ('dom') to refer to the delegates at the 2009 Copenhagen Climate Conference establishes an in-group/out-group distinction. This distinction is, in effect, further secured by the explicit naming of Prime Minister Reinfeldt as one of the 'assholes' in Copenhagen, accompanied by an overt mocking of him via an exaggerated Swedish, nonnative variety of English. On the other hand, the targeting of a fellow Swedish national in this way can also be interpreted as self-targeting, suggesting the existence of another, larger in-group, namely, the in-group of Swedes and their associated speech community. Rocky, his interlocutors and Reinfeldt are mutual members of this in-group, which is further established by Reinfeldt's voiced use of the first-person pronouns 'we' and 'us.' In this way, Rocky, his friends, and Reinfeldt can all be aligned with this in-group based on national identity, and characterized by the common use of a Swedish, nonnative variety of English.

With the exception of Figure 10.1, the source of humor in these strips can be attributed to the use of nonnative English, specifically, the use of a Swedish, nonnative English variety in contrast to standard or native English. As such, there is incongruity resulting from a native/nonnative opposition. Incongruity is generally acknowledged as a basic prerequisite for linguistic humor founded upon linguistic forms:

> At the basis of much linguistic humor are the various types of linguistic units and their interrelationship. The notion of incongruity is crucial to such humor. It involves the disarray of phonological and grammatical elements, the twisting of the relationship between form and meaning, the reinterpretation of familiar words and phrases, and the overall misuse of language. (Apte, 1985: 179)

Each occurrence of English in Figures 10.1 through 10.3 illustrates code switching (as opposed to crossing) for the purpose of communicating with non-Swedish speakers. For this reason, the switches can be further categorized as situational (Blom and Gumperz 1972), since they are direct responses to a situation requiring the use of a specific language, i.e., English. In each example, the use of Swedish, nonnative English is presented as incongruous in its opposition to a standard or native variety. In Figure 10.2, the incongruity develops out of the opposition to the rooster's native-speaker English, while in Figure 10.3, the incongruity can be attributed to the implied opposition of a standard English variety to the run-on structure and orthographical representation of the Swedish, nonnative accent of the Prime Minister (although, in this strip, humor can also be derived from the proposition of seasonally exploiting 'poor' countries).

Critical to the recognition of this incongruity and thus the appreciation of humor is the constant status of Kellerman and the character Rocky as members of an in-group consisting of the Swedish, nonnative English speech community. The mockery resulting from the exploitation of the native/nonnative opposition is acceptable, and ideally even humorous, precisely because it is restricted to other in-group members with regards to its source and target. In this way, it can be considered a form of self-deprecating humor capitalizing on a personal quality that other in-group members might recognize in themselves.

Finally, these examples indicate that situational code switches to English invoke an 'English from above' variety, indicated by institutional usage ('school English'), a more formal register ('how do you do'), and associated with international communication in a high-status domain (the UN's Climate Conference 2009). In this regard, the variety of English employed in native speaker or lingua franca interaction suggests a perceived distance to the language on the part of the speaker, further suggesting that, in these and similar contexts, English is approached as a foreign language.

### 10.4.2 Hip-hop English

In the previous section, I illustrated how the use of Swedish, nonnative English is variably represented in *Rocky*, namely, as translated English and as a nonstandard variety in terms of phonetic, lexical, structural, semantic or pragmatic features. I suggested that the use of Swedish, nonnative English contributes to establishing an in-group identity by highlighting a native/nonnative opposition, the incongruity of which is exploited for humor. Code switching to English was shown to be situational, motivated by interaction with non-Swedish speakers. Situational code switching of this kind thus emphasizes the native/nonnative opposition, resulting in an incongruity which, as was shown, is exploited for humorous purposes for an in-group audience. In the examples presented in this section, the notions of in-group membership and incongruity are revisited, as switches to English are shown to highlight an alignment with a subcultural identity normally associated with the other, namely, a hip-hop identity.

Hip-hop culture is a recurring element in *Rocky*, stemming from Kellerman's own interest in hip-hop music. References to hip-hop are often in the form of singing or reciting lyrics, as shown in Example 4. In this strip, the dialogue is almost entirely in English, and features the lyrics to the song *Gimme the Loot*, by hip-hop artist Notorious B.I.G. Just prior to the time of this strip, Rocky had his license revoked, and is driving illegally with two friends in the car. He is understandably nervous about seeing a police car pull up alongside.

Example 4

(1) Rocky: Damn is he ever staring at me! Is he going to pull me over now when I'm driving without a license?!
(2) Crocodile: Relax. You haven't done anything.
(3) Rocky: He can tell that I'm driving illegally!
(4) Crocodile: **Be cool, fool! He ain't gonna roll up**, all he wants is fuckin donuts!
(5) Dog: **Then why the fuck he keep lookin!?**

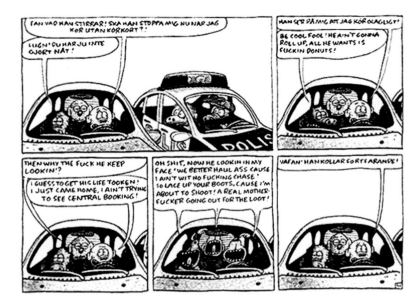

*Figure 10.4* Rocky resists the police. All figures copyright Kellerman/Kartago, 2008. Used by permission.

(6) Rocky: **I guess to get his life tooken! I just came home, I ain't trying to see central booking!**
(7) All: **Oh shit, now he lookin in my face! We better haul ass cause I ain't wit no fucking chase! So lace up your boots, cause I'm about to shoot! A real mother fucker going out for the loot!**
(8) Rocky: Damn it! He's still staring!

The code switch in this strip is triggered by the topic of whether the policeman is staring at Rocky. It is not a communicatively necessary switch, as the situational switches in examples 1 through 3 are. This is instead an example of what Blom and Gumperz (1972) termed 'metaphorical' code switching, relating to 'particular kinds of topics or subject matter rather than to change in social situation.'

It is worth pointing out that the switch to English occurs early in the strip, and persists almost until the end. In other words, nearly the entire strip features hip-hop English. Kellerman's decision to devote a strip of this length, a longer, Sunday edition double strip, to featuring the lyrics of a hip-hop song reflects a deliberate and prolonged move to align himself/the character Rocky, and his/Rocky's friends with a particular in-group, namely, one defined by an interest in and familiarity with hip-hop music. In so doing, he automatically implies an out-group, consisting of those who do not share the same affinity for hip-hop. In this strip, it is the police who visually and conceptually represent the out-group. Two oppositions are thus featured in this strip, Swedish/English and in-group/out-group, from which humor is derived based on incongruity. First, the switch to English is unexpected and thus incongruous in terms of lacking situational motivation. This is not humorous *per se*, but the persistence of the switch, which ultimately features all of the interlocutors simultaneously engaged in a refrain crescendo, may be considered incongruous enough with respect to an expected progression of conversation to be humorous.

Second, the hip-hop in-group/out-group opposition creates an incongruity with respect to expectations of behavior. This strip suggests, for example, that for Rocky and his friends, it is expected behavior to relate ongoing experiences to known hip-hop music in this way, showcasing their expertise, aligning with a hip-hop identity, and thereby reaffirming their in-group membership. Such behavior can, however, be regarded as unexpected from the perspective of an out-group member, and thus would understandably warrant conspicuous, curious, or, in the case of the policemen, even suspicious observation. That Rocky seems surprised that the policemen continue to stare even after the interlocutors' collaborative, impromptu recital confirms the group opposition and creates the necessary incongruity from which to derive another source of humor.

In example 4, the physical representation of the characters enclosed in their respective cars contributes to establishing the group opposition, pitting the two sets of characters against each other. Rocky and his friends are enclosed in their own hip-hop world, while the police are enclosed in another, conflicting, non-hip-hop world. In hip-hop culture, it is not unusual for artists and law enforcers to be presented as having an adversarial relationship. In light of this context, Rocky and his friends can be considered authentic members of the hip-hop community, as they, too, are experiencing harassment (real or imagined) by the police. The physical boundaries depicted in this strip serve to confine the characters and events, thereby emphasizing the opposition, which in turn facilitates an acceptance of Rocky and his friends as ratified representatives of the hip-hop community.

In example 5, Kellerman again trades on the in-group/out-group distinction, this time using it to call into question the authenticity of Rocky's hip-hop in-group membership. Rocky and his friend, a bird, are observing some sheep at a farm; the bird has apparently referred to goats as 'male sheep,' to which Rocky reacts with indignation.

Example 5

(1) Rocky: Goats aren't male sheep, goats are goats! Sheep have their own rams, but they're not goat-rams. Don't you know anything?
(2) Bird: As if you know anything about sheep!
(3) Rocky: Hey, we had 300 sheep when I was little! I know a whole fucking lot about sheep, goats and everything in between. I could make your sweater from **scratch** if I wanted to!
(4) Bird: This sweater isn't made of sheep's wool, it's from Supreme in Tokyo, and not from some local history association on Gotland! /Gotland is a small island off the east coast of Sweden known for sheep raising/
(5) Rocky: What the hell do you think yarn is you stupid ass?!
(6) Bird: Well, it sure the hell isn't wool at any rate, pea-brain!!!
(7) Rocky: Man, I've gotten up at 6 a.m. and fed sheep, I've shoveled a hundred tons of sheep shit, shepherded sheep in a cowboy hat, I've ridden sheep, slept with sheep, sheered sheep, washed and carded wool, spun and colored yarn and knitted many a tightly-pulled pot-holder!

*Figure 10.5* Rocky proves he knows about country life. All figures copyright Kellerman/Kartago, 2008. Used by permission.

(8) Rocky: So **don't step to me talking about** yarn, **nigguh!** (singing) **Watcha** (sic) **know about that? You don't know about that!**

In example 5, the setting is the decidedly un-hip-hop environment of a country field with grazing sheep. Rocky and the bird engage in a heated argument about sheep and goats, prompting Rocky to conduct a diatribe on his experience with sheep as proof of his expertise. His rant is similar to showboating or bragging; it is the kind of shameless self-promotion which figures as another key characteristic of hip-hop discourse (Forman and Neal 2004).

Similar to example 4, the code switch to hip-hop English in the last panel is metaphorical, although motivated by discourse style, i.e. establishing credibility, as opposed to topic. Rocky's first utterance is not a recital of any particular lyrics, but it does feature lexical items associated with the hip-hop vernacular, such as 'step to' and 'nigguh'. His second utterance is reminiscent of hip-hop artist and producer Timbaland's song *Whatcha know about this*, but it is not a verbatim recital.

The pastoral setting combined with an argument about sheep among two Swedish males creates a clear opposition to the conventional hip-hop environment of urban America populated by African American males comparing gangster credibility. The incongruity is equally obvious. Although a member of the hip-hip in-group would recognize Rocky's rant as a discursive trigger and thus expect a switch to hip-hop English, Rocky is actually presented as an out-group member, who only appropriates the code of an in-group to which he aspires. The switch is therefore more accurately labeled a case of crossing. Rocky's failed claim to membership is visually suggested by his flailing arms as an attempt to simulate hip-hop gesturing, as well as the bird's disengaged posture and blasé facial expression. Furthermore, his potential in-group membership is linguistically threatened by his use of Swedish 'garn' instead of 'yarn,' and the (mis?)use of 'nigguh.' The humor thus derives from the incongruous use of hip-hop English as a pragmatic move by Rocky to align himself with an in-group of which he is not recognized as a legitimate member. Kellerman has himself addressed the inherent incongruity of white, middle- to upper-class Swedish males aligning themselves with the African American hip-hop culture: 'When a Swede says something like Jay-Z would say, that's automatically funny. It's still white here, but in Sweden, it's funnier. Most of my friends have grown up on hip-hop, but it's like a joke – we're so not gangsta' (MacDonald 2005).

The (mis)appropriation by Whites of the hip-hop vernacular is an acknowledged and controversial phenomenon (Armstrong 2004). Historically, hip-hop is rooted in African American culture and represented by a distinct vernacular, African-American Language (Smitherman 1997). Authenticity is a recurring theme in hip-hop texts and central to its culture because it 'becomes a node through which flow arguments about who is capable, or not, of legitimately interpreting a culture, and therefore, participating in its most esoteric forms of antecedent oral and aesthetic culture…' (Jones 2006:2). In the past, Whites have managed

*Figure 10.6* Rocky attends a hip-hop music festival. All figures copyright Kellerman/Kartago, 2008. Used by permission.

to participate in hip-hop by appropriating the hip-hop vernacular to confirm their out-group status, mock their whiteness, and call attention to their cultural shortcomings (Fraley 2009). In effect, Rocky is doing precisely this in example 5: he appropriated the hip-hop vernacular (appropriated via lyrics) to express his white European background. It is an incongruity played upon for humor, but which may be misunderstood as a mockery of the hip-hop culture.

The final example features Rocky and his friend, a wolf, at a music festival, where Rocky is due to conduct interviews with some hip-hop artists.

Example 6

(1) Wolf: So, what's the plan for today?
(2) Rocky: The plan is to [**get drunk, get crunk, get fucked up!**]
(3) Wolf: Isn't there anyone you're supposed to interview?
(4) Rocky: No, Kjelis (sic) cancelled her tour because she couldn't find a cat-sitter, so I'm off today!
(5) Wolf: Aren't there any concerts we should see?
(6) Rocky: No, we don't want to see concerts today, today we just want to [**get drunk, get crunk, get fucked up!**]
(7) Rocky: (singing) **I want a girl I can fuck in my hummer truck, apple bottom jeans and a big old butt! Shake that ass fo me! Shake that ass fo me!**
(8) Wolf: You got **fucked up** at least…
(9) Rocky: **Stop the violence in hip hop, waaaiow!**

In this example, the code switch to hip-hop English can be understood as both situational and metaphorical at once. On the one hand, the presence of Rocky and the wolf at a hip-hop music festival invites switches to hip-hop English. In other words, it is the situation in which they find themselves that triggers a code switch. On the other hand, one can consider their presence at the festival to prime them for, as opposed to triggering, code switches to hip-hop English. Instead, it is the subject matter of the wolf's question which invites Rocky's code switch. Despite Rocky's persistent code switching, the wolf refrains from using English until the last panel, switching intrasententially, triggered by Rocky's previous use of the phrase 'fucked up.'

Whether situational or metaphorical, the switch nevertheless reflects group alignment. Membership in the hip-hop in-group is attributed to Rocky and the wolf by virtue of their attendance at the festival.

It is furthermore secured by the explicit mention of Rocky's interview assignment and the naming of Kelis, a well-known hip-hop artist. The first code switch to English is in the form of lyrics to the Eminem song, *Shake That*, recited as an answer to the wolf's question about the plan for the day. Familiarity with the song and the ability to recite its lyrics appropriately with regards to the context are other indicators of hip-hop in-group membership. Nevertheless, there are aspects of Rocky's turns that instead establish the now familiar in-group/out-group opposition. Similar to example 2, Rocky's first and second code switches to hip-hop English, as well as the artist name 'Kelis' are written in nonstandard spelling, suggesting a Swedish, nonnative English pronunciation. The switch in the third panel, however, is written accurately; here, the in-group/out-group opposition is instead created via the presence of a third party, the bird (female) walking by. Despite the accurate recital of lyrics (or, perhaps, because of this), Rocky's code switch is apparently not accepted. This can be attributed to the taboo and sexist content of the switch, which is perhaps made all the more offensive if the bird does not acknowledge Rocky as a legitimate in-group member. The fourth panel suggests this to be the case, as the representation of Rocky in an overturned position, complete with the tell-tale cartoon star signaling dizziness or unconsciousness allows the reader to assume the bird physically assaulted Rocky. The result is that Rocky's out-group member status is confirmed. The wolf appropriates the phrase 'fucked up,' originally used by Rocky in his first two code switches, to assess Rocky's condition, and in so doing, emphasizes the in-group/out-group opposition. 'Fucked up' can have two meanings, referring to either an alcohol- or drug-induced state, or to being physically beaten. By invoking the latter connotation, the wolf also rejects Rocky's overtures towards asserting a hip-hop identity. Recalling Apte's (1985) assessment of the role of incongruity in humor, it can be said that the wolf's appropriation of the phrase 'fucked up' is humorous in its incongruity resulting from 'the twisting of the relationship between form and meaning.' The strip concludes with a final effort by Rocky to assert a hip-hop in-group member identity, now reciting from Boogie Down Productions' *Stop the Violence*.

The fact that nearly all of Rocky's switches to hip-hop English are in the form of verbatim recital of song lyrics gives weight to the powerful influence of music as an example of English from below. It should be acknowledged that consumers of music may memorize and recite lyrics to songs without understanding or reflecting on the meanings. At the same time, it can be proposed that nonnative speakers interested in a musical genre may be particularly keen on learning and understanding lyrics. This may be the case for Kellerman, a hip-hop devotee. Code switches to hip-hop English allow Kellerman/Rocky to showcase a familiarity with hip-hop lyrics via an ability to apply lyrical content to interactional contexts in semantically and pragmatically appropriate ways. This practice of metaphorical code switching suggests an intimacy with the language that would normally be associated with English as a second language. As such, the examples from Rocky ultimately indicate a dichotomy which assigns an approach to English from above as a foreign language, and to English from below as a second language.

## 10.5  Discussion

In this chapter, I have set out to analyze the use of English in the Swedish-language comic strip, *Rocky*. I have first presented evidence of the widespread, cross-domain use of English in Sweden, in order to establish English as a viable language choice in the Swedish speech community, in both high- and low-status domains. The increasing presence of and exposure to English in Sweden reflect influences of English from above and English from below (Preisler 1999), and contribute to a dominance of the English language such that it rivals the use of Swedish in Sweden.

The practice of code switching as the concurrent use of both Swedish and English reflects the status of both languages as viable codes of communication. I have furthermore suggested that English can be considered at once a second language and a foreign language in Sweden, proposing that this distinction may be significant in terms of how Swedish speakers relate to or identify with their use of English. If English is considered a second language, then this can reflect a proprietary stance toward the language, with the rights that this status entails. An approach to English as a foreign language, on the other hand, implies a distance from the language and, accordingly, a lack of rights. I maintain that the labeling is relatively unimportant, but the examples in this chapter suggest an ideological shift in the status of English. On the one hand, English from below in Sweden seems to be assuming the status of a second language while English from above is maintained as a foreign language. This distinction may be significant to the analysis, reception, and interpretation of code switches to English, which ultimately will affect the role of English in Sweden, its further spread, and its effect on Swedish.

That English in Sweden reflects influences from above and below calls to mind the fact that English, like any other language, is itself realized as different varieties or codes, and that native and nonnative speakers alike switch between these codes. The examples from *Rocky* suggest that institutional or 'school' English is used in situational code switching, where the use of English is necessary to communicate with non-Swedish speakers such as native speakers of English or speakers of English as a lingua franca. Metaphorical code switches result in the use of English from below; in the *Rocky* examples, such switches are in the form of English appropriated from hip-hop. I have argued that English has been established in Sweden as a valid choice of language for communication, thus enabling both situational and metaphorical switching. It does not follow that crossing should be disabled, as different varieties or codes of English may not belong to or be accessible by the Swedish nonnative speech community. The use of school English is for this reason not an example of crossing: this is a non-exclusive code. In contrast, the use of hip-hop English does constitute crossing, since it is a code associated with an in-group.

Examples of code switches to English in the Swedish-language comic strip *Rocky* were presented against the background of the language of *Rocky* as manifesting the assimilation of English in Sweden. The examples were then analyzed to reflect two approaches to the use of English: as an indicator of in-group/out-group member identity, and as an indicator of incongruity and, as such, a source of humor in the Swedish comic strip medium. The examples invite a number of interesting conclusions. Situational code switching (Blom and Gumperz 1972) as exemplified in examples 1 through 3 suggest a national Swedish awareness of, and perhaps insecurity about, a native/nonnative opposition with regards to speaking English. In *Rocky*, this opposition is exploited for humor by appealing to an in-group, nonnative identity. As a comedic strategy, this has proven to be successful since, according to Kellerman, his readers 'recognize the situations and how the characters behave. A lot of people just laugh at the characters because they think they're stupid and pathetic, but I don't think they would laugh if they didn't recognize themselves' (Spurgeon 2005).

Metaphorical code switching (Blom and Gumperz 1972), as illustrated in examples 4 through 6, was also shown to create an opposition from which incongruity and humor were derived. Similar to examples 1 through 3, in which code switching created an in-group/out-group opposition based on native speaker status, the examples of metaphorical code switching establish an in-group/out-group opposition based on ratified membership in a hip-hop speech community. These examples illustrate how an expression of cultural alignment, in particular with

US hip-hop culture, can be triggered discursively by conversation topic. Switches to hip-hop English take the form of reciting song lyrics, allowing the speaker to showcase familiarity and expertise which serve to assert a legitimate in-group membership. There is nevertheless an inherent incongruity to the appropriation of the urban, African American hip-hop vernacular by Swedes, which encourages an interpretation of switches to hip-hop English as examples of crossing (Rampton 1995). Again, it is an in-group/out-group opposition which establishes incongruity and an opportunity for a humorous resolution.

The apparent popularity of *Rocky* in terms of production, distribution, marketing, and accolades suggests an appreciation of the characters, their experiences, and, significantly, the discourse and the linguistic expression of humor. *Rocky* thus contributes to bringing the practice of Swedish-English code switching to the Swedish linguistic mainstream. In so doing, *Rocky* ultimately contributes to securing the comic strip medium as a host to linguistic trends and language change in Sweden. Indeed, the influence of *Rocky* on spoken Swedish has already been predicted. In a national newspaper article from 2007 summarizing the findings of a research project on the linguistic future of Sweden, a journalist was prompted to make the following statement (Håkansson 2007): 'Kort sagt kommer vi alla tala som Martin Kellermans Rocky om hundra år och ingen kommer att tycka det är något konstigt med det.' ('In short, we are all going to talk like Martin Kellerman's *Rocky* in one hundred years, and no one will think there's anything weird about it').

## References

Apte, M. L. (1985) *Humor and Laughter: An Anthropological Approach*. Ithaca, NY: Cornell University Press.
Armstrong, E. (2004) Eminem's construction of authenticity. *Popular Music and Society* 27: 335–55.
Auer, J. C. P. (1988) A conversation analytic approach to code-switching and transfer. In M. Heller (ed.) *Codeswitching*, pp. 151–86. Berlin: Mouton de Gruyter.
Berg, E.C., Hult, F.M., and King, K.A. (2001) Shaping the climate for language shift? English in Sweden's elite domains. *World Englishes* 20 (3): 305–19.
Blom, J.-P. and Gumperz, J. J. (1972) Social meaning in linguistic structure: code-switching in Norway. In J. J. Gumperz and D. Hymes (eds.) *Directions in Sociolinguistics*, pp. 407–34. New York: Holt, Rinehart and Winston.
Ferguson, C. (1994) Note on Swedish English. *World Englishes* 13(3): 419–24.
Forman, M. and Neal, M. A. (eds.) (2004) *That's the Joint! The Hip-Hop Studies Reader*. New York: Routledge.
Fraley, T. (2009) I got a natural skill...: hip-hop, authenticity and whiteness. *Howard Journal of Communications* 20 (1): 37–54.

Graddol, D. (1996) Global English, global culture? In S. Goodman and D. Graddol (eds.) *Redesigning English: New Texts, New Identities*, pp. 181–217. London: Routledge.
Grosjean, F. (1982) *Life with Two Languages: An Introduction to Bilingualism*. Cambridge, MA: Harvard University Press.
Gunnarsson, B.-L. (2004) Orders and disorders of enterprise discourse. In C. Gouveia, C. Silvestre, and L. Azuaga (eds.), *Discourse, Communication and the Enterprise. Linguistic Perspectives*, pp. 17–39. Lisbon: University of Lisbon Centre for English Studies.
Håkansson, G. (2007) Rocky talar framtidens svenska. *Dagens Nyheter*. Retrieved on 20 May 2010 from http://www.dn.se/kultur-noje/rocky-talar-framtidens-svenska-1.724341
Haugen, E. (1987) *Blessings of Babel*. Berlin: Mouton de Gruyter.
Hollqvist, H. (1984) *The Use of English in Three Large Swedish Companies*. Uppsala: Studia Anglistica Uppsaliensa 55.
Holm, K. (2006) Antiliksomism. *Språkförsvaret*. Retrieved on 20 May 2010 from http://www.sprakforsvaret.se/sf/index.php?id=15
Hult, F. M. (2003) English on the streets of Sweden: an ecolinguistic view of two cities and a language policy. *Working Papers in Educational Linguistics*. Retrieved on 20 May 2010 from http://www.wpel.net/v19/v19n1_Hult.pdf
Hyltenstam, K. (1999) Svenskan i minoritetsspråksperspektiv. In K. Hyltenstam (ed.) *Sveriges sju inhemska språk – ett minoritetsspråksperspektiv*, pp. 205–40. Lund: Studentlitteratur.
Jones, M. D. (2006) An interview with Michael Eric Dyson. *Callaloo* 29 (3): 786–804.
Josephson, O. (2004) *'Ju'. Ifrågasatta Självklarheter om Svenskan, Engelskan och Alla Andra Språk i Sverige*. Stockholm: Norstedts Ordbok.
Kellerman, M. (2008) *Rocky 10 år. Samlade Serier 1998–2008*. Stockholm: Kartago.
Kinn, G. (2005) November 29. Nu ska Rocky erövra USA. Göteborgsposten/Gränslöst. Retrieved fromhttp://www.gp.se/kulturnöje
Labrie, N. and Quell, C. (1997) Your language, my language or English? The potential language choice in communication among nationals of the European Union. *World Englishes* 16 (1): 3–26.
Li Wei (ed.) (2005) *The Bilingualism Reader*. New York: Routledge.
Ljung, M. (1986) The role of English in Sweden. In W. Viereck (ed.) *English in Contact with Other Languages*, pp. 369–86. Budapest: Akademiai Kiado.
MacDonald, H. (2005) Martin Kellerman: It's a dog's life in Sweden. *Publishers Weekly.com*. Retrieved on 8 April 2010 from http://www.publishersweekly.com/article/420273-Martin_Kellerman_It_s_a_Dog_s_Life_in_Sweden.ph
Myers-Scotton, C. (1988) Code-switching as indexical of social negotiations. In M. Heller (ed.) *Codeswitching*, pp. 187–214. Berlin: Mouton de Gruyter.
Philipson, R. (1992) *Linguistic Imperialism*. Oxford: Oxford University Press.
Preisler, B. (1999) Functions and forms of English in a European EFL country. In T. Bex and R. Watts (eds.) *Standard English: The Widening Debate*, pp. 239–67. London: Routledge.
Rampton, B. (1995) *Crossing: Language and Ethnicity Among Adolescents*. London: Longman.

Sharp, H. (2001) *English in Spoken Swedish: A Corpus Study of Two Discourse Domains*. Stockholm: Almqvist and Wiksell International.

Sharp, H. (2007) Swedish-English language mixing. *World Englishes* 26 (2): 224–40.

Spurgeon, T. (2005) A short interview with Martin Kellerman, a short run from Rocky and a bonus chat with Kim Thompson. *The Comics Reporter*. Retrieved on 20 May 2010 from http://www.comicsreporter.com/index.php/resources/interviews/3434/

Smitherman, G. (1997) 'The chain remain the same': communicative practices in the Hip Hop Nation. *Journal of Black Studies* 28 (1): 3–25.

Teleman, U. (1992) Det svenska riksspråkets utsikter i ett integrerat europa. *Språkvård* 4: 7–16.

Truchot, C. (1997) The spread of English: from France to a more general perspective. *World Englishes* 16 (1): 65–76.

Westman, M. (1996) Har svenska språket en framtid? In L. Moberg and M. Westman (eds.) *Svenska i tusen år: Glimtar ur svenska språkets utveckling*. Stockholm: Norstedts.

# 11
## 'Ah, laddie, did ye really think I'd let a foine broth of a boy such as yerself get splattered...?' Representations of Irish English Speech in the Marvel Universe

*Shane Walshe*

### 11.1 Introduction

Although representations of Irish people in caricatures and cartoons can be dated back to the eighteenth century, they reached their zenith in the nineteenth century, particularly in publications such as *Punch* in Britain and *Puck, Harper's Weekly* and *Yankee Notions* in the United States. These magazines drew on the tradition of the Stage Irish figure which had long been perpetuated by theaters in England and which lampooned the Irish as 'ignorant but harmless drudges, given to drink and emotional excesses, loving a fight, and not above a lie or a bit of minor thievery' (Appel 1971: 367). In addition to presenting these supposedly Irish character traits, such cartoons also propagated stereotypes about Irish people's appearance and speech patterns (Soper 2005). Thus, Irish characters were shown with simian features and with the costumes and props traditionally associated with Stage Irish figures, namely *dudeen* 'clay pipe' and a *shillelagh* 'wooden cudgel.' They were also provided with a repertoire of stock phrases and spoke with a brogue, which was achieved through respellings deemed to reflect their Irish accent. The effect of such portrayals on the readers of those magazines was the same as that identified by Jonathan Swift over a century earlier, namely: '[...] the Irish *brogue* is no sooner discovered, than it makes the deliverer, in the last degree, ridiculous and despised; and from such a mouth, an Englishman expects nothing but bulls, blunders and follies' (Swift 1728: 346).

To gain an impression of what the stereotypical speech in these cartoons may have 'sounded' like, one need only turn to Martin Weimer's analysis of the representation of Ireland and the Irish in *Punch* magazine, which conveniently summarizes the most frequent respellings as being the following: <oi> for <i> (*foine, Oireland*), <i> for <e> (*iligant, gintry*), <a>, <ay>, or <ai> for <ea> (*bastes, trayson, aisy*), with, what Weimer calls, the guttural Irish pronunciation being achieved through the insertion of <h> after consonants, as in 'Ghlory' (1993: 465). He adds that representations of typically Irish English (hereafter IrE) grammatical forms are rare and consist mainly of the use of *them* as a demonstrative pronoun, as in 'thim fowl,' or of adding the tag *is it* to statements to create questions, as in 'An idjut, is it?' (1993: 466). Instead of representing IrE grammar, the most common way of suggesting Irish speech in those cartoons is by resorting to interjections, such as *bejabers*, *begorra* and *bedad*, or to exclamations, such as *arrah*, *accushla*, *musha*, and *faix* or *faith*, all of which emphasize the emotionality and wildness of the Irish (1993: 466–67). It is also common for these Irish figures to swear oaths and to seek the intercession of the saints by using expressions such as '[t]he blessed saints preserve us' or '[t]he holy Saints be with ye' (p. 471). Weimer also singles out expressions such as 'the broth of a boy,' which appears as a typical form of praise from an Irishman, and 'the top uv the mhornin' to ye' or 'more power to yer elbow,' which are supposedly typical greetings (pp. 470–71).

The Stage Irish nature of such expressions was highlighted at the turn of the century by George Bernard Shaw in his satire *John Bull's Other Ireland*. In the play, Doyle, an Irishman, scolds Broadbent, an Englishman, for having been conned into giving money to a Scotsman who had pretended to be Irish by using the stock phrases (Shaw 1907: 81):

BROADBENT: But he spoke – he behaved just like an Irishman.
DOYLE: Like an Irishman!! Man alive, don't you know that all this top-o-the-morning and broth-of-a-boy and more-power-to-your-elbow business is got up in England to fool you, like the Albert Hall concerts of Irish music? No Irishman ever talks like that in Ireland, or ever did, or ever will.

However, despite drawing attention to these expressions, Shaw's play did not signal the death knell for the clichéd representation of Irish speech. Instead, much to the consternation and/or amusement of Irish people, such expressions have continued to appear across a variety of fields, including music, film, television, advertising and, as is the case in this chapter, comics.

Of note for this analysis is the idea that these representations are often British and American constructs, reflecting widespread perceptions of the Irish. Williams, who analyzes the representation of the Irish in popular songs in America, underlines how important it is to remember the origin of these songs: 'Within the Irish community, Irishness – American Irishness – was real. However, the language, images, and symbols that might have expressed that reality to other Americans had been preempted by the clichés of popular culture' (1996: 242). This notion is echoed by Wittke, who adds that the popularity of Irish songs 'helped to reinforce certain features of the Irish stereotype' (1952: 216). Furthermore, he argues that the cumulative effect of exposure to these stereotypes from songs and plays meant that 'American audiences saw these inaccurate stage immigrant characters and heard these dialects so long that many accepted them as completely authentic, although character traits, overplayed for comic effect, obviously emphasize idiosyncrasies and deviations from the general folk pattern' (1952: 232).While some research has already been conducted into the representation of the Irish in various media, e.g., Williams (1996), Walshe (2009a, 2009b) and Negra (2001), no attention has been given to the portrayal of the Irish in modern-day comics. This chapter aims to remedy that gap in the literature by offering an analysis of the representation of Irish figures in Marvel comics. It will examine which features are most frequently used to create the impression of Irish speech and will explore to what extent Stage Irish features are still present.

Together with its rival DC Comics, Marvel Comics has dominated the world of comic books for decades and boasts thousands of characters, who appear in numerous series and crossovers. For the purposes of this study, five Irish characters from the Marvel universe have been chosen and their speech analyzed in a total of 150 comics, published between 1967 and 2009. These Irish characters are: Banshee, Siryn, Black Tom Cassidy, Shamrock and Irish Wolfhound. Since these characters have been drawn and scripted by several, mostly US, artists and writers over the years, an interesting aspect of this analysis will be to see which features are the most pervasive across the entire corpus and to what degree representations of IrE speech in the Marvel universe are consistent.

## 11.2 Irish characters in the Marvel universe

The first of the Irish characters to appear in the Marvel universe was Banshee (Sean Cassidy), a mutant from County Mayo, whose special power is his ability to emit a high-pitched scream, with which he can generate sonic blasts and which enables him to fly. He first appeared in

*Uncanny X-Men #28* (1967). In this early incarnation, penned by Roy Thomas, his Irishness is conveyed more through his appearance than his speech. In keeping with the nineteenth-century caricatures, he is portrayed with simian features and displays a fondness for smoking his clay pipe. His speech, however, does not bear any of the linguistic features described by Weimer, or any other scholars, but rather is reminiscent of the 'ornate, overblown phraseology' (Lee 1978: 60) often used by deities and otherworldly figures in the Marvel universe, such as Thor and the Silver Surfer. Indeed, his description of the citizens of New York as 'these mortal fools' would suggest that in this first appearance, Banshee was less human than the character which readers would become acquainted with eight years later when he joined the X-Men. The latter character, written by Len Wein, and later by Chris Claremont, sounds considerably more 'Irish' than his predecessor, although, only in a Stage Irish way. The first word out of his mouth is the stereotypical *Begorra*, followed by a number of the other features which were also present in the *Punch* cartoons. A series of respellings also suggest that he speaks with an Irish brogue. The sum total of his speech from that historic issue *Giant-Size X-Men #1* (1975) is reproduced below:

(1) Begorra! 'Tis Professor X himself now.
(2) So that's the story, is it? Then sure an' I'll help ye, Professor.
(3) 'Twill be nice to tread the straight an' narrow ... fer a change.
(4) 'Tis a pleasure ta be workin' with ye, laddy.
(5) Saints, laddy–will ye look at the size o' them beasties!
(6) Well, laddy–sure 'n it looks like we've done fer the beasties! We'd best be gettin' on to that temple we spied a touch back.
(7) Faith! 'Tis good t' be seein' ye all again. 'Twas a moment there I had me doubts.
(8) Begorrah! The blinkin' beastie's gettin' stronger now! But how–?

As will be seen, many of these features also occur in the speech of the other Irish characters over the years. One of those figures is Banshee's daughter, Siryn (Theresa Rourke Cassidy). She possesses the same mutant powers as her father, but can also modulate her voice in such a way as to persuade others to do her bidding. She never knew her father growing up and, following her mother's death, was raised by her 'uncle,' Tom Cassidy – Banshee's first cousin and archenemy. She was first introduced to Marvel readers in *Spider-Woman #37* (1981), written by Chris Claremont, where her Irishness was signaled solely by her accent rather than her use of Irish or pseudo-Irish expressions. The opposite approach is evident in her speech in the series *X-Factor* (2005–present), scripted by Peter David. Here, Siryn's accent is not indicated at all, but her Irishness is indicated by her use of religious exclamations, and, for good measure, her red hair, freckles, and alcoholism, all stereotypes associated with the Irish. In this regard, it is also interesting to note that whereas Siryn and Banshee both first appeared with blonde hair, their later characterizations in *Generation X* (1994–2001) and *X-Factor* (2005–present) have often had the red hair and freckles that are typically connected to the Irish, thus confirming Dowling's observation that artists often add 'phenotypic attributes that American audiences would expect' of their comic book characters (2009:185).

Also a mutant is Black Tom Cassidy, who possesses the ability to generate blasts of concussive force or heat by channeling his power through wood, which, in keeping with the Stage Irish tradition, is often in the form of a shillelagh. He was first introduced in *Uncanny X-Men #99* (1976). The representation of his speech is the least consistent of all the Irish characters, varying not only from writer to writer, but also within the work of individual authors. On some occasions, his speech is indicated as being Irish via his use of the same features as Banshee uses, while on others, there is no linguistic evidence of his Irishness. This may be attributable to the fact that, although he grew up in Ireland, he was educated at Oxford and thus may have lost his accent. Moreover, he does not share the phenotype which audiences would typically expect, but instead has dark hair and dark eyes and thus can be characterized as 'Black Irish,' a term often used to describe those 'Irish of Mediterranean appearance' (*OED*).

Shamrock (Molly Fitzgerald) was born in Dunshaughlin, County Meath, where both her father and brother were militant Irish nationalists. When Molly was young, her father called upon ancient spirits to grant his son the power to defeat his enemies. However, that power was given to Molly instead and manifests itself in her having an aura that causes random improbabilities to occur in her favor, giving her what she calls her 'good luck power,' no doubt an allusion to the famed 'luck of the Irish.' She first appeared in the three-part *Marvel Super Hero Contest of Champions* (1982), in which her Irishness was indicated by her Irish accent, as well as her use of *begorrah* and *it*-clefting, a feature which, like all of the grammar features, will be explored in more depth below. She too has red hair and, like Banshee and Siryn, wears a green costume, another allusion to her being from the Emerald Isle.

Finally, Irish Wolfhound (Cuchulain) is a legendary warrior from Celtic mythology, whose powers are his superhuman strength and stamina.

*Figure 11.1* Shamrock from Marvel.com. All images copyright Marvel Comics. Used by permission.

He first appeared in *Guardians of the Galaxy Annual #3* (1990) and, like the early incarnation of Banshee, generally speaks in a formal and archaic manner. In keeping with his pre-Christian mythological background, he does not use the same religious oaths or exclamations as the characters in *Punch* do, but rather exclaims the names of pagan or mythological figures. However, he does adhere to another cultural stereotype in that he certainly fits the bill of the belligerent yet romantic Irishman.

## 11.3 Methodology

### 11.3.1 Establishing the corpus

The most convenient and certainly the most cost-effective way of gaining access to such a dauntingly large back catalogue of comic books was to subscribe to Marvel's own digital comic database online and to then search the over 10,000 comics available there for issues featuring the Irish characters. This search was greatly aided by the website's own guide to 'significant issues' in which the characters appear. Significant issues are those in which characters are introduced, develop love interests, lose their powers, die, or simply those in which important back story is revealed. While not all of the most significant comics are included in the Marvel database, this source proved to very effective, particularly for issues featuring the Cassidys, as it provided a large and representative sample of appearances. Unfortunately, but understandably, comics featuring the minor characters of Shamrock and Irish Wolfhound were not (yet) available in the database and thus were sourced through a comic store.

In total, the corpus amounts to 150 comic books compiled from 28 different series by 28 different writers or writing teams. Within those 150 comics, several of the Irish figures appear in the same issues, and thus the number of appearances per character breaks down as follows: Banshee (80), Siryn (59), Black Tom (22), Shamrock (8) and Irish Wolfhound (2).[1]

Every utterance for each Irish character was transcribed, regardless of whether it contained IrE features, with the total number of words in the corpus amounting to approximately 25,500, which can be broken down into approximate number of words per speaker: Banshee (11,900), Siryn (7,700), Black Tom (3,100), Shamrock (2,100) and Irish Wolfhound (700). This measure gives the reader a better indication of the size of the corpus than the number of appearances does, as in some cases these appearances amounted to little more than a few sentences per issue. When it came to counting the linguistic features, however, rather than adding up the total number of occurrences in the corpus, each feature was noted as either occurring or not occurring in a particular comic. This measure was taken to show the distribution of features over the entire corpus, thus preventing a high number of appearances of a feature in individual comics from distorting the general impression of IrE speech created for the whole corpus. For example, although *laddy* is a very conspicuous feature in *Giant-Size X-Men #1*, appearing three times in Banshee's first six dialogues, such frequency is misleading as the term actually only appears a total of 17 times in the entire corpus, and, more importantly, in only nine different comics.

It should be noted that, unlike the representations of Irish speech in *Punch*, *Puck* and so on, the intention behind the portrayals in the Marvel corpus does not appear to be for the sake of humor, but rather to lend the characters who inhabit the Marvel universe a greater degree of verisimilitude. While it may be unusual to claim verisimilitude or realism in a world where people can fly or walk through walls, Marvel legend Stan Lee explains it as follows: 'Perhaps the most important element in the so-called Marvel style is the fact that we have to stress realism in every panel. I can imagine the scoffers among you leaping to their feet and exclaiming "How can he speak of realism when Marvel specializes in fantasy?" The trick is to create a fantastic premise and then envelop it with as much credibility as possible' (Lee in Daniels 1991: 9). This credibility is attempted via Marvel's making its superhuman characters as human as possible, with everyday fears and worries, and via setting its stories in real world locales, such as New York. An extension of this quest for verisimilitude is Marvel's representation of accents and dialects.

### 11.3.2 Establishing how speech is portrayed

In her article on the representation of Jewish characters in comics, Dowling notes that although the X-Men are from all over the world, 'with few exceptions their "speech patterns" are "standardized" middle American' (2009:192). Even so, it is the speech of these 'few exceptions' which is of interest to us. Before looking in more detail at the representations of Irish speech in the Marvel universe, it is important to examine how nonstandard speech in general is conveyed in these comics.[2] This will help to contextualize the strategies that are used in the case of Irish characters.

Within the 150 comics that make up the corpus, both foreign languages and varieties of English are spoken and two conspicuous patterns emerge with regard to how they are marked as being linguistically different or other.[3] The first is that, in the case of foreign languages, the representations can involve occasional code switching, the use of respellings to indicate accent, or both, while in the case of dialects of English, respellings to indicate accent are used together with the occasional insertion of dialectal words. The second noticeable pattern is that code switches and dialect words tend predominantly to take the form of affirmatives, vocatives, and exclamations. This is the case irrespective of whether the character's native tongue is a foreign language or an English dialect, as can be seen, for example, in Nightcrawler's use of German terms (*ja, mein Freund,* and *verdammt!*) or Wolfbane's use of Scottish expressions (*aye, lassie,* and *och!*). As we shall see, the same strategies also apply to the representation of Irish speech in the comics and thus the features are grouped according to these categories below. In the interest of conserving space, I have chosen examples which include a number of features and can be returned to repeatedly. Therefore, rather than digressing to refer to additional features which are encountered along the way, I shall cross reference them at a later stage. I shall begin with affirmatives.

## 11.4 Data and analysis

### 11.4.1 Affirmatives

The most common method of suggesting that a character is Irish is their use of *aye* rather than *yes* as a means of expressing affirmation:

> (9) **Aye.** It's been good havin' the lot o' ye here these past few days–after that battle with Proteus, we all needed a rest–I'm sorry to see ye go, Cyclops.
> (Banshee, *Uncanny X-Men #129*)

This feature, which is also common in dialectal usage in Britain (*OED*), occurred in 32 of the 150 comics. While that may seem to be quite a small number, it is helpful to contextualize it, and indeed all of the findings, by contrasting its rate of occurrence with that of *yes*, a word one would expect to occur frequently in any corpus of spoken English. Unlike *aye*, *yes* appears in only 19 issues. Admittedly, to those familiar with IrE, the low occurrence of *yes* may not be quite so surprising, as Irish people have a tendency to avoid using *yes* (and *no*), particularly in response to *yes/no* questions. This phenomenon has its root in the fact that Irish (Gaelic) which influences IrE 'has no exact equivalents of the affirmative and negative particles *yes* and *no*' (Filppula 1999:160). Instead, speakers of Irish and, by extension, of English in Ireland tend to resort to other strategies, such as repeating the verb of the question in their answer (Is that your comic? *It is*) or, to a lesser degree, of using a form of *do*, sometimes preceded by the words *indeed* or *that* ([Do you read comics?] *Indeed I do/That I do*). The latter phenomenon occurs four times in the corpus and can be seen in the following exchange:

> (10) [You owe me somethin', Irish.[4] Remember the Farouk affair?]
> **That I do.** Very well, Deadpool. But if anythin' happens to her...
> (Banshee, *Deadpool #1*)

### 11.4.2 Vocatives

The category of vocatives consists of *lad, laddy/laddie, boyo* and *bucko* as terms of address for male characters, as well as *lass, lassie* and *darlin'* for female ones. In addition to being used in the vocative, some of these words are occasionally employed when referring to other characters. Each will be looked at in turn.

*Lad.* One of the most frequent features in the corpus is *lad*, which, like *aye*, appears in 32 comics. The term *lad(s)* is used by Irish characters from the very beginning, even in the comics by Roy Thomas, who generally eschewed any indications of Irish dialect, or indeed accent, and whose portrayal of Irish speech was otherwise atypical for the corpus. It should be noted that *lad* is predominantly used when Irish characters address males who are younger than they are, which is almost always true of Banshee, one of the oldest X-Men:

> (11) Gently does it, **lad**. All right, Peter. Set her down.
> 
> (Banshee, *Uncanny X-Men #148*)

However, despite its popularity, it should be noted that the use of *lad* (singular) in the vocative form, as it mostly appears in the comics, has Stage Irish connotations (Walshe 2009a: 141). The vocative plural form, in contrast, is used very frequently in contemporary IrE and can even apply to females (Walshe 2009a: 142). An example of the plural form can be seen below where Shamrock orders some schoolboys to fetch Irish Wolfhound.

> (12) You **lads** go fetch himself! I'll deal with this mechanical beastie until he arrives!
> 
> (Shamrock, *Guardians of the Galaxy #51*)

*Laddy/Laddie.* As with the vocative use of *lad*, the use of *laddy/laddie* as a term of address has strong Stage Irish connotations, and indeed is more typical of Scottish English (hereafter ScE) than of IrE (*OED*). Nonetheless, that does not prevent it from occurring in nine different comics, each time in the vocative. Some examples of the *laddy* spelling were already evident in Banshee's appearance in *Giant-Size X-Men #1* above (examples 4, 5 and 6). Below is an example of the other, more common, spelling:

> (13) **Laddie** – are ye all right – Don't try to get up, boyo – just rest easy an' ye'll be foine...
> 
> (Banshee, *Uncanny X-Men #96*)

*Boyo* and *bucko.* Like *lad*, *boyo*, which appeared in the previous example, is used both when addressing and when speaking about other male characters and occurs in 24 separate comics. The *OED* describes it as colloquial and dialectal and acknowledges its being chiefly an Irish feature, meaning *boy*, which is used 'especially as a jovial form of address' (*OED*):

> (14) Cain, me **boyo**–is that you?
>
> (Black Tom, *Deadpool #1*)[5]

It should be noted that the suffix *-o*, which is present in *boyo*, has its origin in Irish and denotes affection (Dolan 2006: 33). The suffix is also present in the term *bucko*, which occurs four times in the corpus. Dolan defines *bucko* as 'a young fellow' and claims that it is a combination of the Old English *buc*, meaning a male deer, with the suffix *-o* (2006: 38). The *OED* cites *bucko* as being nautical slang, meaning 'a blustering, swaggering, or domineering fellow,' a description that is certainly fitting in the example below, in which Black Tom refers to the unstoppable character known as the Juggernaut:

> (15) C'mon **bucko**–where are ye? Ye can survive this ... ye have to!
>
> (Black Tom, *Amazing Spider-Man #230*)

*Lass.* According to the *OED*, *lass* is the ordinary word for girl in 'northern and north midland dialects.' Although *lass* is not attested to Ireland in that particular definition, it does occur in 14 different comics, twice as a descriptor and 12 times as a vocative, as in the following:

> (16) Don't worry, **lass**; help's on th' way! My sonic scream combined wi' Cyclops' optic blasts'll have ye free in–Saints preserve us–what now?!
>
> (Banshee, *Uncanny X-Men #110*)

*Lassie.* Like *laddy/laddie*, *lassie* is a term which is more reminiscent of ScE than it is of IrE, so much so that alone Shamrock's use of the word seems to have been reason enough for an *Irish Times* journalist to describe her as having 'an inexplicable Scottish accent' (Freyne 2011). The Scottishness of this term is also confirmed by the *OED*. Nonetheless, *lassie* is used by Irish characters in three different comics. It is also used mockingly by Captain America when speaking to Shamrock. In each case, it is used in the vocative, as follows:

> (17) Moira, me darlin' – ye're a foine figure of a woman – an' a brave one t'boot – an' I'm thinkin' I like ye a lot – but here an' now, **lassie** – ye're playin' way outa yer league.
>
> (Banshee, *Uncanny X-Men #96*)

*Darlin'*. The term *darlin'* is, of course, by no means exclusive to Ireland. However, its use is quite conspicuous in the corpus, and would seem to imply that for the writers it is typical of Irish speech. It is used 13 times as a vocative by Banshee and Black Tom, and not only when speaking affectionately to lovers or relatives.[6] Banshee also uses the term when speaking to female members of the X-Men, and even when talking to Magneto, the X-Men's arch nemesis, whereas Tom uses it when speaking to Siryn and Juggernaut:

(18) That's what ye think, Maggie-**darlin'**!

(Banshee, *Uncanny X-Men #113*)

(19) There 'tis, me **darlin's**. The source of untold wealth.

(Black Tom, *Spider-Woman #37*)

### 11.4.3 Exclamations and oaths

As was evident in Nightcrawler's speech, exclamations and oaths associated with a particular nation or region are commonly used to identify a character as being from there. This also applies to the Irish characters in the corpus.

*'Saints' and other religious expressions.* In comparison with other varieties of English, IrE employs a greater range of religious language and with greater frequency (see Amador Moreno 2010: 69–70), with religious terms being used to express shock, excitement, surprise, impatience, anger, and so on (see Farr and Murphy 2009). Just as Weimer had observed in *Punch*, the Irish characters in the Marvel universe tend to use exclamations featuring the word *saints*. Examples include *saints* on its own, as well as more elaborate versions such as *Saints above*, *Blessed Saints*, *Saints preserve us*, *Saints preserve us all* and *Saints be praised*.[7] Variations of exclamations featuring *saints* occurred in 17 comics and can be seen in examples (5) and (16) above and in (20) below.

(20) What's that–?! **Saints above**–a net!! I didn't hear 'em comin' over the sound o' me sonic scream. And the bloody net's too close to dodge!

(Banshee, *Uncanny X-Men #116*)

Other religious exclamations are spread throughout the comics. Examples include *Jesus, Mary and Joseph*; *Holy Mother o' God!!*; *Mary, Mother of God*; *sweet mother o' mercy*; *Lord almighty*; *Lord above*; *good Lord*; *Lord in Heaven*; *Lord rest her soul*; *in the name of our Lord and Savior*; *Glory be…; in Heaven's name; by Heaven; by the Eternal–!* and the inventive *Lord and Lady preserve me* and *St. Brigid's Cross!* Given such a large variety of religious expressions in the corpus, Siryn could very well be speaking for the other Irish characters in the Marvel universe when she says in confession: 'Oh. Well, uh … I took the Lord's name in vain a lot' (*X-Factor #28*).

Irish Wolfhound, as mentioned earlier, is an exception regarding such religious oaths, as he is a figure from pre-Christian Celtic mythology. However, this does not prevent him from having Irish oaths of his own. These include: *By the Red Branch!*, a reference to the mythological Ulster army, the Red Branch Knights; *Daughters of Calatin!*, an allusion to the druid Calatin, whom Cuchulain killed, and whose three daughters swore to avenge their father's death; and *Boru's Harp!*, a reference to the harp of Brian Boru, the last High King of Ireland.

### 11.4.4 Grammar

As was the case in the representation of the Irish in *Punch*, typical IrE grammar features were not very common in the corpus.[8] However, this may be a blessing, as although the writers correctly identified particular structure types as being those which are used in IrE, their execution of these structures in their own writing was often problematic. For example, particular structures are either used incorrectly (the *after*-perfect and the progressive form), or they are used correctly but so frequently as to draw negative attention to themselves (*it-* clefting). Given the low frequency of IrE grammar structures in the corpus, such incorrect use or overuse is bound to stand out all the more and is therefore worthy of consideration. I shall look at each feature in turn.

*The after-perfect.* The *after*-perfect is one of the most frequently discussed grammatical features of IrE and it is thus no surprise that it is also 'the most widely used to portray Irish characters in fiction' (Amador Moreno 2010: 38). It is used to describe an action which took place right before the time of speaking or right before the time of reference and typically takes the form *be + after + verb + -ing*, which is a direct English translation of an Irish substratal structure. Thus, the Irish sentence *Tá mé tar éis mo dhinnéar a ithe* becomes the IrE 'I'm after eating my dinner' (Dolan 2006: 3), which in Standard English (hereafter StE) would be 'I have just eaten my dinner.' However, this type of structure is very frequently misunderstood by those who are unfamiliar with IrE, as they usually understand the *after* to imply a desire or intention to do something, and thus interpret the example above as 'I want to eat my dinner' (see Harris 1993). If people who are unfamiliar with IrE are likely to misinterpret such *after*-perfect sentences as signifying intention, then they are also likely, when creating their own IrE structures, to imagine that sentences expressing intention also need to have the form *be* + *after* + *verb* + *-ing*. This would explain the following erroneous example:

(21) **I'll be after joining** you, Major Astro!
(Irish Wolfhound, *Guardians of the Galaxy #51*)

While *after*-perfects can occasionally appear in the future tense in IrE (Hickey 2007:198), erroneous constructions such as the one above have invariably resulted in sentences featuring *will* + *be* + *after* + *verb* + *-ing* being regarded as 'Stage Irish' (Bartley 1954: 130).

*It-clefting*. This syntactic form is the division of a single clause into two subclauses, each with its own verb. The first subclause is introduced by *it* and a present or past tense form of the copula, namely *it is*, *it's*, *'tis*, *is it?*, *it was*, *'twas*, *was it?*, and then followed by the element to be fronted. The second subclause follows and resembles a relative clause beginning with *that*, although the relative pronoun itself is very frequently omitted in IrE (Harris 1993: 173). Although *it*-clefting can also be found in StE, it does not occur there as frequently as in IrE (Filppula 1999: 248–49).

While the structure does not occur very frequently in general in the Marvel corpus, it is very prominent in Shamrock's speech. For example, in *Guardians of the Galaxy #51*, *it*-clefting is used three times in the first three pages alone, which strikes this reader as an example of overkill.

(22) **It's grateful they were** for our help and leadership. We set them to work constructing real homes. Pride and dignity were easy flames to rekindle.
(Shamrock, *Guardians of the Galaxy #51*)

(23) **It's great plans we have** for this community we call 'Dochas' which in my native tongue means 'hope.'
(Shamrock, *Guardians of the Galaxy #51*)

(24) **It's help you'll get**, Klaus, … but from Shamrock!
(Shamrock, *Guardians of the Galaxy #51*)

Research by Taniguchi (1972) shows just how flexible the Irish English use of *it*-clefting is. It can be used to introduce many parts of speech: nouns, pronouns, adjectives, past participles, present participles, bare infinitives, *to*-infinitives, prepositional phrases, and adverbs. While example (22) may seem unusual to some readers, '[a]ccording to Quirk *et al*. (1985: 1385), informal IrE is the only variety which allows either subject complement nouns or adjectives in focus position' (quoted in Filppula 1999: 251).

*Nonstandard use of the progressive.* Generally, IrE allows a more liberal use of progressive forms than StE does, enabling them to occur together with both dynamic verbs and stative verbs, particularly those of perception and cognition, such as *hear, want, wonder, know, think* and *believe* (Harris 1993: 164). However, some of the uses in the corpus are particularly jarring, even to Irish ears. As with the erroneous *after*-perfect above, these problematic examples of the progressive stand out because of their use of the future *will* form. In examples (25) and (26), the future *will* form is used where the conditional and present tenses, respectively, would be considered standard:

(25) Wait. You didn't call. Maybe we should be, oh, waiting by the phone? Is that where **you'll be wanting** us, then? Like a dateless colleen on a Friday night?

(Siryn, *X-Factor* #9)

(26) **I'll be knowin'** how t' do me own job, thank ye very much – I been doin' it nigh on forty years. Now, fer the last time, close that flamin' door!!

(Postmaster, *Uncanny X-Men* #99)[9]

Furthermore, although *will* can be used in imperatives ('Will you shut up!'), which examples (27) and (28) effectively are, and although imperatives can feature the progressive in IrE (see example 32), the combination of the two does not work. This is because imperatives in the progressive generally only work in the negative form in IrE (Walshe 2009a: 70). Thus, while sentences such as 'Will you don't be lookin' a' that' or 'Will ye don't be shuttin' the door' would be acceptable in IrE, examples (27) and (28) sound, at best, antiquated or Stage Irish:

(27) **Will ye be lookin'** a' that–The head honcho, Dr. Steven Lang, hisself ... come t'pay us all a visit.

(Banshee, *Uncanny X-men* #98)

(28) **Will ye be shuttin'** the door, ye flamin' idjit–!

(Postmaster, *Uncanny X-Men* #99)

## 11.4.5 Phonology

As with the typical lexicogrammatical features above, when it came to examining the features which were used to reflect the phonology of IrE, each feature was noted as either occurring or not occurring in a particular issue. It should be noted, however, that although the corpus consists of 150 comics, it would be misleading to present the findings for phonology out of 150, as some writers make no attempt at all to represent accent and instead rely solely on Irish expressions to create the impression of Irishness. This is the case for 46 issues. Thus rather than looking at occurrences out of 150, it is more accurate to treat them out of 104.

*Vowel and consonant substitutions.* Whereas traditional representations of Irish accents in literary dialect have tended to focus on substitutions of vowels and on occasional consonant changes, as in *Punch*, this was not the case in the comics examined. In fact, the findings were very surprising. There is not a single example of <e> being rewritten as <i>, nor of <ea> being respelled as <a>, <ay> or <ai>, and there are only three realizations of <oi> for <i>, namely the one in the title of this chapter and those in examples (13) and (17) above, all of which are lexically bound to the word *foine*.

Even if one were to look beyond the type of respellings discovered by Weimer in *Punch* and were to consult Taniguchi's (1972) findings regarding the representation of IrE in plays and novels, the search would not be much more successful. The respellings Taniguchi mentions, which are also attested elsewhere (Sullivan 1980), are also only present a handful of times, if at all. For example, there is no evidence of <s> being respelled as <sh>, as in <shtop> rather than <stop>, or of <j> being used instead of <d>, as in <projuce> rather than <produce> (Taniguchi 1972: 239–41). What is even more surprising is that there are only three substitutions of <t> or <d> for <th> in all the comics, even though this is almost always a salient feature in fictional representations of Irish accents (Amador Moreno 2010: 77). The examples are in (29), (30) and (47) below:

> (29) Hur-ry, hur-ry, hur-ry! Come one, come all t' the greatest little show on earth! We got **t'rills** an' chills, sights t' bedazzle the eye an' freeze the heart!
>
> (Banshee, *Uncanny X-Men #111*)
>
> (30) I'm not sure I agree **wit'** what ye're sayin', Charles – but ye're the boss.
>
> (Banshee, *Uncanny X-Men #101*)

The respelling of <o> as <ou>, which Taniguchi mentions, also only occurs once, in the word *oul(d)*, although there is also evidence of it being respelled as <au>. However, it should be noted that <auld> is the Scottish spelling and rhymes with *called*, whereas the Irish form is *ould* and rhymes with *howled*.

(31) I'm just interested in the sexual habits of Madrox the Multiple Man. Don't try to tell me yeh're not, Monet, yeh dirty **oul** slag!
(Siryn, *X-Men #128*)

(32) Don't be congratulatin' yerselves too soon, lads! The Banshee's a tough son o' the **auld** sod, and me jaw's had the time to recover!'
(Banshee, *Captain America #172*)

The only substitutions *per se* that occur with any frequency in the corpus, and which the reader will no doubt already have noticed are the diphthong in *my* being realized as the vowel in *me*, and the substitution of the alveolar nasal for the velar nasal, with <-ing> being respelled as <-in>. The substitution of *me* for *my*, like the substitutions in *foine* and *oul* above, is also lexically bound and occurs in 45 of the 104 issues in which accent is indicated. On three occasions, this change can also be found in the related word *myself*:

(33) I wouldn't have known of it **meself**, if I hadn't seen her go t' pieces when we fought **me** cousin, Black Tom, and Juggernaut.
(Banshee, *Uncanny X-Men #110*)

The respelling of <-ing> as <-in> is more common and occurs in 78 of the 104 comics. As one would expect, the progressive forms of verbs constitute the majority of such occurrences, boosted somewhat by the tendency of Irish characters to use the progressive form in contexts where it would not be used in StE (see above). However, this phenomenon is not limited to verbs, but also occurs in nouns such as *darlin'*, *nothin'*, *somethin'*, *mornin'* and adjectives such as *interestin'*, *fascinatin'*, and *bleedin'*.[10] Evidence of this realization of <-ing> as <-in> will already have been apparent above. However, some additional examples include:

(34) **Somethin'** that imp said about **programmin'** – if he's some kind o' robot, I'm **bettin'** my sonic scream'll scramble his circuits some. I'm **bettin'** me life. There he is! **Usin'** me scream as an airborne sonar led me right to him!
(Banshee, *Uncanny X-Men #108*)

Apart from these examples, respellings involving vowel and consonant substitutions are rare, with the writers relying instead on other features to suggest Irish phonology, namely elision, reduction, and weak forms. While one could correctly argue that these features are typical of all vernacular English and thus by no means exclusively Irish, this is not reflected in the comic books in the corpus, where these features occur considerably more often in the speech of Irish characters than in that of others. A reason for this may be that such representations are meant to reflect the common perception of Irish people speaking quickly and unclearly. For example, Walshe (2010), a perceptual dialectological study in which 30 native speakers of English from five countries were asked to describe linguistic features which they associate with Irish speech, found that Irish speakers' pace and lack of clarity were mentioned time and again.

*Elision*

<an'> or <'n'> for <and>
While elision of *and* is also common in spoken StE, its occurrence is more frequent among Marvel's Irish characters than any others. It occurs in 59 of the 104 issues which feature Irish accents and can involve the elision of both the initial vowel and the final consonant or just the omission of the final consonant. Both strategies are evident in the following example:

> (35) **An'** on that note, me boyos, Moira **'n'** I'll be movin' on ourselves ... we're t'be showin' each other the sights o' New York.
> (Banshee, *Uncanny X-Men #98*)

<t'> for <to>
The previous example also features another instance of elision, namely a respelling which signifies the weakening of /tuː/ to /tə/. This phenomenon is present in almost half of the issues examined, occurring 48 times. Again, the reader will have already noticed several examples, but the following illustrates the feature particularly well:

> (36) I don't think any of the great powers are goin' **t'** sit still f'r such an ultimatum. Some idiot somewhere's sure **t'** challenge him an' then, I fear, we'll see some fireworks. Sorry **t'** intrude, Dr. MacTaggart, Ma'am. But ye've been closed down in your lab f'r so long, without word or even a bite **t'** eat, I was startin' **t'** worry.
> (Banshee, *X-Men #1*)

<o'> for <of>

The weakening of <of> to <o'> was already visible in the expression *top o' the mornin' to ye* which appeared in Shaw's *John Bull's Other Island*. While that particular expression is used only once in this corpus, the substitution of <o'> for <of> occurs regularly, with 43 occurrences. Below is an example:

> (37) I've got to slide a wall **o'** sound across the base **o'** the island–counter energy with energy–block Magnum's beam ... then reflect it back on itself.
>
> (Banshee, *Uncanny X-Men #119*)

<'tis> for <it is>

Elision of *it is* to <'tis> rather than <it's> is very common in representations of Irish speech and also appears in 17 comics. This phenomenon, which is perhaps better categorized as a form of contraction, is archaic and was common in English up until the seventeenth century (Peitsara 2004: 77). Its presence in IrE can be explained by the fact that it was brought to Ireland during the Cromwellian conquest. According to Bliss (1979: 20), 'the English spoken in most parts of Ireland today is descended from the English of Cromwell's planters, and since the early part of the eighteenth century no other type of English has been spoken in any part of Ireland except in Ulster.'

Elision or contraction of the past tense and future forms of the copula also occur in the corpus, with *it was* being realized as <'twas> and *it will* being realized as <'twill>. Examples of the various forms are evident in (1), (3), (4) and (7) from *Giant-Size X-Men #1* above, while below is another example from Shamrock, once again displaying her affinity for *it*-clefting.

> (38) **'Tis** lucky I am that me costume blends with the foliage–otherwise Captain America might have seen me–before I could snatch this vine from out of his grasp!
>
> (Shamrock, *Marvel Superhero Contest of Champions #3*)

Although Shamrock's use of the structure above is correct, the following example illustrates the problem of writers becoming too fond of a structure and using it unquestioningly in contexts where it is not appropriate:

> (39) Jimmy...? **'Tis** it really ye?
>
> (Siryn, *X-Force #57*)

In this case, the writer includes two subject pronouns where only one is permitted. The sentence thus essentially reads as 'It is it really ye?' and is ungrammatical, even in IrE.

*Weak forms*

<ye> or <yeh> for <you>
The use of the weak form <ye> or <yeh> instead of <you> is very common in the corpus, just as it was in *Punch*. The <ye> spelling occurs in 65 of the 104 comics in which accent is represented, whereas <yeh> appears in ten issues, all of which are in the *Generation X* and *New X-Men* series. It is important that this <ye> spelling not be confused with the second person plural form *ye*, which is very common in IrE. In fact, although plural forms of the second person plural (*ye, youse, yiz/yis*) are common in IrE (Hickey 2007), they are very rare in the comics, with no attempt being made to recreate the pronunciation of the plural *ye* or *youse* and only three occurrences of *yiz/yis*, albeit always written as <ye's>. An example is:

(40) Look out, all o' **ye's**–the beastie's loose again–!

(Banshee, *Uncanny X-Men #96*)

The weak form also extends to other words containing *you*, such as *you're, your, yourself* and *yourselves*. The respellings <yer>, <ye're> and <y'r> occur in 30 comics, while <yehr> appears six times. The reflexive form <yerself> also occurs six times, with the plural <yerselves> only making one appearance, in example (32). Examples of the various forms include:

(41) An' **ye** – Major Maple Leaf, or whate'er **ye** call **yerself** – **ye're** the man responsible!

(Banshee, *Uncanny X-Men #109*)

(42) Is that a joke **ye're** crackin', Peter Rasputin? Will wonders never cease? **Yer** torment's almost over, though – because, my friends – we have arrived.

(Banshee, *Uncanny X-Men #101*)

(43) Polite? **Yeh** want to talk about polite? The last time **yeh** graced us with **yehr** presence **yeh** very nearly murdered the children.

(Banshee, *Generation X #68*)

While some of the spellings above do succeed in reflecting Irish pronunciation, the tendency of some writers to simply substitute a weak form in any (and every) context is problematic. This can best be seen in examples (41) above and (44) below, where the use of the unstressed form in stressed positions is incorrect.

(44) Moira–! **Ye** an' Storm find Nightcrawler and Colossus, warn 'em–! 'cause if the sentinels are back...

(Banshee, *Uncanny X-Men #98*)

<fer> for <for>
A similar pattern of respelling to that seen in *your* and *you're* above is seen in *for*, the weak form of which is respelled as <fer> or <f'r> in 26 comics. This change also is evident in words such as *forgetting*, *forgotten*, and *forever*. Examples of this phenomenon include:

(45) I do want him t'pay **f'r** his crimes ... but not by dyin'. Sorry t'stick ye with this, Daniel. Tonight seems ta be a night **f'r** callin' in old favors.

(Banshee, *Deadpool #2*)

(46) Can ye now? That's mighty big talk, I'm thinkin'; but then, talk is what ye're most famous for. Could it be that ye're **fergettin'** that Banshee–

(Banshee, *Uncanny X-Men #104*)

The standard spelling of *for* in the last example may just be an oversight on the part of the writer, but can just as readily be explained by it occurring in a stressed position.

## 11.5  Discussion

As illustrated above, the representation of Irish speech in the Marvel universe involves a combination of supposedly typical IrE lexicogrammatical features, as well as a system of respellings and contractions to indicate an Irish accent. (See Table 11.1 for a list of lexicogrammatical features and the number of issues they occur in.)

Terms and expressions which had been deemed most characteristic of Irish speech in nineteenth-century caricatures (*begorrah*, *faith*, *top o'*

*Table 11.1* The distribution of supposedly typical Irish English features in the corpus

| Feature | Number of issues | Feature | Number of issues |
|---|---|---|---|
| aye | 32 | bucko | 4 |
| lad | 32 | grand | 4 |
| boyo | 24 | bleedin' | 4 |
| lass | 14 | begorra(h) | 3 |
| darlin' | 13 | faith | 2 |
| laddy/laddie | 9 | top o' the mornin' | 1 |
| wee | 6 | broth of a boy | 1 |
| lassie | 4 | musha | 1 |

*the mornin' to ye, broth of a boy, musha*) still occur in Marvel comics in the late twentieth and early twenty-first century, albeit very sparingly. Instead, the features which are most common are those which are more often associated with Scottish English. This is certainly true of *laddie* and *lassie* (*OED*), but also applies to *lass, wee* and *aye*.[11] More importantly, similarities to, and potential confusion with, ScE are not restricted to these lexical items. For example, *beastie*, meaning a small animal, is used in four different comics, by four different teams of writers (examples of which can be seen in (5), (6), (8), (12) and (40) above), yet *beastie* is a Scottish term and not an Irish one (*OED*). Similarly, Black Tom's utterance 'Laird, laird! Why does the boy have to be so headstrong?' (*Amazing Spider-Man #229*), which is intended as a religious exclamation, is also problematic – on two counts. First, *laird* is a Scottish term, not an Irish one, and, second, although *laird* does mean *lord*, it is used to describe a member of the landed gentry and not the son of God or God Himself (*OED*).

In another example, Shamrock uses the Scottishism *bonny* together in a sentence with *lassie*: 'That's right, lassies! Ain't a single head o' hair I couldn't make into a work o' art! 'course that was before that bloody accident that shot me superhero career t' high heaven ... now who's ready t' become a bonny goddess?' (*Excalibur #108*). Even if one were to justify this usage as an example of accommodation, since one of the people Shamrock is talking to is Wolfsbane, a Scottish mutant, such an argument does not justify the many examples of Irish characters using Scottish forms of negation (see Lenz 1999: 51–3), as in *cannae, dinnae, didnae, nae,* and so on. (See Figure 11.2.) These erroneous uses of Scottishisms in an Irish context are conspicuous and not restricted to one writer, but rather occur in the works of several.

*Figure 11.2* Siryn in *X-Force* #58. All images copyright Marvel Comics. Used by permission.

Examples of such errors include:

(47) Stickin' ye nose into my transistors is **nae** what I thought ye meant – when ye offered me a hand wit' converting the comm board.

(Siryn, *Uncanny X-Men* #294)

(48) I **didnae** want to see him suffer, so I arranged for him t'be doctored while in prison.

(Banshee, *Deadpool* #2)

(49) I **dinnae** know. She was **nae** here when I awoke–

(Siryn, *X-Force* #58)

(50) Ye **cannae** come, Nate. But the way I hear it from Moira MacTaggert – see it in yer face – ye know all about the nature of sacrifice already...

(Siryn, *X-Men* #18)

(51) Whether me power stays or goes 'tis of little importance ... this ship is still full of the virus. We **canna** leave it be.

(Banshee, *Marvel Comics Presents #24*)

In addition to linguistic errors, the characters occasionally make references that are more characteristic of Scottish (popular) culture rather than Irish. For example, in *X-Men #1* Banshee mentions one team taking the 'high road,' an allusion to the traditional Scottish song *The Bonnie Banks o' Loch Lomond*, while in *X-Factor #7* Siryn believes that the DVD that Banshee has left her is *Brigadoon* (Vincente Minnelli's stereotypical portrayal of Scottishness), as 'he's always swearing he'll find a way to make me watch that damned movie.' It may be reading too much into the choice of film to suggest that the writer is confusing Ireland with Scotland, but given the previous evidence, it is difficult to pass by this cultural reference without comment. Indeed, in light of all these examples, I believe it is safe to say that Broadbent in Shaw's play is not the only person to mistakenly believe that something of Scottish origin is Irish.

## 11.6 Conclusions

This first analysis of Irish characters in contemporary comics has revealed some very telling insights into the way that Irish speech is conveyed. Not only is there still occasional evidence of the same stereotypical Stage Irish features which were used over a century ago in periodicals such as *Puck* and *Punch*, but there is also a tendency for writers to combine (or confuse) features of IrE speech with those of ScE. Given the nature of these findings in Marvel comics, future projects should explore whether or to what extent such patterns are evident in the work of other publishing houses, starting perhaps with Marvel's main rival DC and their Irish characters, Jack O' Lantern and the Gay Ghost.

### Notes

1. This refers to appearances in which the character speaks. Cases where a character appears but does not speak (see Shamrock's appearance in *Rom #65*) have been discounted from the corpus, for obvious reasons.
2. It is important to note that speech does not refer to the spoken word alone, that is, to the words which appear in speech balloons, but also to those which appear in thought balloons. This is particularly important in the context of Banshee and Siryn, as much of their 'speech' occurs in the form

of thought balloons, as they use their sonic screams to keep themselves in the air when they are flying and thus cannot speak at the same time. Since these thoughts also contain (supposedly) Irish features and are marked for accent, they are a useful source of information.
3. *Otherness* is defined in contrast to the norm, which can be understood to be standard American English, which bears no indication of accent via respellings.
4. The Irish characters' ethnicity is also underlined by the nicknames given to them by others. Banshee is frequently called 'Irish,' as in example (10), while Irish Wolfhound receives more derogatory treatment, being referred to as 'Big Mick,' 'Gaelic Gorilla,' 'Spud,' and 'Mr. Potato Head' (*Guardians of the Galaxy #3*). 'Mick' is (or was) a common name in Ireland and, like 'Paddy,' is often used as a pejorative by non-Irish people. 'Spud' and 'Potato Head' refer to the staple of the Irish diet, at least in pre-Famine times, while the reference to 'Gorilla,' once again, alludes to the association of the Irish with simians.
5. *Boyo* is used by Black Tom to such an extent that Deadpool imitates him, saying: 'Say goodbye 'boyo'…' (*Deadpool #4*).
6. The Gaelic term of endearment *accushla*, which was listed by Weimer as occurring frequently in *Punch*, is also present in the Marvel corpus: 'Moira? What ails you, accushla? Have ye been cryin'?' (Banshee, *X-Men #1*). Like so much IrE vocabulary, such terms of endearment were popularized in Irish songs (Williams 1996: 34–35).
7. Interestingly, during Banshee's first few appearances, as written by Roy Thomas, he never uses religious oaths, but rather uses the word *sirens* where *saints* would later be used, as in 'Sirens preserve us' and 'Sirens be praised!'
8. The lack of IrE grammar features in the corpus could be due to the rules of the Comics Code Authority (1954 and 1971), which state that 'although slang and colloquialisms are acceptable, excessive use should be discouraged and, wherever possible, good grammar shall be employed' (Nyberg 1998: 173).
9. Examples (26) and (28) are from the Postmaster, a peripheral Irish character in the corpus. As is also often the case in Stage Irish drama, peripheral characters tend to speak with a stronger dialect than main characters.
10. *Bleedin'* is a swearword, which means the same thing as *bloody*. Both are common in the British Isles (including Ireland).
11. Even if *wee* and *aye* can also be found in Northern Ireland, the use of *aye* there is not particularly Irish but, according to Kirk and Kallen (2010: 205), is rather a means of expressing affiliation with Scottishness or the new Ulster-Scottishness.

## References

Amador Moreno, C. P. (2010) *An Introduction to Irish English*. London: Equinox.
Appel, J. J. (1971) From shanties to lace curtains: the Irish image in Puck, 1876–1910. *Comparative Studies in Society and History* 13 (4): 365–75.
Bartley, J. O. (1942) The development of a stock character: the Stage Irishman to 1800. *The Modern Language Review* 37 (4): 438–47.
Bliss, A. (1979) *Spoken English in Ireland 1600–1740*. Dublin: The Dolmen Press.

Daniels, L. (1991) *Marvel: Five Fabulous Decades of the World's Greatest Comics*. New York: Harry N. Abrams.

Dolan, T. P. (ed.) (2006) *A Dictionary of Hiberno-English*. 2nd edition. Dublin: Gill and Macmillan.

Dowling, J. (2009) 'Oy Gevalt!' A peek at the development of Jewish super heroines. In A. Ndalianis (ed.) *The Contemporary Comic Book Superhero*, pp. 184–202. New York: Routledge.

Farr, F. and Murphy, B. (2009) Religious references in contemporary Irish English: 'For the love of God almighty...I'm a holy terror for turf.' *Intercultural Pragmatics* 6 (4): 535–59.

Filppula, M. (1999) *The Grammar of Irish English: Language in Hibernian Style*. London: Routledge.

Freyne, P. (2011) Just in time – A new breed of Irish superhero. *The Irish Times*. 18 March. Retrieved on 11 August 2011 at http://www.irishtimes.com/news paper/features /2011/0318/1224292498323.html

Harris, J. (1993) The grammar of Irish English. In J. Milroy and L. Milroy (eds.) *Real English: The Grammar of English Dialects in the British Isles*, pp. 139–86. London and New York: Longman.

Hickey, R. (2007) *Irish English: History and Present-Day Forms*. Cambridge: Cambridge University Press.

Kirk, J. M. and Kallen, J. L. (2010) How Scottish is Irish Standard English? In R. McColl Millar (ed.) *Northern Lights, Northern Words. Selected Papers from the FRLSU Conference, Kirkwall 2009*, pp. 178–213. Aberdeen: Forum for Research on the Languages of Scotland and Ireland.

Lee, S. (1978) *Marvel's Greatest Superhero Battles*. New York: Fireside.

Lenz, K. (1999) *Die schottische Sprache im modernen Drama*. Heidelberg: Universitätsverlag C. Winter.

Negra, D. (2001) Consuming Ireland: Lucky Charms cereal, Irish Spring soap and 1-800-Shamrock. *Cultural Studies* 15(2): 76–97.

Nyberg, A. K. (1998) *Seal of Approval: The History of the Comics Code*. Jackson: University Press of Mississippi.

Peitsara, K. (2004) Variants of contraction: the case of it's and 'tis. *ICAME Journal* 28: 77–94.

Quirk, R., Greenbaum, S., Leech, G., and Svartvik, J. (1985) *A Comprehensive Grammar of the English Language*. London: Longman.

Shaw, G. B. (1907) *John Bull's Other Island*. London: Constable and Company Ltd.

Soper, K. (2005) From swarthy ape to sympathetic everyman and subversive trickster: the development of Irish caricature in American comic strips between 1890 and 1920. *Journal of American Studies* 39(2): 257–96.

Sullivan, J. P. (1980) The validity of literary dialect: evidence from the theatrical portrayal of Hiberno-English forms. *Language in Society* 9(2): 195–219.

Swift, J. (1728) On barbarous denominations in Ireland. In T. Scott (ed.) (1905) *The Prose Works of Jonathan Swift D.D. Volume 7 (Historical and Political Tracts)*, pp. 340–50. London: George Bell and Sons.

Taniguchi, J. (1972) *A Grammatical Analysis of Artistic Representation of Irish English with a Brief Discussion of Sounds and Spelling*. Tokyo: Shinozaki Shorin.

Walshe, S. (2009a) *Irish English as Represented in Film*. Frankfurt: Peter Lang.

Walshe, S. (2009b) 'So, it's our syntax you're criticizin' then?'– Irish English speech in *The Simpsons* and *Family Guy*. Paper presented at *Myth and Reality: Language, Literature and Culture in Modern Ireland*, Dalarna, Sweden, 29–30 October.

Walshe, S. (2010) Folk perceptions of Irishness: separating the Irish from the 'Oirish.' Paper presented at *New Perspectives on Irish English*. University College Dublin. 11–13 March.

Weimer, M. (1993) *Das Bild der Iren und Irlands im Punch 1841–1921*. Frankfurt: Peter Lang.

Williams, W. H. A. (1996) *'Twas Only an Irishman's Dream. The Image of Ireland and the Irish in Popular Song Lyrics, 1800–1920*. Urbana and Chicago: University of Illinois Press.

Wittke, C. (1952) 'The immigrant theme on the American stage.' *The Mississippi Valley Historical Review* 39 (2): 211–32.

# Conclusion

*Frank Bramlett*

Through my critique of McCloud and Eisner in the introductory chapter, I argued that their metaphor 'language of comics' helps give comics artists and scholars common ground on which to build discussions of the way images and texts go together. At the same time, the metaphor stands as a barrier to a sufficient engagement with comics from a linguistic viewpoint. As the chapters in this collection demonstrate, comics scholars must eschew the notions of the 'vocabulary of comics' or the 'grammar of comics' so that a more productive engagement with the language *in* comics may proceed. In this way, the linguistic codes at play in comics are freed and the distinction can be used more fruitfully to characterize and explore the text and image as well as the linkages that bind them one to the other. Further, linguistic scholarship can be used productively to examine nonlinguistic domains (like visual representations of action).

I do not mean to use this book to establish disciplinary boundaries; quite the contrary, I hope that this volume helps to extend the breadth and depth of what we know about comics. In this regard, the possibilities are numerous. For instance, an examination of language in comics requires that we develop a wide variety of robust research programs.

*Sources of linguistic data.* Comic books of course must be considered sources of linguistic data and objects of linguistic analysis. An opportunity for expansion, however, is web comics. Hundreds of web comics are available in English; some have been published for years. The potential for web comics scholarship has not yet been tapped in a significant way in linguistics.

*Data collection, preparation, and analysis.* The chapters in this collection range from case studies to a reliance on significant corpora. The preparation of data for empirical analysis is both time- and labor-intensive, and the need for scholars to pursue these methodological approaches will continue to grow. Linguistics can provide a rich source of research methods regarding the collection and preparation of comics data as well as subsequent analysis, encompassing corpus linguistics, critical discourse analysis (CDA), translation studies, and gesture studies.

*More languages.* How might diverse languages and cultures play a role in comics and graphic novels? Chapters in this collection cover Dutch, English, French, Spanish, Swedish, and Turkish. Comics research in Arabic, Chinese, Diné (Navajo), Greek, Korean, Portuguese, and Swahili, for example, could also encourage comparative linguistic studies, essential for verifying and expanding linguistic theory.

*Manuscript studies or text linguistics.* The analysis of long passages of (uninterrupted) text in the comic arts will demand that new ways of accounting for language in the context of the visual be developed. Multimodal analysis holds great promise for this kind of approach. Just as important, scholars must be willing to challenge received notions regarding comics and ask research questions to verify or perhaps overturn assumptions that appeal intuitively. For example, how might linguistics expand our understanding of the gutter as a marker of space, of time, and of narrative structure?

*Literacy and language processing.* How do (multilingual) readers come to (multilingual) comics? How do they perceive language codes and characters? How do they find humor or stereotype or racism or compassion articulated in comics? Cognitive linguistics and psycholinguistics can help us understand how readers process the verbal-visual blend. Devoting time to experimental/clinical research may reveal the neural and cognitive resources that readers draw upon.

*Rates of linguistic change.* The speed at which languages change depends on many factors: internal linguistic forces as well as external social and cultural factors. Given that comics have been published for more than a century, it is possible to get a sense of how language and especially language ideologies fluctuate from one generation to the next. Comics can provide a dependable source for data. Do web comics represent an opportunity for deeper and wider freedom for artists or potentially new kinds of constraints? How might web comics serve as witness to linguistic variation, and especially to linguistic change?

*Representation of nonstandard or nonprestige varieties.* Given the internet as a site of relatively less restrained freedom of expression, comics artists may feel freer to wield codes that they prefer rather than constrained to employ more generalized or standardized codes. In other words, if a comics artist chooses to write in a local, regional dialect or in a low-prestige language rather than a standardized or high-prestige language, this provides ever-greater opportunity to investigate linguistic diversity around the world.

*Language preservation, language revival.* Scholars predict that many endangered languages are heading rapidly toward extinction. How might comics, especially the use of dialogue and the use of orthography, help us understand more about minority or endangered languages and, in turn, the people who speak those languages and the cultures they have created? How might comics artists play a role in preventing the death of endangered or minority languages or in helping to bring languages back out of extinction?

Scholars who work in linguistics and comics studies have an opportunity to revise and expand current disciplinary boundaries, marrying approaches and disciplines that may have seemed too disparate or intractable to be blended usefully. However, we also have the chance to erase those boundaries and promote a tenacious intellectual engagement in linguistics and comics scholarship. I hope that this volume serves as a stepping stone along that path.

# Index of Language Varieties

**Artificial languages/ constructed languages**

Poulpe 156
Schtroumpf 156–57
Yorthopia 156

**Natural languages**

AAE *see* African American English
Amharic 146, 153–54
Arabic 89, 121, 134, 146, 151–52, 154, 159, 292
Azeri 134

Belgian French: slang expressions 159
Bulgarian 134

Chinese 89, 145, 150, 152–55, 158, 292

Diné (Navajo) 292
Dutch 10, 102, 163, 174, 176, 292: Belgian/Flemish Dutch 10, 163–65, 167, 176, 181; Belgian Standard Dutch 165, 174, 179; central Brabantian dialect group 164; Colloquial Belgian Dutch (CBD) 164–65, 173–74, 176, 179–80; colloquial Dutch 165; Dutch Dutch 167, 181; Dutch Standard Dutch 163, 165, 167, 176, 178–79; Standard Dutch 163–64, 169, 173–74, 176–79, 181

English 9–11, 46, 63, 88, 93, 98, 100, 102, 128, 133, 135–36, 142–44, 146–53, 157–58, 176, 184, 187–89, 192, 196, 198, 210–11, 216–19, 223–25, 227, 233, 235–36, 239–51, 253, 257–61, 271–72, 276, 281–82, 291–92; African American English (AAE) 187–89, 191–94, 196–200; American English (General) 4, 10, 163, 187, 190, 198, 219–20, 226: standard 288; Black English *see* African American English; British English 10, 147, 163; California English 217; Chicano English 216–17, 235; hip-hop English 241, 245, 252–53, 255–61; Irish English (IrE) 11, 197, 264–65, 277, 285; native English 247, 250–51; nonnative English 241, 245, 247, 249–51; Old English 274; 'school' English 246–47, 251, 260; Scottish English (ScE) 273, 285; Swedish English 248, 250–51, 258 (nonnative variety of English); Standard English 251, 276; vernacular English 281

Farsi 121
French 9–10, 143–44, 146–58, 164, 292

Gallic 150–51; pseudo-Gallic 151–52
German 10, 145, 150, 152, 154, 271
Greek 134, 292
Gypsy (Romani) 151

Hebrew 105, 146, 150–54, 157

Irish (Gaelic) 267, 272, 274: *see also* Stage Irish in Subject Index: brogue 264, 267; pseudo-Irish 267; Stage Irish 264–66, 268, 273, 277, 287–88 (note 9)
Italian 151

Japanese 98, 184, 187, 190–91, 194, 196, 203; Standard Japanese 194

Korean 292
Kurdish 134

Latin 150–52, 157; pseudo-Latin 151

Navajo (see Diné)

Persian 134
Polish 150, 152, 154
Portuguese 292

Russian 10, 145–46, 150–53, 155

Spanish 10, 146, 150–53, 155, 157, 210–11, 216–19, 235, 292; Chicano Spanish 234; Mexican Spanish 234; Mock Spanish 210, 216, 220–21, 223, 225–28, 23–33, 235; Standard Spanish 234

Spanglish 10, 211, 216, 218, 223, 225–26, 235

Swahili 292

Swedish 10–11, 239–43, 245–46, 259, 261, 292

Turkish 124, 127, 132, 134–36, 138, 292; Ottoman Turkish 121

Yiddish 150, 152, 157

# Subject Index

accent 4, 101, 233, 268, 279–80, 283, 288 (note 3): American English 220, 226; fake Spanish 210; 'foreign' 233; 'ignorant country (hick) accent' 233; imitation of 220; Irish 264, 267–68, 279, 281, 284; marked for 288 (note 2); nonnative 250; stereotypical 221, 227; Mock Spanish 226–27; representation of 271; Scottish 274; spellings to indicate 271; Spanish 232; Swedish 247
acquisition 93, 98, 110–11, 217
adjective: as property 109; English 147; word order 148; *Schtroumpf* 156; Dutch 167
adverb: in Irish English 278; no case marker with 175; *Schtroumpf* 156
affirmative (grammar) 271–72
*after*-perfect (Irish English) 276–77
alphabet 2, 184
allomorph 174; *see also* morphology
alveolar 191, 193, 280; *see also* nasal
anime 10, 183–85, 187, 190–96, 199–200 (note 1), 201 (notes 2 and 3)
applied linguistics 94
approximant (phonetics) 248
artificial language 155–56; *see also* Language Index
assimilation 214, 219: of English 240–41, 245, 260
audience 61–62, 68, 89, 122, 136, 186–87, 189, 191, 197, 226, 235, 242, 245, 268: American 266, 268; and code switching 134; appeal to 59; Dutch 169; in Turkey 120; in-audience 251; international 240; Japanese 197–98; sociocultural 144; survey 8; target 43; US 198; white 197–98

authentic 197: dialect 266; identity 197; language 188, 197; members 254; Spanish 219; *see also* sign
authenticity 54–55, 189, 254, 256: caption 48, 50–51, 55–56; idea of 188; linguistic 189; of language 55; of the dialogue 47
balloons 5–7, 106, 115, 128, 132, 135, 137, 180, 204 (note), 287 (note 2): balloonics 5; language in balloons 179; motion lines and speech balloons 104; text balloon 165
*bandes dessinées* (BDs) 9–10, 142–44, 146, 149–52, 155–58
bilingual: abilities 245; code switching 10, 184, 190–93; in English and Spanish 217; linguistic repertoire 245; signs 150; speakers 216–17, 244; speech communities 134, 245; strip 158
bilingualism: and language contact 244; education and 233; issue of 214; stable 219
*British National Corpus* 46
brogue *see* Irish

calque: French from English 158; lexical 148; syntactic 147
caricature 62, 184, 222, 264: nineteenth-century 267, 284; racial 186; racist 187
cartoon: analyses of 41, 46 ff, 64 ff, 128–38, 222; artists 122–23; caricatures and 264; characters 240; discourse 68; editorial 37, 59, 63, 88, 96, 211, 221; *Far Side* 8, 42 ff; fears exploited in 64 ff; gag 37, 55; humor in 37, 39, 82; image 183;

Japanese 59; medium 122; metaphor in 61; multimodal 39, 55; *New Yorker* 63; political 8–9, 59–60 ff, 85, 88, 96, 119, 122–26, 212; Punch 267; research methodology 42–43; semantic content of 54 ff; swine flu 87
cartoonist: agenda 60; artists 211; creation 135; devices used by 61; Mexican political 212; political 60, 120, 235; rhetoric 85; skill 62; Turkish political 120, 122; US 76; use of balloon text 165
censorship 9, 122
Chicano 211, 214–15, 217–18, 221, 228, 234–35: activist 217; anti-Chicano rhetoric 231; art 213; barrio 211, 223; community 213, 216–17, 229, 233, 235; culture and language 218; identity 10, 210, 216, 225; leftist ideology 229; movement 213, 215; post-Chicano Discourse 210, 212–13; satire 212; sociopolitical 211, 235; speech 227; Spanglish-speaking 210, 225; students 234; zoot-suiter 215
clitic: enclitic 192; proclitic 192
code (linguistic, semiotic, visual) 2, 5, 8–10, 94, 119, 123, 125, 127, 129, 136, 139, 149, 158, 183–84, 186–87, 196, 198–99, 211, 240, 246, 256, 260, 291–92
code choice 9, 112, 184
code crossing 11, 244–45, 261
code switch 10, 134–38, 151–55, 184, 190–93, 196, 210–11, 214, 216, 218–19, 223–24, 235, 240–47, 250–51, 255, 257–61, 271: intersentential 223, 244; intrasentential 134, 223–24, 244, 257; metaphorical 253, 255, 257, 259–60; situational 251, 253, 257, 260
cognitive: architecture 100; cognitive/social construct 2; deficits 112; denotation and connotation 138;

development 110; encoding 103; entities 119; linguistics 8–9, 13, 19, 34, 94–95, 292; metaphor 8–9; models 5; neuroscience 94; psychology 94; representation 31; research 4; resources 125, 292; science 94, 112; structures of knowledge 38; system 110; understanding of a language 96
comic book 2, 11, 28, 159 (note 1), 165, 173, 180, 200, 210, 213–14, 216, 268, 281, 291
comic strip 4, 10, 142, 210–13, 216, 222, 235, 239–41, 259–61
comics 1, 3–4, 8, 28, 95, 113, 165, 167, 169–70, 172, 180–81, 187, 200, 266, 270, 291: action comics 8; black comics 200; language in comics 2, 3, 5, 291; language of comics 1–2, 291; linguistics and 4; protocomics 113; superhero comics 8, 11, 187; web comics 291
comics research 4
comics scholarship 1, 8, 187
community 60, 122, 125, 136: African American/black community 187, 192; Chicano community 213, 217, 229, 235; hip-hop community 254; Irish community 266; Jewish community 146, 153; Latino community 231; Mexican American community 213–14, 221
conceptual metaphor *see* metaphor
consonant 102, 265: final 281; substitution 279, 281; *see also* vowel
construction grammar *see* grammar
convergence 10, 22, 157: construction (of panels) 109–10; of heritages 198; of lines 22; with Dutch 10, 165, 167, 181; *see also* divergence
conversation 2, 4, 114 (note 1), 130, 132, 134, 137, 164, 190, 191, 203 (notes), 205 (notes), 212, 247, 253: conversation-driven

## Subject Index

conversation – *contined*
aspect 240; everyday 174; partner 191; topic 261; telephone 246
conversational: 47, 50; exchange 244; interaction 190, 192; turn 132, 192
corpus 9, 46–47, 144, 150, 165, 169, 170, 173, 176, 178, 180–81, 266, 269–71, 273–85, 287 (note 1)–88 (notes 6, 8, 9): *British National Corpus* 46; of Flemish comics 10; *Corpus Gesproken Nederlands* 165; *Jommeke* corpus 172; Marvel (digital database) 270; Marvel (printed comic books) 11; of spoken English 272; of spoken language 165
*Corpus Gesproken Nederlands* 165
corpus linguistics 10, 292
country: bring revolutionary change to the 231; English-speaking communities in each 88; fears for each 80; imaginary 156; Islamization of the 119; 'it's a nice country' 247; life 255; Middle Eastern 121; non-English speaking 242; of origin 63; organized by 64; *Schtroumpf* 156; swine flu cartoons by 74, 81; Tintin's 155; unity of the 221; with high rate of vegetarianism 69
critical discourse analysis 292
cross-cultural: analysis 59, 64; comparisons 110; perspective 89
cultural: affirmation 136; alignment 11, 240, 260; allusions 63–64, 67–68, 71; appropriation 11; aspects 219; authority 189; awareness 66; backdrop 189; background 61; bridge 198; category 97; codes 94, 134; conformity 219; consciousness 62; consumption 149; context 126; context of communication 125; diversity 216: experience 15, 212; factors 292; group 226; hegemonic codes of behavior 127;
heritage 143; hybridity 218; ideas 216; identity 151, 189, 198; imperialism 121; influences 121; institution of comics 113; knowledge 111; linguistics 119; linguocultural codes 136–38; list phenomenon 233; manifestations 98; models 119, 198, 210, 212–15; models of textual imagery 119; nodes 127; objects of comics 104; orientations 143; reference 233, 287; perspective 127; polemics 126; protectionism 135; references 67–68; setting 126; shared habits 188; shared knowledge 60–61; shortcomings 257; spaces 150; stable entities 119; stereotype 269; symbology 62; translation 120; values 4; variability 93
culture 11, 60, 63, 71, 89, 93, 97, 100, 104, 111–13, 119, 126, 138, 144–45, 158, 215, 218–19, 235, 256, 293: African American 11, 188, 200, 256; Anglo 214, 226; Asian American 188; Chicano 218; East Asian 200; fears of 79; francophone youth 10; global 200; hegemonic 243; hip-hop 188, 252, 254–59; Indian 67; internet 135; Irish people in popular 11; Japanese 187, 198; juvenile 142; language 122; language and 127, 292; Latino 210, 217, 219, 222; legitimate 218; mainstream 142; Mexican 226; news 90; of superhero comics 186; popular 11, 60, 142, 198, 266; popular media 243; samurai warrior 10, 183, 185, 198; Scottish (popular) 287; Spanish 232; Swedish 239, 241–42; Turkish 120; US 200; US hip-hop 261; Western(-ized) 120; white majority 200; youth 143, 158

deixis  41, 63
dialect  4, 101, 164: Arabic  121; characteristics of  164; Dutch  164; English  216, 271; 'eye dialect'  4; features  11; graphic  101; in Flanders  164; Irish  266, 273; literary  279; literature  187; local  179; native  216; nonstandard  234; northern and north midland  274; perceptions of  112; reading dialect  192, 196; regional  292; replacing  165; representation of  271; research  4; sociodialect  3; stronger  288 (note 9); usage of  196, 272; words  271; written  187–88
dialectology: perceptual  112, 281
dialogue  28, 47, 135, 137, 165, 180–81, 240, 252, 270, 293; *see also* spoken
discourse  102, 125, 130, 188, 191–92, 215, 228, 235, 261: about the social order  60; academic  41; anti-Mexican  73; capital 'D' Discourse  210, 212, 214–15, 223, 229, 235; cartoon  68; coherence and cohesion in  41; context  55, 110; contributions  200; cultural and political  120; effect  192; features  11, 194; French  149; H1N1 85, 90; hip-hop  255; imagined  51, 54; linguistic  127; little 'd' discourse  212; mainstream  85; opinion news  60; political  211, 235; political cartoons  59; post-Chicano  210–13; private  125; public  125; rhetorical  73; shared  60–61; spoken  15; style  5, 255; tendencies  193; theories of verbal  95; visual  73, 184
discourse analysis  10: critical discourse analysis  292
discourse studies  106
divergence  22: from Dutch  10,165, 167–68, 179; of lines  22; *see also* convergence
domain  17, 19, 114: abstract  18; behavioral  93; conceptual  17,

95; high-status  242–44, 251, 259; humor research  37; interpersonal  243; low-status  243–44, 251, 259; manual  96, 102; mental  17; non-linguistic  112, 291; non-spatial  17; of cognition  93; of experience  18, 34; of industry  243; of trade  243; of writing  93; public  243; semantic  38; verbal  96, 102, 114; visual  114; visual-graphic  97
Dutch: editions  178; evolution of  163; in Flanders  179–80; in the Netherlands  163; market  167, 180–81; readers  169, 180; readership  181; speaking  10; speaking part of Belgium  165, 180; spoken in Belgium  163; television  163; speakers of  164; standardized variant of  164, 167; variant of  177, 180; vocabulary  167; young Belgian speakers of  165; *see also* Language Index

emotion  15, 96, 190, 219
emotional: availability  194; chaos  7; engagement  24; excess  264; moments  223; state  7
emotionality  265
English: as a foreign language  245, 251, 259: as a second language  244, 259; assimilation of  260; dialects of  271; English-only proponents  214; English-speaking world  241; 'from above'  243, 251, 259; 'from below' 243, 259; influence of  243; native speakers of  281; role of  244, 259; status of  244–45, 259; terms in French  153; translated  251; translation of  276; usage  245; use of  243, 245, 248, 259–60; varieties of  248–49, 251, 271, 275; *see also* Language Index
ethnicity  10, 78, 185, 210, 288 (note 4)
exclamation  271, 275–76

French: Academy 144; authors 104;
*bandes dessinées* (BDs) 143–44, 152,
156–57; calques 158; comics 9,
145–47; discourse 149; forms
and styles 158; French /r/ 157;
French-speaking countries 144;
French-speaking locales 142;
French-speaking part of Belgium
165; French-speaking youth 157;
influence upon Flemish Dutch
164; linguists 143; name 147;
practice of 158; 'pure' French 158;
speech 150; translation 153, 155;
utterances 158; *see also* Language
Index
fricative (phonetics) 248
function 3–4, 9, 13: conceptual 32;
conceptual structuring 15;
framing (of English) 149;
humorous 240; narrative 107,
194; of grammatical forms 14; of
the plural (Dutch) 176; of race
in Japan 189; of a standardized
variant 179; utterance 248
function words: (Dutch) 164, 167

gender: ideology 193, 215; non-
conforming 137; stereotyped
role 201 (note 3)
genre 47: action comics 34;
cartoons 37; comics 92;
humor 55–56; identification 48,
50–52, 54–55; musical 188,
259; narrative 94; political
cartoons 60; story telling 54;
superhero 5, 13–14, 31, 198
German: borrowings 154; code
switchings 154; officers 146;
terms 271; visitors 151; *see also*
Language Index
gestalt 16, 239
Gestalt 32, 99
gesture 19, 92, 110: articulatory 99;
co-speech 93; hand 196; in
spoken discourse 15; posture
and 7; research 114 (note 3);
studies 292
grammar 107, 119, 177, 188: and
vocabulary 2; construction 96,
103, 106–07; features 268, 276;
generative 94, 96, 107, 111;
of *ass* words 193; regionally
marked 173; rules 127
grammar of comics 15, 93, 97:
closure 95, 106; sequence of
panels 92; sequential art 1;
visual 109; visual linguistics 98
grammatical: aspects 170;
categorization 109;
competence 248; deviations 247;
elements 110, 250; errors 149,
154, 236 (note 4); features 10–11,
168, 172–73, 176–79; forms 14,
265; lexicogrammatical
features 279; model 110;
patterns 109; perspective on code
switching 134; regionalisms 173;
rules 156, 223; sentences 97;
structure 225; variance 177
grapheme 99, 101–02, 114 (note 2)

Hebrew: markers 153: names 151;
term 150, 153; *see also* Language
Index
hedge (pragmatics) 248
hegemony 5, 129: of whiteness 186
hero 11, 25, 29, 146, 183, 231; *see
also* superhero
hip-hop 10–11, 183, 187–88,
252–53, 256–57, 260; *see also*
Language Index
hip-hop: artist 252, 255, 257–58;
devotee 259; environment 255;
gesturing 256; identity 252–53,
258; music 252–53, 257;
performances 189; song 253;
speech community 260;
vernacular 240, 255–57, 261;
world 254; *see also* community,
culture, discourse, English
Hispanic 229, 236 (note 8)
humor 8, 22, 37, 39, 41, 62, 82, 122,
165, 177, 212, 229, 251, 256, 258,
260, 270, 292: adventure and 168;
creation of 37, 39–42, 46, 48,
50, 54–56; effects of 41; genres
of 55; in cartoons 39, 137; in
comics 292; linguistic 250;

linguistics of   38; metafunctions in   42; perceptions of   38; political   212; research   8, 37, 39, 41–42; role of humor   37; source of   54, 194, 240–41, 250, 254, 260; theories of   38, 40; verbal humor   38, 211

icon   1, 95, 124, 222
iconic   5, 102
ideational metafunction *see* systemic functional linguistics
identity   70, 112, 120, 129, 132, 139, 184, 189, 195–96, 198, 210–11, 213, 219, 233, 241: authentic   197; character   183–84, 189; cultural   189, 198; ethnic   10, 199, 219, 223; fixed   131; formation   184, 189; in-group   240, 249, 251, 258, 260; kits   210; language and   11, 195, 216–17, 223; markers   143, 179, 219; modernist   131; national   121–22, 151, 250; participant   183; political   138; politics   10; promotion   59; regional   180; shifts of   131; social   199; socially situated   212
ideological: developments   136; forces   60; shift   244, 259
ideology   3, 5, 60–61, 88, 90, 121, 134, 157, 187, 197, 199, 214–15, 228: conservative   215; English-only   233; gender role   193; heteromasculinist   5; language/linguistic   199, 200, 210–11, 215–16, 221–23, 232, 235, 292; nationalist   5; political   133, 210, 235; social   199; white/Anglo   187
image   1–3, 7, 14, 17–18, 22, 27–29, 32, 34, 74, 97, 105, 111, 122, 130, 137, 184, 211, 266: and text   9, 37, 39–41, 110, 211, 214, 291; cartoon   131, 183; comics   14–15; humoristic   125; in the mind   16; juxtaposition of   14; of blackness   185–88, 197; of comics   8; schemata in the   39; semiotics of   123; sequence

of   13, 95, 108 (caption), 114 (note 1); sequential   93, 95–98, 104, 107–11, 113; static   13–14, 20, 22, 114 (note 1); still   21, 26; symbolic   126; underground comix   184; verbal   3; visual   60; words and   2, 7, 120, 122, 139; writing and   113
image schemas   8, 13, 15–18, 20, 23, 32–34
image schematic structure   18, 20–22, 27, 30–34
impact flash   15, 19, 25–27, 29–34
interaction: between modalities   110; conversational   190, 196; force-dynamic   31; interpersonal   195; lingua franca   251; of languages   134, 143; perceptual   16; social   39–40, 47–48, 51, 54–56, 210, 240, 243; text and image   39, 113, 211, 214
interactional: contexts   259; styles   188
interactional metafunction *see* systemic functional linguistics
interactional sociolinguistics   112
interjections   147–48, 151–52, 157, 167, 223, 265
Irish: accent   264, 268, 273, 279, 281, 284; 'Black Irish'   268; brogue   264; characters   264, 266–69, 271, 273–76, 280–81, 285, 287–88 (notes   4, 9); dialect   273; English translation of   276; expressions   267, 279; features   288 (note 2); Irishman   265, 269; Irishness   266–68, 279; *Irish Times*   274; people   11, 264–66, 268, 272, 281; phonology   281; pronunciation   265, 284; sentence   276; songs   266, 288 (note 6); speech   11, 265–6, 270–71, 273, 275, 281–82, 284, 287; stereotype   266; *see also* Language Index
Irish English: grammar   265; grammar features   288 (note 8); grammar structures   276; grammatical features   276;

302  Subject Index

Irish English: grammar – *continued*
grammatical forms   265;
lexicogrammatical features   284;
oaths   276; *see also* Language Index
*it*-clefting   268, 276–78, 282
Italian: borrowings   153;
names   151; *see also* Language Index

Japanese: American and Japanese heritage   198; and American emblems   104; audience   197, 198; cartoons   59; children   111; code switch to   196; comics   98, 110–11; culture   187, 198; identity   10; ideologies   187; influence   186; Japanese-Korean relations   59; manga   96, 101; racial identity   198; samurai warrior culture   183; speaking community   196; syllabary   184; use of   193; Visual Language   98, 101; *see also* Language Index
jargon   4

Korean: Japanese-Korean relations   59; Korean American male identity   198–99; Korean American speakers of English   198; *see also* Language Index

language   1, 5, 14–15, 19, 38, 40, 94, 96, 120, 126, 130, 136, 138, 144, 158, 169, 174, 183, 188, 198, 217, 244, 266, 292; *see also* comics; identity; ideology; Language Index
language: acquisition   111, 217; and media   59, 63; and visual rhetoric   63; as sociocultural phenomenon   9; authentic   197; change/evolution   163, 178, 180, 261, 292; choice   88, 210–11, 219, 221, 223, 235, 259–60; contact   143–45, 153, 157, 217, 244; donor   220; endangered   293; first   234; foreign   150–52, 155, 241, 244–45, 251, 259, 271; human behavior   92; interlanguage   155, 159 (note 5); killer   242; metaphor in   19; minority   242, 293; mock   226; national   242; natural   102; of blackness   187; of thought   17; official   10, 134, 242; patterns   18, 96; politics of   10, 179, 181; practices   143, 200; preservation   293; prestige   292; processing   292; regional   173, 177–78; religious   275; representations of   6; revival   293; second   234, 241, 244–45, 259; sign language   2, 93, 97, 110, 114 (note 1); shift   217; spoken   97, 114 (note 1), 163, 165; standard   164; study of language   2; system   123–24; target   220; variable   138; variants   10, 164, 167; variation   111, 112, 163; variety   167, 183, 188, 217; verbal   97–99, 102; visual   9, 93, 97–114; visual representations of   3; written   110, 184; *see also* comics; identity; ideology; Language Index
language code *see* code
language use   40, 55, 119–20, 122, 126, 134, 150, 166, 169–70, 177, 180–81, 185, 190, 200, 212
Latin: characters   152, 154; names   151; pseudo-Latin names   151, 156; *see also* Language Index
Latin American   3; characters   155; culture   218
Latino   214–15, 218–19, 221, 227–29, 231, 233, 235
Latino/Latina: author-artists   216; buying power   227; characters   223; comic book/comic strip   210, 213–14; community   214, 216, 231; culture   217–18; (Chicano) culture   219, 226; holidays and culture   210, 222, 227; identities   10, 210; immigrants   225; political and social activists   218; readers   221, 228; visual and written works   213;

vote 230; native-born 215; non-Latinos 229; young 231
lexicogrammatical features 279 *see also* grammar
linearity 99
lingua franca 143–44, 241, 251, 260
linguistic: ability 217; analysis 210, 240–41, 291; appropriation 220; aptitude 248; aspect 143; authenticity 189; behavior 11, 240; borders 169; change 157, 292; choice 138–39, 183, 199, 235; components 127; constraint 219; content 128; creativity 156; data 291; development 157; devices 3; elements 41, 56, 63, 151; entity 125; error 287; evidence 268; 'exaggeration and simplification' 183; extralinguistic factors 136, 139; features 4–5, 40–41, 51, 150, 187, 190, 197, 267, 270, 281; forms 125, 196, 250; framework 120; heritage 143; hybridity 218; identity 217; inquiry 93–94, 96; knowledge 88, 97; landscape 211, 218, 223; mainstream 261; marker 151; material 3; metaphor 61; methods 96; models 63, 120; norms 11, 87, 240; performance 183, 189, 197; production 189, 191, 194, 196, 199; references 120; repertoire 244, 245; research/scholarship 8, 37, 41, 94, 107, 291; rule 129; sciences 113; sign 130; stereotype 5, 11, 184, 226; structure 95, 127, 134, 196; subject 131, 133; system 1, 96–97, 104, 112, 188, 196, 243–44; text 88–89; theory 96, 292; typology 112; variation 138, 180, 292; variety 215; *see also* ideology
linguistics 59, 92–93, 96, 98, 106, 110, 112, 183, 291–92: and comics 4, 96–97, 240, 293; applied linguistics 4; comparative 111, 292; computational 112; corpus linguistics 10, 292; cultural 119;

neurolinguistics 112; sociolinguistics 112, 128, 134, 184; text 292; tools of 109; visual 98, 111–12; *see also* ideology
literature: and culture 67; *bandes dessinées* (BD) 157; dialect 187–88; French 158; nineteenth-century 187; present-day 142; race in 188; written 143
literary 1: allusions 63–64, 67, 71; dialect 279; Japanese representations 189; register 165

manga 4, 10, 93, 96, 101, 183–85, 187, 190–94, 196, 199, 201 (note 1)
meaning: construction 15, 34, 138; correspondences to 109; creation 19; development of 136; graphic 95; orders 138; patterns 96; pragmatic 40; meaning recognition 125; transmission of 132; unitized 103
media 214, 221, 266: alarmist tendencies in 89, array of 99; art and 94; censorship 120; coverage 71; diversity of images 221; Flemish 164; hype 71, 89; images 221; images of blackness in 188; in Turkey 137; internet 200; language and 9, 63; mass 231; panic-ridden 71; popular culture 243; portrayals in 215; propaganda 167; representations of blackness 189; representations of race in 188; represented by 221; scholars of 59; source 227; trust of 70; twentieth-century 187; twenty-first century 189; visual 14
metafunction *see* systemic functional linguistics
metafunctions in language *see* systemic functional linguistics
metaphor/metaphoric 59–64, 73–80, 88–89, 95–96, 125: analysis 89;

## 304  Subject Index

metaphor/metaphoric – *continued*
  code switch  253, 255, 257,
    259–60; component  89;
    conceptual metaphor  8,
    13, 15–23, 27, 32, 34, 95,
    106; function of English  4;
    mapping  27; multimodal  96;
    supercategory  74; theory  61–63;
    use of  124; verbal  19; visual  19,
    25; *see also* cognitive
metonymy  5, 19, 106
Mock Spanish *see* Language Index
modality  40, 98, 102: graphic  98,
    101, 105; manual  114 (note 2);
    verbal  93, 109; visual-graphic  98,
    113
morpheme  94, 102–06:
    allomorph  174; English
    possessive  225
morphological  104–06, 176, 193,
    219
morphology  93, 95, 98, 102–05,
    109, 167
morphosyntactic  134, 164
model: cognitive  5; cultural  119,
    198, 210, 212–15; English  148;
    grammatical  110; ideational
    metafunction  40; language
    interactions  143; linguistic  63,
    120; national identity  121;
    plurilingualism  143;
    statistical  112; topoi  68;
    visual  123
monolingual speakers  216–17
motion line  15, 19, 23–25, 34, 104
multimodality  18, 93, 98, 110

narrative/narration  7, 14, 143,
    145–46, 148, 156–57, 190, 193, 199:
    arc  6; boxes  6; breakdown  3;
    captions  5–6; development
    of  123; devices  229;
    differences  185; 'erotics'  4;
    forms  94; framing  150,
    157; function  107, 194–95;
    genres  94; intertextual  130;
    meta-narrative  213; patterns  3;
    pictorial  3; power  14; sand  98,
    104, 110, 112, 114 (note 1);
    sequence  107; structure  108
    (caption), 292; techniques  211,
    235
nasal (phonetics): alveolar  191, 193,
    280; velar  191, 193, 280
nonstandard: dialect  234;
    English  240; speech  271;
    spelling  4, 258; use of progressive
    aspect  278; variety  216, 251,
    292; verb form  217; *see also*
    standard
non-visual: event  27; nature of
    speech  2; *see also* visual
noun  148, 156, 167, 176, 179,
    225, 277, 280: proper  235;
    semantic class  109; subject
    complement  278

onomatopoeia *see* sound effects

panel  2–3, 6–7, 15, 20–22, 25,
    28–31, 33–34, 94, 99, 104, 106–10:
    four-panel jokes  14; juxtaposition
    of  14, 95; panel-to-panel  14;
    sequence of  92; single-panel  8,
    13, 20, 132, 211; transitions  95,
    107, 110; with motion lines  25;
    within-panel sense  28–29;
    words = panels  93, 98
*parler jeune*  143–44, 149, 158
    (note 1)
perceptual dialectology  112, 281
performance *see* hip-hop; linguistic;
    verbal
phoneme  99, 101
phonetic: element  227;
    feature  217, 233, 236 (note 4),
    251; indicator  195;
    representation  247; spelling  4
phonetics  98–100, 135
phonology  99
pictorial: expressions  19;
    forms  126; graphics  133;
    information  18;
    narratives  3; non-pictorial
    icons  124; protagonist  130;
    representation  137; runes  5
picture  1, 37, 39, 42–43, 47, 92, 157,
    211: caption and  37, 41, 51, 55;

in the mind's eye   16; mental   25; properties of   19; words and   19
Pocho: politics   210–11, 213, 216, 235; Pocho Hour of Power   213, 235 (note 2); Spanglish-speaking   215
Polish terms   150; *see also* Language Index
politeness: and honorification   41; and pronoun usage   174, 179; contexts   174
political cartoon *see* cartoon
political ideology *see* ideology
politics   64, 71, 120, 213, 222, 228: identity   10; Indian   65; of language   10; opinions   37; presidential   210, 222, 235; *see also* Pocho
possessive: English possessive formation   225; morpheme   225; pronoun   174; Spanish possessive formation   225
power   5, 27, 122, 127, 189, 220; and control   5; authority and   189; holders   212; Latino's buying power   227; 'more power to yer elbow'   265; narrative   14; of visual texts   59; relations   127
power (of characters, heroes)   185, 266, 268, 270, 287 (example   51): 'good luck power'   268; mutant   267; through wood   268
pragmatic: abilities   15; appropriate English   248; coherence   125; continuum   220; deviations from idiomatic English   247; features   251; felicitous use   248; impact   196; meaning   40; move   249, 256; usage of dialect   196
preposition   167, 174: preposition *de* (of)   225; prepositional phrases   278
press: English-language   242; printed   243
pronoun: Dutch demonstrative   174; Dutch personal   164, 169–70, 174, 178; Dutch possessive   174; English personal   191, 201 (note 3), 250;

Irish English   265, 277; Irish English relative   277; subject   283; Turkish pronominal   9
pronounce: pronouncing post-vocalic [r]   191; Spanish place names   220; Spanish words   220; unpronounced   127; with a Spanish accent   232
pronunciation   191, 194: changing   232; differences in   10, 163; English   227; guttural Irish   265; Irish   284; mispronunciation   248; mutilation of   186; native   233; nonnative   258; of French /r/   157; of plural *ye* or *youse*   283; prescriptive norms   4; style   191
psycholinguistics   9, 292

quantitative: analysis   5; techniques   10; *see also* statistics

race   186: and ethnicity   10; ideologies of   215; in Japan   189; in literature   188; representing   198; social realities of   10
racism   10, 218–19, 292
reading dialect   192, 196
reduplication   105
reference   67–68, 95, 102, 120, 128, 136, 142, 233, 252, 276, 287
referent   62, 124
referential   126
register   112, 143, 169, 183, 193, 196: casual   190; exaggerated   3; formal   194, 251; General American English   191; informal   219; literary   165; of English   184; of oral storyteller   165; register variation   4; socially situated   183; use of   191; varieties of   143; youth register   149
respelling   264–65, 267, 271, 279–81, 283–84, 288 (note 3); *see also* spelling
rhetoric   71, 85, 225; anti-Chicano   231; *see also* visual

rhetorical: devices 3, 60;
   discourse 73; methodologies 63;
   presentation 59; rhyming 192,
   194; strategies 60, 87;
   techniques 211, 235
ribbon path 15, 19–26, 29–34
runes 5; *see also* pictorial
Russia 150
Russian: characters 154, 159;
   immigrants 155; Jewish Russian
   painter 146, 153; Jews 154;
   names 151; non-Jewish 153;
   signs 150; *see also* Language Index

sand narrative *see* narrative/narration
satire 37, 212, 214, 265
schema (schemata) 38–39,
   109: acquisition of 111;
   conflicting 56; content 39;
   force-dynamic 33; graphic 111;
   semantic 38; *see also* image
   schema
schematic: encodings 102;
   gestalt 16; information 102;
   patterns 103, 107–08; quality 15;
   structure 34; style 143
script (cognition) 38–39: and
   frame 38; incongruity 39, 41;
   semantic 38
script (writing): Arabic 154
semantic(s) 40, 93, 95, 98, 106,
   109–10, 123: aspects 105; Chicano
   features 217; content 38–43,
   54, 56; derogation 220;
   deviations 247; domains 38;
   features 106, 251; incongruity 8,
   55, 248; nuances 148;
   phenomena 106; phrases 233;
   properties 109; prosody 40,
   43; role 95; *Schtroumpf* 157;
   scripts 38; sets 47; system
   123–24; visual language 107
semiological 94, 107
semiotic 2–3, 123: analysis 89;
   approach 119; content 120;
   field 139; multimodal 39;
   order 120; process 2;
   system 119, 123–24; types 95

semiotics 4, 95, 105, 123: of the
   image 123; visual 62–63
sequence 28–29, 31, 33–34, 94,
   97, 107, 109: action 29, 31,
   33; actions in 20; BD (*bandes
   dessinées*) 146; English 147;
   multimodal 96; narrative 107;
   of events 8, 26, 30–31, 33; of
   happenings 28; of images 13,
   114 (note 1); of movements 30; of
   modalities 97; of panels 92; of
   utterances 248
sequential: art 1–2; images 93,
   95–98, 104, 106–07, 109–11, 113;
   language 15; movements 30;
   patterns 98; role 109;
   sounds 97; structure 96, 107;
   units 93
sexuality 197: heterosexual
   women 5; homosexuality 138;
   same-sex sexual attraction 5
sign 104, 106, 114 (note 1)–115
   (note 3): arbitrary 5;
   authenticating 188; bound 104;
   language 4; linguistic 130;
   liquid encoded 125;
   morphological 104, 106;
   motivated 5; stigmatized 188; *see
   also* linguistic
sign language *see* language
signification 136, 138, 189
signifier 123–25, 129, 131, 138
slang 4, 149, 159 (note 4), 192, 248,
   274, 288 (note 8)
sociocultural: artifact 113;
   audiences 144; code 5, 10,
   183, 186, 194; construct 187;
   context 113–14,
   137; experience 125;
   implications 184; landscape 9;
   phenomenon 9–10, 96;
   settings 112; understanding 96
sound 13, 27–28, 31–32, 101, 114
   (note 2): combinations of 100; in
   language 97, 102; in manga 4;
   percussive 27, 34; phonemes 99;
   quality and magnitude of 18;
   sequential 97; speech and 14;

visible 4; vowel 32; *see also* consonant; nasal; fricative
sound effects (onomatopoeia) 13, 18–19, 26–28, 34, 184: absence of 29; in a panel 29
soundless comics 13
SOURCE-PATH-GOAL image schema 8, 16, 20–22, 31, 34
space (in comics) 14, 24–26: and time 23, 28; conceptualized 23; gutter 292; juncture of 22; of a cartoon 62; of the panel 20, 29; locations in 20
Spanish: accent 232; authentic features 219; barrio 225; 'borrowed' words 220; characters 155; code switching 235; culture 232; for heritage speakers 233; heritage language learners 222–23; immigrant 155; importance of 219; 'incorporation' of language materials 220; interviews conducted in 219; lexical items 225; Mock 219–20; Mock accent 226–27; Mock stereotype 227; native fluency 220; nonnative speaker of 234; proper nouns and place names 235; refugees 146; speaker's accent 220; Spanish-speaking population 220; surname 228; switching from English to Spanish 224; switches to English 223; term 150; words and expressions 151; *see also* Language Index
Spanglish: Spanglish-speaking Chicano 225; speaker 218; *see also* language index
speech balloon *see* balloon
speech community 244, 250: [Black] 188; English-speaking 196; hip-hop 260; Japanese 196; nonnative English 251, 260; Swedish 242, 259; Swedish-English bilingual 245

spelling 169, 273, 279–80, 283–84: change 192; conventions 191; deviations 4; nonstandard 4, 258; phonetic 4; standard 246, 250, 284; Swedish-spelled English 246, 248, 250; *see also* respelling
spoken: Chicano English 216; dialect 216; dialogue 13, 165; discourse 15; English 217, 271, 281–82; Irish English 11; language 97–99, 114 (note 1), 149, 163, 165; language in balloons 179; reality 187; register 3; Spanglish 218, 225; Spanish 217; Swedish 261; words 2, 177, 287; *see also* language
standard: approach to comics 29; code 292; contractions 190; Japanese style 101; language 164, 292; linguistic forms 125; spelling 246, 250, 284; traits 189; variant 164, 179; variety 163–64; *see also* nonstandard; Language Index
statistical; modeling 112; significance 172; test 172
statistically significant 167, 171–72, 193
statistics 170, 180; *see also* quantitative
stereotype 62, 113, 184, 186–88, 199–200, 221, 228, 292: accent 221, 227, 223; black 187, 189; category 5; characters 31; Chicano 235; cultural 269; expectations 189; fallacies 229; ideology and 197; Irish 264, 266, 268; language 227; Latinos 220; linguistic 5, 11, 184; Mock Spanish 226–27, 235; negative 155, 184, 215, 226; of the minstrel 198; portrayal 215, 287; racial 186, 201 (note 3), 219; rhetorical presentation of 59; social realities of 10; speech 197, 226, 264; Stage Irish 287; US 70, visual 185; word 267

stress (phonology): stressed 284; unstressed 284; unstressed variant 174
strip: American comic 226; bilingual 158; cartoon 210; characters 11; comic 4, 210–11, 213, 216, 221, 225, 232, 235–36 (note 4), 239–40, 246; dialogue-driven 240; English-language 142, 145; Latino 210, 214; medium 261; newspaper 166; Sunday-edition double 253; Swedish 239–40, 259–60
style (linguistic) 127, 143, 183–84, 219: discourse 5, 255; French 158; interactional 188; pronunciation 191
style (non-linguistic) 131: abstract 95; artistic element 3; clothing 62; drawing 183; fashion 71; graphic 98; hair 194; Japanese 101; lifestyle 89, 137; Marvel 270; media presentation 229; narrative 146; representational 95; shirt 227
subcultural identity 243, 252
superhero 31, 187, 285: black 185; comics 5, 8, 13, 20, 22, 96, 186–87, 198; genre 5, 14, 31; Marvel comics 11; stories 34
suppletion (morphology) 104, 105
Swedish: comic strip 239; comic strip medium 240, 260; culture 239, 241–42; effect on 259; English translations of 246; evolution of 242; fate of 242; in Sweden 259; language (television) channels 242; language usage 245; males 255–56; nonnative speech community 260; nonnative variety of English 248; non-Swedish speakers 250–51, 260; reading public 11, 240; PEN (Poets, Essays, and Novelists) 239; phonetic representations of accent 247; society 241; speakers of English 248;

speech community 242, 259; Swedish-English bilingual speech community 245; Swedish-English code switching 261; Swedish-language comic strip 260; Swedish-spelled English 246; websites 243; see also Language Index
syllabary 2, 184
symbol 1, 15, 19, 28, 30, 34, 62, 92, 95, 119, 143, 212, 231, 266: appropriation of 62; conceptual 33; cultural 142; graphic 92; image 126; motion 30; movement 33; nonconventional 123; system of 120, 122, 138–39; see also visual
symbology 62
synesthesia 4, 18
syntax 108: of Schtroumpf language 157; variation in 163
systemic functional linguistics 8, 37–38: communicative metafunction 40; conceptual metafunction 40; ideational metafunction 39–40, 48, 54–55; interactional metafunction 8; interpersonal metafunction 8, 38, 40–43, 47–48, 50, 54–55; metafunction 37, 39–40, 46; poetic metafunction 40; textual metafunction 8, 40–41, 46–48, 51, 54–56

tense (verbs) 40, 129, 133: future in Dutch 175, 178; future in Irish English 277–78, 282; in Turkish 133; past 22; past in Irish English 277, 282; past in Turkish 129; present in Dutch 174; present in Irish English 277–78; present progressive in English 133
tension: between codes 199; between words and images 3, 7; political in Turkey 119, 122, 126, 138
text 94, 138: antebellum 187; balloon 165; dialogue as 28; English 243; hip-hop 256;

keywords in  46; linguistic  88–89; linguistics  292; multimodal  95; of Talmudic law  153; production principles  3; sound rendered as  13, 26–27; text-based jokes  39, 55; visual texts  59, 63; *see also* image
textual: connections  48, 51, 55–56; content  119; cue  28; flux  130; imagery  119; metaphors  62–63; realities  184
textual metafunction *see* systemic functional linguistics
theory: conspiracy  69–70, 85; incongruity  38–40, 56; linguistic  292; metaphor  61–63; of comics  94, 96; of graphic expression  96; of language  94; of verbal humor  38; of visual language  9, 112; psychic release  38; relevance  212; superiority  38
*topoi/topos*  9, 59, 63–64 ff, 79, 89
trajector (cognitive linguistics)  20–21, 25, 33
translation  119, 120, 131, 153, 155–56, 243; Chinese  151; English  129, 276; studies  292
Turkish: artist  128; cartoonist  122; culture  120; expression  128; grammar  9, 135; newspaper  122; political cartoonists  120, 123; political cartoons  119–20, 122, 124–27, 130–31, 133, 136, 138; Prime Minister  131, 135; pronominals  9; replacement words  135; structural integration of English and  136; 'Turkish Language Society'  134; use of  134; *see also* Language Index

utterance  190–91, 194–95, 202–07, 248, 255, 270, 285: French  158; meaning of  40; written  2

verbal  87: abstraction  123; allusion  130; camp  5; communication  97, 139; concepts  120; content  120; counterpart  62; discontinuity  129; discourse  95; domain  96, 114; encoding  119; expression  126; form  93, 95, 98, 105, 110; humor  38, 211; image  3; information  123, 125–26, 128; jokes  37; language  93, 98–99, 102, 104, 107, 112; meaning  123; modality  93, 109; narrative  3; non-verbal representation  8; performance  197; representation  123; system  164; verbal-auditory channel  99; *see also* image; visual
verb: auxiliary  133; be + after + verb + -ing  276–77; changes in  167; construction  178; deviant construction  175; dynamic  278; final group  175–76; finite  176; nonstandard form  217; past  176; progressive form  280; repeating the verb  272; *Schtroumpf*  156; stative  278; tense of  133
verbal-visual: blend  3, 7, 292; construct  3; modality  110
'Verb-ing the TIME Away' *see* construction grammar
velar  191, 193, 280; *see also* nasal
visual/nonvisual  8, 126, 292: and written works  213; art  211; codes  8; cognitive/social construct  2; communication  63; conventions  8, 15; cues  8, 14, 18; design  7; devices  29; discourse  73, 184; domain  97, 114; elements  2–3, 61; experience  16; flash  27; form  111; format  122; frame  132, 133; grammar  109; lexicon  104; linguistics  111, 112; media  14; metaphor  19, 25, 61–63; morphology  104; motion  19–20; perception  16, 27, 100, 105; perspective  15, 20, 22, 25; protagonist  131; relationship with verbal  2–3, 187–89; representation  2–3, 14, 18, 21, 27, 183, 291; rhetoric  59, 62–63; stereotype  185; strategies  60; symbol  15, 18, 31, 33–34; text  59, 63; *see also* language

visual-graphic: aspects 102;
   domain 97; form 109;
   modality 98, 113; vocabulary 92
visual language *see* language
visual text *see* text
vocabulary 4, 130, 163, 167,
   169–70, 288 (note 6): grammar
   and 119, 127; of *bandes
   dessinées* 143; of comics 1–2,
   95, 291; pronunciation and 10;
   regional 167, 169, 171; visual 92
vocative 271–75
vowel 280: central 32;
   diphthong 191; initial 281; long
   or short 102; monophthong 191;
   substitutions 279, 281; *see also*
   consonant

word 89, 102, 104–05: and
   image 2–3, 7, 18, 95, 120, 122,
   130, 139; and language 15; and
   lyrics 231; and panels 93, 98;
   and picture 1, 19; Arabic 154;
   *ass* 193–94; balloons 106, 115
   (note 3); beginning of 100;
   blend 218; borrowed 220;
   bubbles 104; Chinese 154;
   colloquial 181; content 167;
   count 46; dialectal 271;
   English 135, 151, 225, 241;
   ethnic 225; form 194;
   frequency 46, 170; fulcrum 223;
   function 164, 167; games 158;
   keyword 46–47; length 46;
   level 192; loanword 243;
   meanings 157; Mock
   Spanish 235; onomatopoetic 28;
   order 148, 176, 225; play 156,
   227; regional 169–70, 177;
   replacement 135; size 102;
   slang 159 (note 4); Spanish 151,
   210, 220, 226; spoken 2, 287
   (note 2); stem 174; swear 190,
   288 (note 10); usage 136, 274;
   written 124
writing 93, 113, 130, 185, 212: and
   drawing 92; and images 113;
   process 92; student 59
writing system 2: Arabic 152, 154;
   Chinese 152, 154; Cyrillic 152;
   Belgian 163; Dutch 163;
   Egyptian hieroglyphs 152;
   Hebrew 152; Latin 152; left-to-
   right 89; non-Western 89
written: calligraphy 184;
   comics 93, 98, 113;
   component 63; dialect 187;
   dialogues 165; documents 121;
   in English 88, 210, 216, 223;
   interjections 152; Farsi 121;
   language 93, 110, 179, 184;
   literature 143; nonstandard
   spelling 258; rarely written
   pronoun 174; sound effects 27;
   texts 46, 177; utterances 2;
   word 124; works 213